AF361663

The Antioxidant Potential of Fermented Foods: Challenges and Future Trends

The Antioxidant Potential of Fermented Foods: Challenges and Future Trends

Guest Editors

Michela Verni
Carlo Giuseppe Rizzello

Basel • Beijing • Wuhan • Barcelona • Belgrade • Novi Sad • Cluj • Manchester

Guest Editors

Michela Verni
Department of
Environmental Biology
Sapienza University of Rome
Rome
Italy

Carlo Giuseppe Rizzello
Department of
Environmental Biology
Sapienza University of Rome
Rome
Italy

Editorial Office
MDPI AG
Grosspeteranlage 5
4052 Basel, Switzerland

This is a reprint of the Special Issue, published open access by the journal *Fermentation* (ISSN 2311-5637), freely accessible at: www.mdpi.com/journal/fermentation/special_issues/antioxidant_foods.

For citation purposes, cite each article independently as indicated on the article page online and using the guide below:

Lastname, A.A.; Lastname, B.B. Article Title. *Journal Name* **Year**, *Volume Number*, Page Range.

ISBN 978-3-7258-2844-9 (Hbk)
ISBN 978-3-7258-2843-2 (PDF)
https://doi.org/10.3390/books978-3-7258-2843-2

Contents

Editorial

The Antioxidant Potential of Fermented Foods: Challenges and Future Trends

Michela Verni * and **Carlo Giuseppe Rizzello**

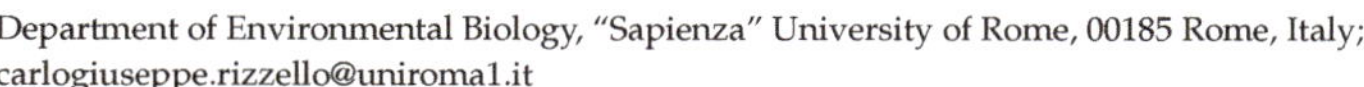

Department of Environmental Biology, "Sapienza" University of Rome, 00185 Rome, Italy; carlogiuseppe.rizzello@uniroma1.it
* Correspondence: michela.verni@uniroma1.it; Tel.: +39-06-4991-2395

check for updates

Citation: Verni, M.; Rizzello, C.G. The Antioxidant Potential of Fermented Foods: Challenges and Future Trends. *Fermentation* **2023**, *9*, 790. https://doi.org/10.3390/fermentation9090790

Received: 17 August 2023
Accepted: 25 August 2023
Published: 27 August 2023

The major roles of antioxidant compounds in preserving food shelf-life, as well as in providing health-promoting benefits, combined with the increasing concern regarding synthetic antioxidants, is progressively leading the scientific community to focus on natural antioxidants. Polyphenols, bioactive peptides, amino acids, and vitamins are among the most common antioxidant compounds naturally present in foods. Nevertheless, enabling further improvements to food antioxidant activity in vitro, which could potentially reflect on that in vivo, is a topic of the utmost significance. The bioconversion elicited by the use of microbial enzymes, and/or fermentation with selected starters, can be considered a tool for enhancing the activity of bioactive compounds by facilitating their release or changing their structural conformation. Indeed, fermentation is known to affect food features and, although it began as a means to extend food shelf-life, over the last few decades, research shifted to the investigation of its health benefits, among which are those provided by antioxidant compounds [1]. In this framework, this Special Issue aimed to cover the most recent advances in the use of fermentation as a means to enhance food antioxidants' potential.

Many of the papers published relied on the use of in vitro assays to determine how fermentation, either spontaneous [2–4] or with selected starters [5–11], can improve the antioxidant potential of cereals [9], legumes [6,9], milling by-products [8], and other vegetable matrices [2–5,7,10]. In most of the cases, the activity of lactic acid bacteria (LAB) and/or yeasts led to a higher bio-accessibility of phenolic compounds which, in turn, determined a higher antioxidant activity [2,6–11], often measured as DPPH (2,2-difenil-1-picrylidrazyl) radical scavenging activity, as well as a higher anti-inflammatory and antiplatelet potential [5].

Even though, due to their limitations, the use of in vitro tests has generated controversies over the years, they are still of great importance in the selection of potential antioxidant compounds or studying conspicuous sets of microorganisms/matrices [1], hence they are often coupled to other methods including those comprising cellular models. This is the case of the research of Polo et al. [12], who studied the effect of fermentation with several LAB starters on ice cream. The authors found that ice cream fermented with *Lacticaseibacillus casei* F14 counteracted the accumulation of proinflammatory interleukin mediators IL-8 and IL-6 and reactive oxygen species in Caco-2 cell culture, thus showing antiradical and anti-inflammatory features [12].

Nevertheless, in vitro and ex vivo assays can only be predictive tools of the antioxidant activity in vivo, and human trials are time-consuming and heavily regulated by ethical committees. Some papers collected in this Special Issue [13–16] relied on animal studies to further explain the in vivo roles of fermented foods. Indeed, the consumption of yogurt or LAB used as probiotic by rats experiencing oxidative stress resulted in a significant decrease in triglycerides, total cholesterol, low-density lipoprotein, and a remarkable increase in high-density lipoprotein [13–15], thus demonstrating the feasibility of fermentation for enhancing the functionality of foods. However, the elucidation of new bioconversion pathways, the study of antioxidant bioavailability and bio-accessibility, as well as their functions in in vivo digestion, are areas that still needs exploring.

Author Contributions: M.V. and C.G.R. equally contributed to the proposal, editorial work of this Special Issue, and to the writing of the Editorial. All authors have read and agreed to the published version of the manuscript.

Conflicts of Interest: The authors declare no conflict of interest.

References

1. Verni, M.; Verardo, V.; Rizzello, C.G. How Fermentation Affects the Antioxidant Properties of Cereals and Legumes. *Foods* **2019**, *8*, 362. [CrossRef] [PubMed]
2. Cai, K.; Dou, R.; Lin, X.; Hu, X.; Wang, Z.; Liu, S.; Li, C.; Li, W. Changes in Phenolic Profiles and Inhibition Potential of Macrophage Foam Cell Formation during Noni (*Morinda citrifolia* Linn.) Fruit Juice Fermentation. *Fermentation* **2022**, *8*, 201. [CrossRef]
3. Palachum, W.; Klangbud, W.; Chisti, Y. Spray-Dried Nipa Palm Vinegar Powder: Production and Evaluation of Physicochemical, Nutritional, Sensory, and Storage Aspects. *Fermentation* **2022**, *8*, 272. [CrossRef]
4. Várady, M.; Tauchen, J.; Klouček, P.; Popelka, P. Effects of Total Dissolved Solids, Extraction Yield, Grinding, and Method of Preparation on Antioxidant Activity in Fermented Specialty Coffee. *Fermentation* **2022**, *8*, 375. [CrossRef]
5. Glenn-Davi, K.; Hurley, A.; Brennan, E.; Coughlan, J.; Shiels, K.; Moran, D.; Saha, S.; Zabetakis, I.; Tsoupras, A. Fermentation Enhances the Anti-Inflammatory and Anti-Platelet Properties of Both Bovine Dairy and Plant-Derived Dairy Alternatives. *Fermentation* **2022**, *8*, 292. [CrossRef]
6. Leksono, B.; Cahyanto, M.; Rahayu, E.; Yanti, R.; Utami, T. Enhancement of Antioxidant Activities in Black Soy Milk through Isoflavone Aglycone Production during Indigenous Lactic Acid Bacteria Fermentation. *Fermentation* **2022**, *8*, 326. [CrossRef]
7. Budak, H. Determination of Changes in Volatile Aroma Components, Antioxidant Activity and Bioactive Compounds in the Production Process of Jujube (*Ziziphus jujuba* Mill.) Vinegar Produced by Traditional Methods. *Fermentation* **2022**, *8*, 606. [CrossRef]
8. Siroli, L.; Giordani, B.; Rossi, S.; Gottardi, D.; McMahon, H.; Augustyniak, A.; Menon, A.; Vannini, L.; Vitali, B.; Patrignani, F.; et al. Antioxidant and Functional Features of Pre-Fermented Ingredients Obtained by the Fermentation of Milling By-Products. *Fermentation* **2022**, *8*, 722. [CrossRef]
9. Kwon, H.; Choi, J.; Kim, S.; Kim, E.; Uhm, J.; Kim, B.; Lee, J.; Kim, Y.; Hwang, K. Optimization of Solid-Phase Lactobacillus Fermentation Conditions to Increase γ-Aminobutyric Acid (GABA) Content in Selected Substrates. *Fermentation* **2023**, *9*, 22. [CrossRef]
10. Liu, N.; An, X.; Wang, Y.; Qi, J. Metabolomics Analysis Reveals the Effect of Fermentation to Secondary Metabolites of Chenopodium album L. Based on UHPLC-QQQ-MS. *Fermentation* **2023**, *9*, 100. [CrossRef]
11. Akmal, U.; Ghori, I.; Elasbali, A.; Alharbi, B.; Farid, A.; Alamri, A.; Muzammal, M.; Asdaq, S.; Naiel, M.; Ghazanfar, S. Probiotic and Antioxidant Potential of the *Lactobacillus* Spp. Isolated from Artisanal Fermented Pickles. *Fermentation* **2022**, *8*, 328. [CrossRef]
12. Polo, A.; Tlais, A.; Filannino, P.; Da Ros, A.; Arora, K.; Cantatore, V.; Vincentini, O.; Nicolodi, A.; Nicolodi, R.; Gobbetti, M.; et al. Novel Fermented Ice Cream Formulations with Improved Antiradical and Anti-Inflammatory Features. *Fermentation* **2023**, *9*, 117. [CrossRef]
13. Shahein, M.; Atwaa, E.; Babalghith, A.; Alrashdi, B.; Radwan, H.; Umair, M.; Abdalmegeed, D.; Mahfouz, H.; Dahran, N.; Cacciotti, I.; et al. Impact of Incorporating the Aqueous Extract of Hawthorn (*C. oxyanatha*) Leaves on Yogurt Properties and Its Therapeutic Effects against Oxidative Stress Induced by Carbon Tetrachloride in Rats. *Fermentation* **2022**, *8*, 200. [CrossRef]
14. Shahein, M.; Atwaa, E.; Alrashdi, B.; Ramadan, M.; Abd El-Sattar, E.; Siam, A.; Alblihed, M.; Elmahallawy, E. Effect of Fermented Camel Milk Containing Pumpkin Seed Milk on the Oxidative Stress Induced by Carbon Tetrachloride in Experimental Rats. *Fermentation* **2022**, *8*, 223. [CrossRef]
15. Qu, H.; Zhang, L.; Liu, X.; Liu, Y.; Mao, K.; Shen, G.; Wa, Y.; Chen, D.; Huang, Y.; Chen, X.; et al. Modulatory Effect of *Limosilactobacillus fermentum* grx08 on the Anti-Oxidative Stress Capacity of Liver, Heart, and Kidney in High-Fat Diet Rats. *Fermentation* **2022**, *8*, 594. [CrossRef]
16. Chen, Q.; Mao, J.; Wang, Y.; Yin, N.; Liu, N.; Zheng, Y.; An, X.; Qi, J.; Wang, R.; Yang, Y. Fermented Wheat Bran Polysaccharides Improved Intestinal Health of Zebrafish in Terms of Intestinal Motility and Barrier Function. *Fermentation* **2023**, *9*, 293. [CrossRef]

fermentation

Article

Fermented Wheat Bran Polysaccharides Improved Intestinal Health of Zebrafish in Terms of Intestinal Motility and Barrier Function

Qiuyan Chen [1,2,†], Jinju Mao [1,2,†], Yuan Wang [1,2,*], Na Yin [1,2], Na Liu [1,2], Yue Zheng [1,2], Xiaoping An [1,2], Jingwei Qi [1,2], Ruifang Wang [1,2] and Yanping Yang [1,2]

1   College of Animal Science, Inner Mongolia Agricultural University, Hohhot 010018, China
2   Inner Mongolia Herbivorous Livestock Feed Engineering Technology Research Center, Inner Mongolia Agricultural University, Hohhot 010018, China
*   Correspondence: wangyuan@imau.edu.cn
†   These authors contributed equally to this work.

Abstract: Intestinal barrier dysfunction and gut microbiota disorders have been associated with various intestinal and extraintestinal diseases. Fermented wheat bran polysaccharides (FWBP) are promising natural products for enhancing the growth performance and antioxidant function of zebrafish. The present study was conducted, in order to investigate the effects of FWBP on the intestinal motility and barrier function of zebrafish, which could provide evidence for the further potential of using FWBP as a functional food ingredient in the consideration of gut health. In Experiment 1, the normal or loperamide hydrochloride-induced constipation zebrafish larvae were treated with three concentrations of FWBP (10, 20, 40 µg/mL). In Experiment 2, 180 one month-old healthy zebrafish were randomly divided into three groups (six replicates/group and 10 zebrafish/tank) and fed with a basal diet, 0.05% FWBP, or 0.10% FWBP for eight weeks. The results showed that FWBP treatment for 6 h can reduce the fluorescence intensity and alleviate constipation, thereby promoting the gastrointestinal motility of zebrafish. When compared with control group, zebrafish fed diets containing FWBP showed an increased villus height ($p < 0.05$), an up-regulated mRNA expression of the tight junction protein 1α, muc2.1, muc5.1, matrix metalloproteinases 9 and defensin1 ($p < 0.05$), an increased abundance of the phylum Firmicutes ($p < 0.05$), and a decreased abundance of the phylum Proteobacteria, family Aeromonadaceae, and genus *Aeromonas* ($p < 0.05$). In addition, 0.05% FWBP supplementation up-regulated the intestinal mRNA expression of IL-10 and Occludin1 ($p < 0.05$), enhanced the Shannon and Chao1 indexes ($p < 0.05$), and increased the abundance of Bacteroidota and Actinobacteriota at the phylum level ($p < 0.05$). Additionally, 0.1% FWBP supplementation significantly improved the villus height to crypt depth ratio ($p < 0.05$) and increased the mRNA expression of IL-17 ($p < 0.05$). These findings reveal that FWBP can promote the intestinal motility and enhance the intestinal barrier function, thus improving the intestinal health of zebrafish.

Keywords: fermented wheat bran polysaccharides; intestinal motility; intestinal barrier; intestinal health; zebrafish

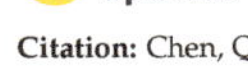

Citation: Chen, Q.; Mao, J.; Wang, Y.; Yin, N.; Liu, N.; Zheng, Y.; An, X.; Qi, J.; Wang, R.; Yang, Y. Fermented Wheat Bran Polysaccharides Improved Intestinal Health of Zebrafish in Terms of Intestinal Motility and Barrier Function. *Fermentation* **2023**, *9*, 293. https://doi.org/10.3390/fermentation9030293

Academic Editors: Michela Verni and Carlo Giuseppe Rizzello

Received: 18 February 2023
Revised: 12 March 2023
Accepted: 15 March 2023
Published: 17 March 2023

1. Introduction

As an important organ of the digestive and immune systems in human body, the intestine is in charge of nutrient digestion and the absorption of food [1,2]. In normal digestion, food is transited through the gastrointestinal tract by rhythmic contractions. Gastrointestinal dysmotility could lead to spasms or paralysis, thereby causing gastrointestinal diseases and morbidity [3]. Furthermore, intestinal constipation may change the gut microbiome, potentially contributing to the impairment of gastrointestinal functions. As the key determinant of gut health, the intestine is colonized by trillions of microbes comprising the gut microbiota, which participates in nutrient metabolism, regulation of immune system and energy regulation [4]. Numerous observations have indicated that the changes in the composition of gut microbiota are associated with various intestinal diseases [5]. Therefore, the promotion of gastrointestinal motility and the stability of the gut microbiota are essential ecological characteristics, given their importance to gut health. Previous studies showed that various polysaccharides possess effects regarding the regulation of microbiota, enhancing immunity, and improving health status [6].

Wheat bran, as one available by-product of wheat processing, contains many high-value components: proteins, non-starch polysaccharides, enzymes and vitamins. In recent years, wheat bran polysaccharides (WBP) have attracted particular attention for their demonstrated beneficial effects, derived from their immunomodulatory, antioxidant, antihyperlipidemia, and antitumor activities [7–10]. However, wheat bran contains non-starch polysaccharides, which are not easily digested or absorbed by fish. Microbial fermentation technology can change the structure of non-starch polysaccharides. Recently, a novel polysaccharide component was isolated from fermented wheat bran in our laboratory. It is a 21.19 kDa hetero-polysaccharides (the total polysaccharide content was 96.96%), and it is mainly composed of glucose, xylose, arabinose, galactose and mannose [11]. Through our previous study, *S. cerevisiae* and *B. subtilis* were selected to fermented wheat bran, which could obtain the highest yield of fermented wheat bran polysaccharides (FWBP) [12]. In addition, our previous study showed that FWBP exhibited significant antioxidant activity in vitro, and exhibited stronger effects on growth performance and antioxidant function in juvenile zebrafish than WBP [13]. However, the influence of FWBP on gut health in zebrafish is poorly investigated.

Zebrafish are fresh water fish, and has been widely studied for various purposes, such as in ecotoxicology, immunity, and neurophysiology studies. When compared with murine models, which have high costs, long life cycles, and complex operation, zebrafish models' popularity is due in part to the fact that they have a shorter life cycle, easy and low-cost breeding, a high presence of human orthologous genes, and the availability of a large array [14]. In addition, the intestine of zebrafish is quite similar to that of humans [13,15,16]. In this aspect, zebrafish provide a useful platform for studying host–microbe interactions. Therefore, in the current study, we assessed the effectivity of FWBP on intestinal motility in zebrafish larvae. Furthermore, we sought to understand as to whether diets containing FWBP can alter the composition of the gut microbiota and modulate the gut immune response in zebrafish.

2. Materials and Methods

2.1. Preparation of Fermented Wheat Bran Polysaccharides

Wheat bran was fermented, and the FWBP was extracted from fermented wheat bran as described in our previous study [11]. Wheat bran was fermented according to previous methods, and the inoculum was prepared by mixing activated *S. cerevisiae* and *B. subtilis* in a ratio of 3.3:6.7, with a final concentration of 1×10^8 CFU/mL. The WB was inoculated with 10.4% (v/v) inoculum. Then, sterile, distilled water was added to achieve a 1:1.16 material: water ratio. The substrate was fermented at 36 °C for 47 h and dried at 45 °C for 48 h to obtain fermented WB. The fermented WB was ground and stored at 4 °C for the polysaccharide extraction.

2.2. Experiment 1: Effect of FWBP on Promoting Intestinal Peristalsis and Alleviating Constipation in Zebrafish Larvae

2.2.1. Intestinal Peristalsis-Promoting Effect in Zebrafish Larvae

Zebrafish larvae at 5 dpf were randomly selected and placed in a 6-well microplate, with 3 mL 10 µg/L Nile red, for 16 h. Then, the zebrafish were washed with fresh embryo media to remove the dye, and FWBP was added at concentrations of 0 (Control), 10, 20, 40 and 30 µg/mL domperidone (DOM) for 6 h. At the end of the treatment, the zebrafish were rinsed with fresh embryo media, and we then randomly selected 20 zebrafish from each concentration. The intestinal fluorescence intensity of the zebrafish was measured under a fluorescence microscope [3]. Then, according to the results of the intestinal fluorescence intensity of zebrafish, the promotion effects of DOM and FWBP were calculated as:

$$\text{promotion effect} = [(\text{DOM}/\text{FWBP group} - \text{Control group})/\text{Control group}] \times 100\%$$

2.2.2. Alleviating the Constipating Effect of FWBP in Zebrafish Larvae

Further study was carried out by establishing the constipation model of zebrafish. The constipation zebrafish model was established with a concentration of 10 μg/mL loperamide hydrochloride (LH). Then, we followed the steps outlined in Section 2.2.1. According to the results of the intestinal fluorescence intensity of zebrafish, the alleviating effects of DOM and FWBP were calculated as:

$$\text{alleviating effect} = [(\text{DOM}/\text{FWBP group} - \text{LH group})/\text{LH group}] \times 100\%$$

2.3. Experiment 2: Effects of FWBP on the Immune Activity, Intestinal Morphology and Gut MicroBiota of Zebrafish

2.3.1. Animals and Experimental Diets

A total of 180 one-month old zebrafish (66.0 ± 0.7 mg) were randomly selected from 18 aquariums (3 L), at the rate of 10 fish per aquarium, and were adapted to a recirculating system for 14 days. After the nursery period, the zebrafish were weighed and divided into three treatments: (1) the basal diet (Control), (2) the basal diet supplemented with 0.05% FWBP (0.05% FWBP), and (3) the basal diet supplemented with 0.1% FWBP (0.1% FWBP). Six replicate tanks were randomly assigned per treatment group. The composition of the commercial diet was 38.9% crude protein, 15.1% crude fat, 93.6% dry matter, and 11% ash. Zebrafish were fed to apparent satiation for 8 weeks. In this study, the zebrafish were fed four times a day until apparent satiety. During the experimental period, water was exchanged automatically and the basic physicochemical parameters of the water, including the temperature, pH, and amount of dissolved oxygen, were maintained at 28 ± 1 °C, 7.2 ± 0.52, and 7.28 ± 0.39 mg/L, respectively.

2.3.2. Histological of the Intestines

The intestinal samples were prepared for histological analyses, according to routine laboratory procedures. At the end of the feeding trial, segments of the middle intestine were surgically removed; three fish per tank (n = 18/group) were selected and preserved in freshly prepared 4% paraformaldehyde solution. Following fixation, the fixed intestine tissues were dehydrated in gradient ethanol, hyalinized in xylene, and embedded in wax. Embedded midguts were sectioned at 4–5 μm and stained with hematoxylin and eosin (HE), using a standard protocol. Then, we observed the sections using a microscope [17].

2.3.3. RT-PCR Analysis

The total RNA of intestines from zebrafish was extracted using TRIzol Reagent (Invitrogen, Carlsbad, CA, USA), according to the manufacturer's instructions. To construct cDNA, the Super-Script III First-Strand Synthesis System (Invitrogen, Carlsbad, CA, USA) was used. The primers for gene expression detection are shown in Table 1. The real-time PCR method was used to determine the relative expression of genes, as described in our previous work [12]. β-actin was selected as the reference gene, and was used to normalize the gene expression levels. For each gene, the mRNA expression levels of the target genes were calculated using the $2^{-\Delta\Delta Ct}$ method, and data for each target transcript were normalized to the control zebrafish (1.0) [18].

Table 1. The primers for gene expression used in this study.

Gene	Primer Sequences (5′→3′)	Product Size (bp)	Accession Number
IL-10	F:ACGCTTCTTCTTTGCGACTG R:TTGGGGTTGTGGAGTGCTT	295	NM_001020785.2
IL-17	F:CCCATCCATCCAATCAACAA R:ACCTCAACGCCGTCTATCAG	134	NM_001020787.1
TJPα	F:CCATTGAGACAGGAGTAAGCATT R:ACATCACCAGAGGACTCAACAGA	153	XM_009303250.3
Occludin1	F:GATGTGGAGGACTGGGTCAATA R:GCCGCTGCTAATAGGGACTG	183	NM_212832.2
Mucin2.1	F:GCCGCTGCTAATAGGGACTG R:CGACAGTTTTCGATTTACGTG	210	XM_021470771.1
Mucin5.1	F:AATAATCTTGCCTGCCCAGAGT R:CGACATTGATTTCAGTGATGTTCA	190	XM_021470622.1
MMP9	F:GCCTGCCAAATCAAGGAGTT R:CGTTCACCATTGCCTGAGAT	101	NM_213123.1
defensin1	F:GCATCCTTTCCCTGGAGTT R:AGCCTAATGGTCCGAAGTAAA	91	NM_001081553.1

IL-10: Interleukin 10; IL-17: Interleukin 17; TJPα: tight junction protein α; MMP9: Matrix metalloproteinases 9.

2.3.4. 16S rRNA Gene Sequencing Analyses of Zebrafish Gut Microbiota

Zebrafish were fed FWBP diets for eight weeks, then, four hours after the last feeding session, the digesta samples were collected from three biological replicates for each group [19], with six pooled samples per treatment as biological replicates. The gut microbiota from the experimental zebrafish was analyzed using bacterial 16S rRNA gene sequencing. Total genomic DNA from the intestinal content was extracted according to the manufacturer's instructions. Then, 16S rRNA genes from the genomic DNA samples were amplified by PCR using specific primers for the V4 region of bacterial 16S rRNA. The PCR amplification was performed with the following protocol: 30 s of initial denaturation at 98 °C; 98 °C for 10 s; 50 °C for 30 s; 72 °C for 30 s; repeated for 30 cycles; a final elongation step at 72 °C for 10 min. All procedures were conducted using the Novo-gene Bioinformatics Technology Co., Ltd. (Beijing, China). The sequencing library was quantified by Qubit and qPCR, and the barcoded V4 PCR amplicons were sequenced using a NovaSeq6000 platform, according to the manufacturer's protocol [20].

2.4. Statistical Analysis

All results for different treatment groups are shown as mean $\pm$ standard error (SE). The data were analyzed by one-way ANOVA in SAS. For treatments that showed significant differences, means were compared using the Tukey's test. The level of significance was set at $p < 0.05$.

3. Results

3.1. Experiment 1: Effect of FWBP on Promoting Intestinal Peristalsis and Alleviating Constipation in Zebrafish

FWBP was shown to promote intestinal peristalsis in the zebrafish, with the results shown in Figure 1. The intestinal fluorescence intensity of zebrafish in the morpholine group, as well as the FWBP groups with different concentrations, is significantly lower than that of the control group ($p < 0.05$). The intestinal motility in zebrafish, promoted by different concentrations of FWBP, increases and then decreases with increasing concentrations of FWBP. The intestinal motility promotion effect of FWBP in zebrafish at 20 μg/mL was 25.98%.

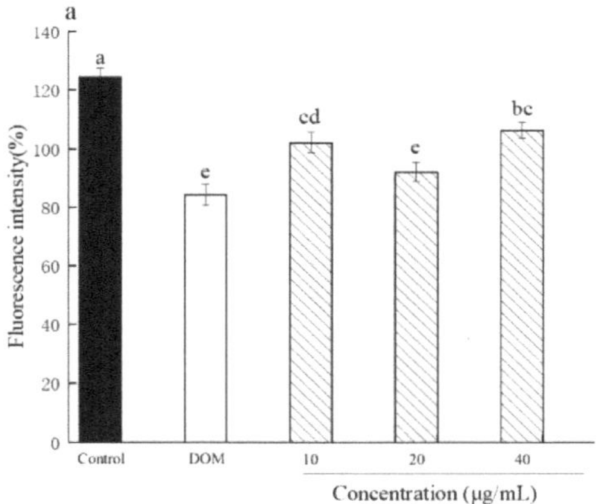
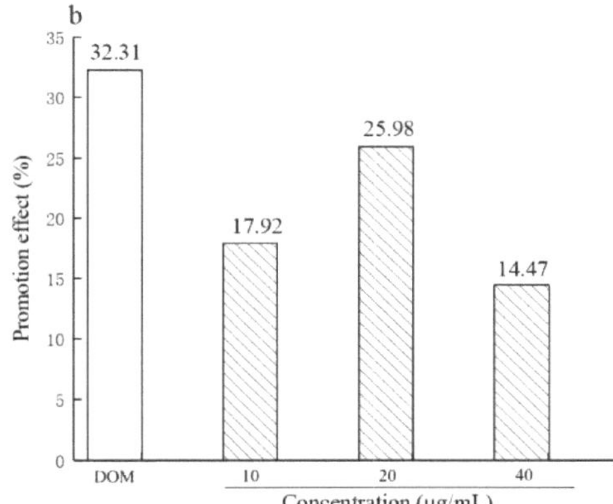

Figure 1. Effects of FWBP on intestinal peristalsis in zebrafish ($p < 0.05$). (**a**) Fluorescence intensity, %; (**b**) Intestinal motility promotion effect, %. DOM: Domperidone. Different lowercase superscript letters denote statistically significant differences among different groups ($p < 0.05$).

The effect of FWBP on the relief of constipation in zebrafish is shown in Figure 2. The fluorescence intensity of the LH group was significantly higher than that of the control group ($p < 0.05$), indicating that the constipation model was successfully established. However, zebrafish groups treated with morpholine, as well as 20 and 40 μg/mL FWBP, showed dramatically decreased fluorescence intensity. This result indicates that FWBP relieves intestinal constipation in zebrafish.

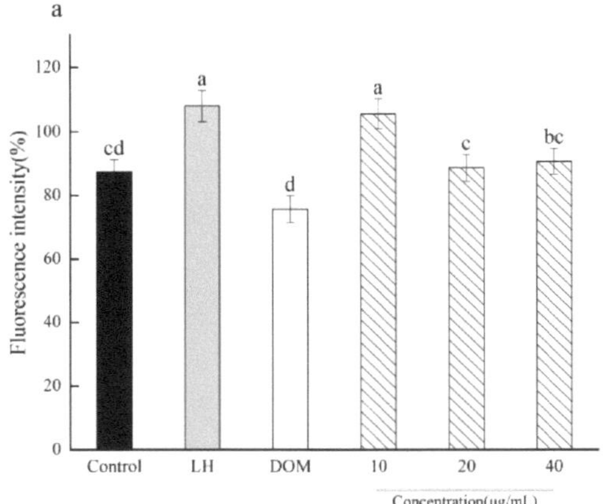
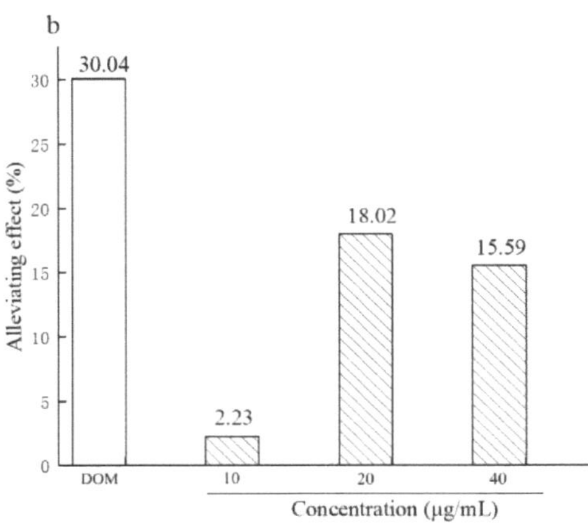

Figure 2. Alleviating effects of FWBP on intestinal motility inhibition in zebrafish ($p < 0.05$). (**a**) Fluorescence intensity, %; (**b**) Alleviating effect, %. LH: loperamide hydrochloride; DOM: Domperidone. Different lowercase superscript letters denote statistically significant differences among different groups ($p < 0.05$).

3.2. Experiment 2 Effects of FWBP on Immune Activity, Intestinal Morphology and Gut Microbiota of Zebrafish

3.2.1. Effect of FWBP on Intestinal Morphology

The effects of FWBP on the intestinal morphology of zebrafish are presented in Figures 3 and 4, and, compared with control group, the FWBP groups showed a significant improvement in the height of the villi ($p < 0.05$).

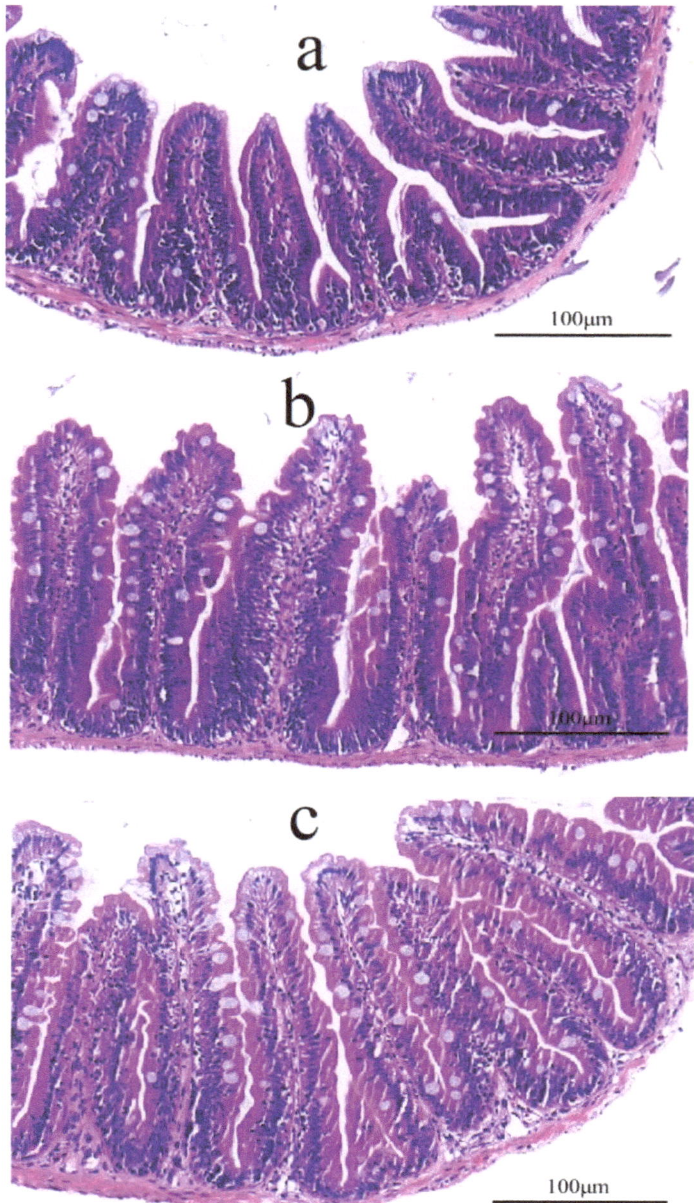

Figure 3. Representative HE-stained intestine sections from zebrafish fed with different diets ($p < 0.05$). (**a**) Control; (**b**) 0.05% FWBP; (**c**) 0.1% FWBP. FWBP: Fermented wheat bran polysaccharides. Scale bars = 100 μm.

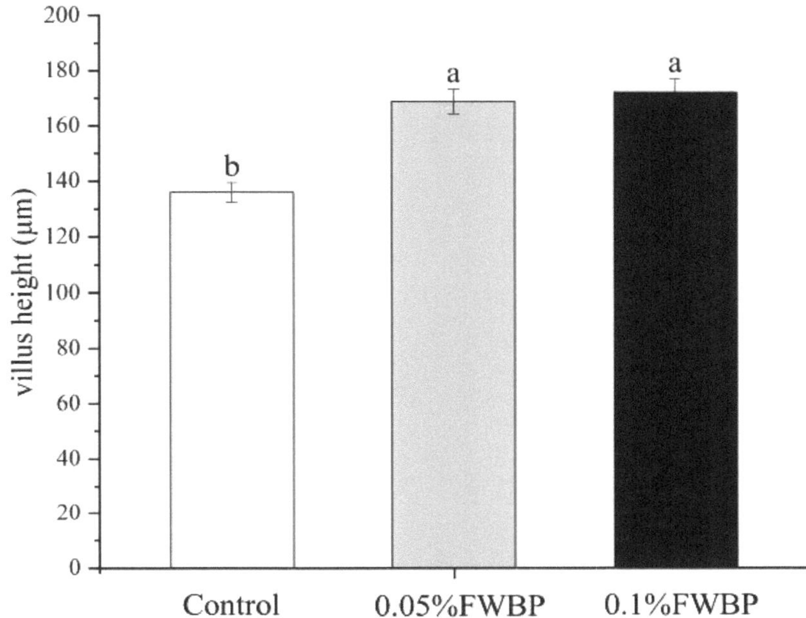

Figure 4. The effects of FWBP on villus height and crypt depth in the intestine. FWBP: Fermented wheat bran polysaccharides. Different lowercase superscript letters denote statistically significant differences among different groups ($p < 0.05$).

3.2.2. Expression of Intestinal Inflammation-Related Genes and Intestinal Mucosal Barrier-Related Genes

As shown in Figure 5, after zebrafish were fed diets with the supplementation of FWBP, the mRNA expression of interleukin 10 (IL-10) in the intestines of zebrafish in the 0.05% FWBP group was significantly increased, comparted to the control ($p < 0.05$). Zebrafish in the 0.1% FWBP group showed significantly increased IL-17 mRNA expression when compared with the control group ($p < 0.05$). The mRNA expression of intestinal mucosal barrier-related genes was determined, and the results are presented in Figure 6. When compared with the control group, FWBP groups showed up-regulated mRNA expression of tight junction protein α(TJP1α), muc2.1, muc5.1, matrix metalloproteinases 9 (MMP9), and defensin1 ($p < 0.05$). The 0.05% FWBP group showed higher Occludin1 mRNA expression than in the control and 0.1% FWBP groups ($p < 0.05$).

3.2.3. Effect of FWBP on Gut Microbiota
Diversity of Gut Microbiota

The Venn diagrams (Figure 7) show that a total number of 404 operational taxonomic units (OTUs) is shared by the three groups, and the numbers of unique OTUs in the control, 0.05% FWBP, and 0.1% FWBP groups were 214, 298, and 264, respectively. In the analysis of α-diversity (Table 2), the Shannon and Chao1 indexes were significantly enhanced in the 0.05% FWBP supplementation group ($p < 0.05$).

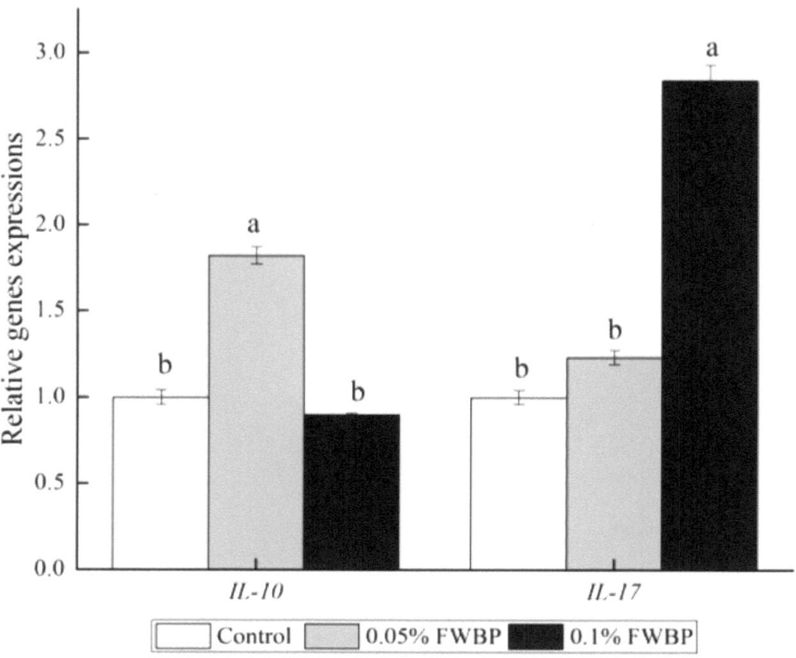

Figure 5. Gut inflammation-related gene expression in zebrafish fed with FWBP. FWBP: Fermented wheat bran polysaccharides. IL-10: Interleukin 10; IL-17: Interleukin 17. Different lowercase superscript letters denote statistically significant differences among different groups ($p < 0.05$).

Composition of Intestinal Microbiota

In regards to the intestinal microbial composition, at the phylum level (Figure 8), Fusobacteriota, Proteobacteria, Bacteroidota, Actinobacteriota, and Firmicutes were the predominant bacterial phyla in the gut of the zebrafish. At the family and genus level, Fusobacteriaceae and Aeromonadaceae, as well as *Cetobacterium* and *Aeromonas*, were the dominant families and genera, respectively (Figures 9 and 10). Taxonomic profiling showed a diverse gut microbiota community at the phylum, family and genus level. The significantly decreased abundance of Proteobacteria and the significantly increased abundance of Firmicutes at the phylum level were observed in 0.05% FWBP and 0.1% FWBP groups, in comparison with control group ($p < 0.05$). In addition, the 0.05% FWBP group showed a significantly increased relative abundance of Bacteroidota and Actinobacteriota ($p < 0.05$). At the family and genus level (Figures 9 and 10), by comparison with control group, a significantly decreased abundance of the family Aeromonadaceae and genus *Aeromonas* was observed in the 0.05% FWBP and 0.1% FWBP groups ($p < 0.05$).

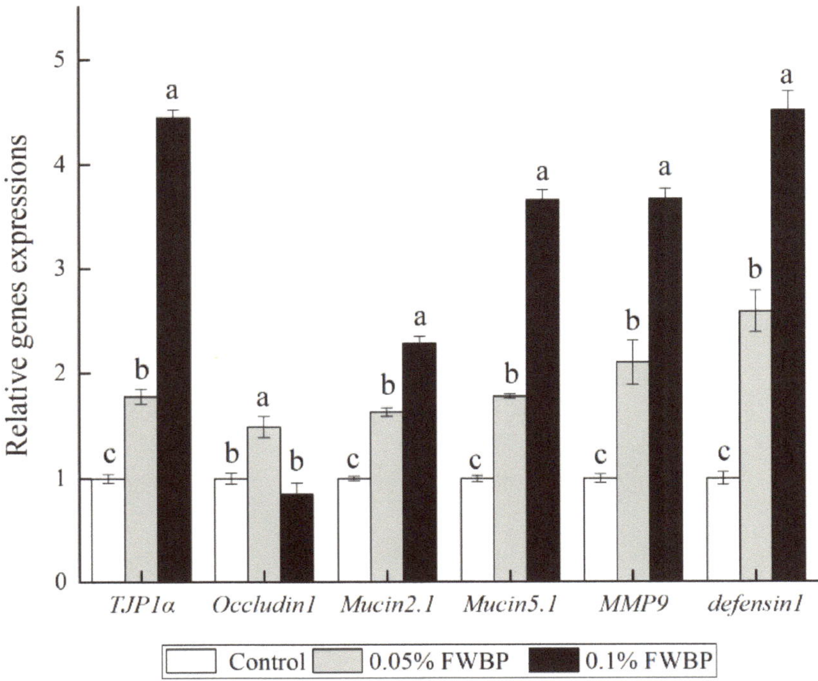

Figure 6. Gut mucosal barrier-related gene expression in zebrafish fed FWBP. FWBP: Fermented wheat bran polysaccharides. TJPα: Tight junction protein α; MMP9: Matrix metalloproteinases 9. Different lowercase superscript letters denote statistically significant differences among different groups ($p < 0.05$).

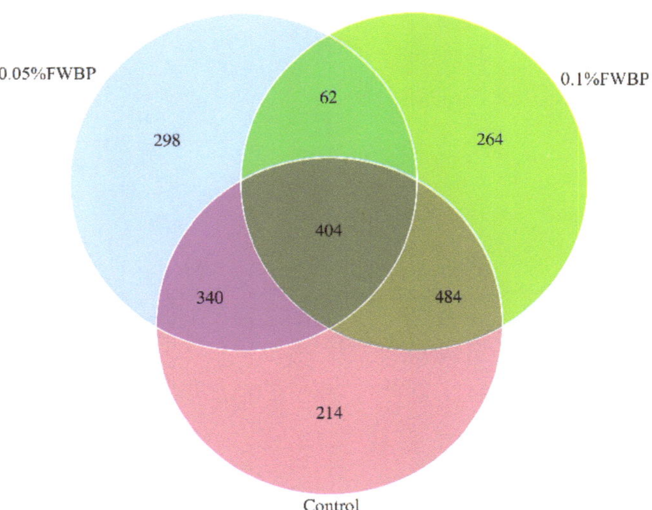

Figure 7. Venn diagram of OTUs among the dietary groups. FWBP: Fermented wheat bran polysaccharides.

Table 2. The effects of FWBP on α-diversity in the gut.

Groups	Observed-Species	Goods-Coverage	PD-Whole-Tree	Shannon	Simpson	Chao1
control	341.67 ± 98.96	0.999 ± 0.0001	24.99 ± 4.02	1.76 ± 0.09 [b]	0.61 ± 0.03	274.12 ± 74.44 [b]
0.05% FWBP	422.67 ± 78.65	0.999 ± 0.0001	41.93 ± 5.65	3.53 ± 0.51 [a]	0.71 ± 0.09	565.43 ± 32.47 [a]
0.1% FWBP	342.66 ± 88.91	0.999 ± 0.0002	31.51 ± 6.29	1.63 ± 0.31 [b]	0.57 ± 0.13	270.33 ± 46.10 [b]

FWBP: Fermented wheat bran polysaccharides. Data are expressed as the mean ± SE. Different lowercase superscript letters denote statistically significant differences among different groups ($p < 0.05$).

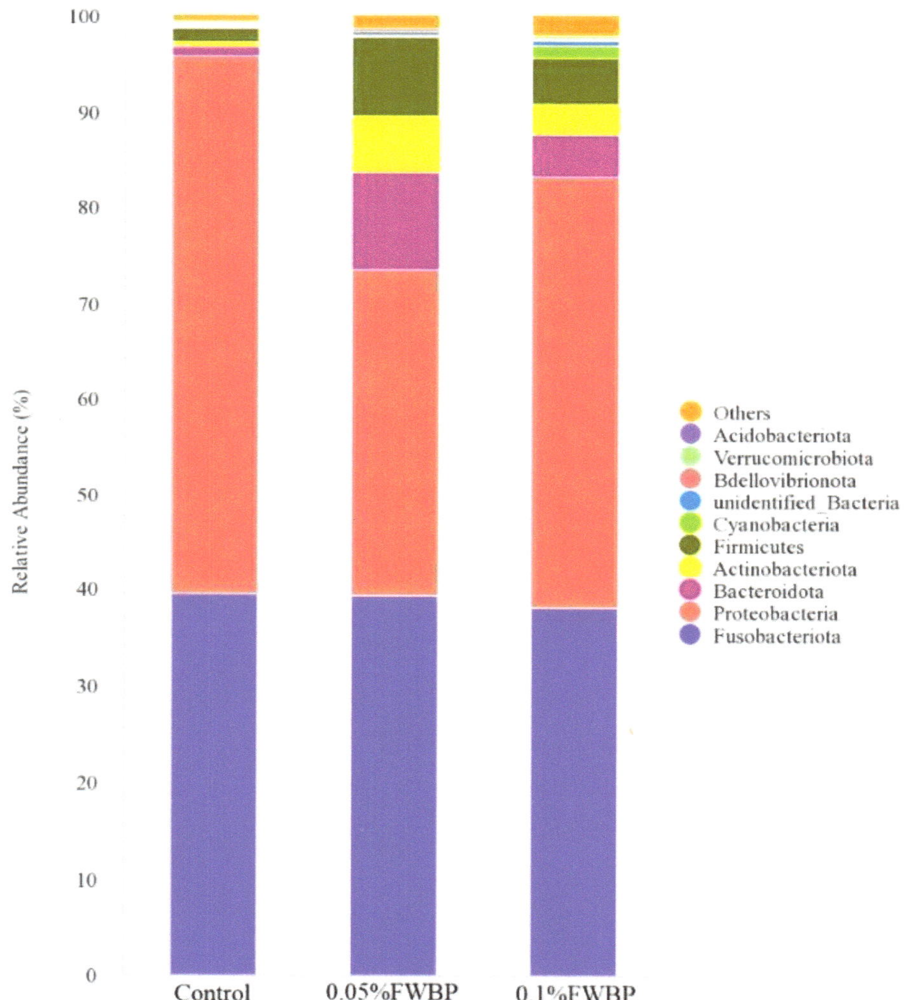

Figure 8. *Cont.*

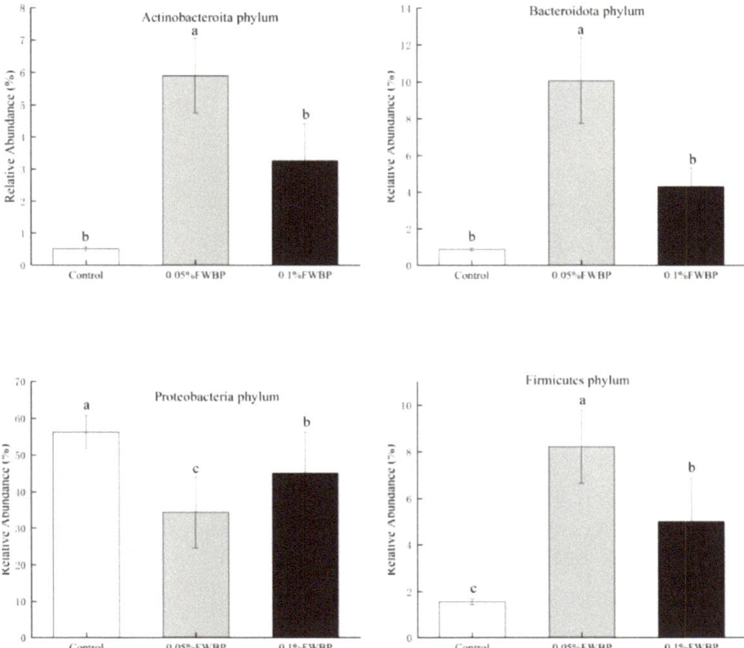

Figure 8. Microbial community in zebrafish fed with FWBP, at the phylum level. FWBP: Fermented wheat bran polysaccharides. Different lowercase superscript letters denote statistically significant differences among different groups ($p < 0.05$).

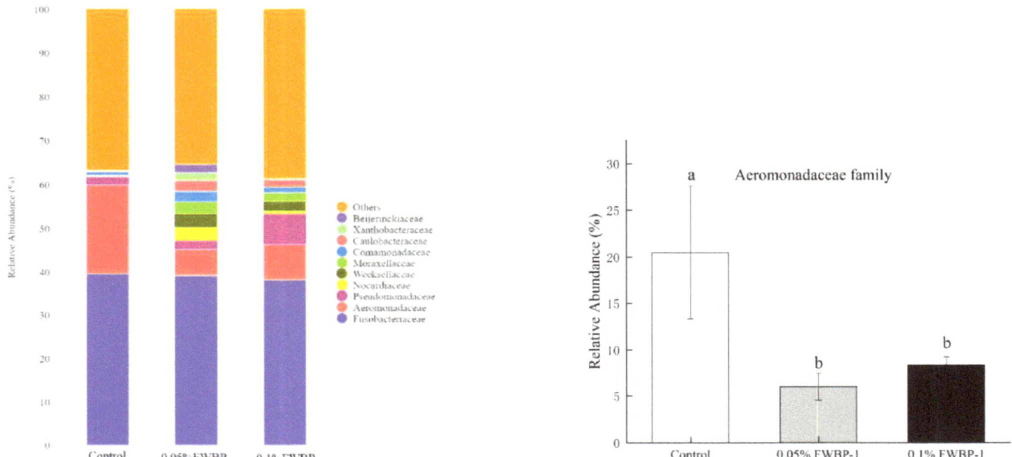

Figure 9. Microbial community in zebrafish fed with FWBP at the family level. FWBP: Fermented wheat bran polysaccharides. Different lowercase superscript letters denote statistically significant differences among different groups ($p < 0.05$).

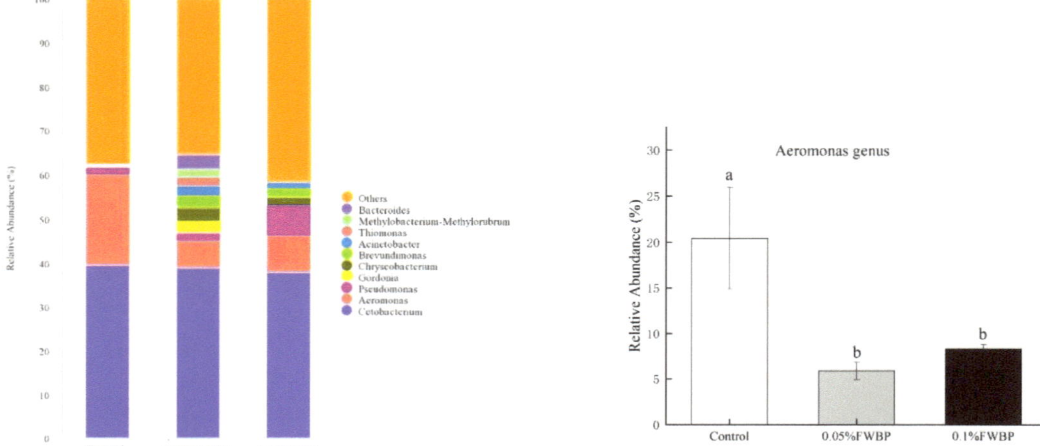

Figure 10. Microbial community in zebrafish fed with FWBP at the genus level. FWBP: Fermented wheat bran polysaccharides. Different lowercase superscript letters denote statistically significant differences among different groups ($p < 0.05$).

4. Discussion

Wheat bran, as a rich source of dietary fiber, contains many non-starch polysaccharides, including cellulose, and other non-cellulose polysaccharides, such as arabinoxylan, β- glucan, glucomannan, araban, and so on [21]. However, the gut tract of zebrafish cannot directly digest the non-starch polysaccharide. Therefore, the current study used microbial fermentation technology to break down wheat bran by microorganisms [22]. Studies have shown that wheat bran could be altered by microbial fermentation, and active substances, such as soluble polysaccharides, were significantly increased [23,24]. Based on our previous study, the results found that using *S. cerevisiae* and *B. subtilis* could change the structural characteristics of wheat bran and increase the yield of FWBP when compared to using a single strain. Because *S. cerevisiae* can secrete many enzymes and assimilate five-carbon sugars, *B. subtilis* can produce α-amylase, cellulase, β-glucanase, phytase, pectinase, xylanase, and so on [25]. In addition, our previous research found that the antioxidant activity of WBP can be improved by fermentation with *S. cerevisiae* and *B. subtilis*, both in vitro and in zebrafish models [11]. Previously, our study results showed that the FWBP group had better growth performance, higher antioxidant-associated gene expression, and a more positive effect on gut microbiota than the WBP group in zebrafish [13].

Intestinal motility disorders are an important cause of morbidity. The primary clinical manifestation of this effect is constipation, with the potential to produce intestinal obstruction, intestinal infarction, and a paralytic ileus, leading to death in sporadic cases [26]. The zebrafish is a potentially valuable model for gastrointestinal studies, due to its transparency, low cost, and ease of screening [27]. Studies have shown that loperamide hydrochloride can induce an increase in intestinal fluorescence intensity, which indicates that loperamide hydrochloride can be used to establish the zebrafish model of constipation [27]. Researchers have found that wheat bran polysaccharide induces cytokine expression via the toll-like receptor 4-mediated p38 MAPK signaling pathway, and prevents cyclophosphamide-induced immunosuppression in mice [7]. To determine whether FWBP can be used as a natural product for the treatment of intestinal motility disorders, we investigated FWBP's promotion of intestinal peristalsis in normal zebrafish and in zebrafish with loperamide hydrochloride-induced constipation. In the present study, the results showed that FWBP had a strong promoting effect on the intestinal motility of normal zebrafish. Furthermore, the fluorescence intensity was significantly altered in the zebrafish treated with loperamide hydrochloride when compared with the normal group. After treatment with FWBP, the fluorescence intensity was dramatically decreased, indicating the amelioration of loperamide hydrochloride-induced constipation in zebrafish. This initial data suggested that the FWBP effectively promoted intestinal peristalsis.

Intestines with a large surface area acting as a barrier are acknowledged as the first line of resistance against external harmful substances' penetration into the intestine, associated with structural integrity, immunological status and microbiota homeostasis [28,29]. The villus height is a common indicator used to evaluate the function of the gut; an increase in villus height reflects an improvement in the digestion and absorption ability of the gut, and thereby positively affects the utilization of nutrients [2,30]. Additionally, complete intestinal barrier function is closely related to integrated intestinal morphology [31]. Polysaccharides have been shown to promote intestinal health in fish. However, there are very few studies regarding the effects of polysaccharides derived from wheat bran on intestinal function in zebrafish. In this work, the effect of FWBP on the histological structure of zebrafish gut was studied. Although the crypt depth did not show any alteration between experimental groups, villus height significantly increased in the FWBP-treated versus the control group. In addition, 0.1% FWBP significantly increased the villus height to crypt depth ratio. Our results were in partial accordance with Zahran et al., who observed that Nile tilapia fed with 1500 mg/kg *Astragalus* polysaccharides had an increased villus height in the anterior intestine [32].

Intestinal barrier function is considered to be the most important line of defense against external stimuli, which is composed of an immune barrier, mechanical barrier, chemical barrier, and biological barrier [33]. Inflammatory factors are a series of proteins secreted by endothelial cells, lymphocytes, monocytes, and fibroblasts that play an important role in regulating inflammatory processes, and are important facets of the intestinal immune barrier [34]. Our results displayed that the 0.05% FWBP group had anti-inflammatory effects on the intestine of zebrafish, via significantly increasing the expression of IL-10. It can be concluded that supplementation of appropriate amounts of FWBP in diets could improve the zebrafish intestinal immune system. In present study, the expression of the pro-inflammatory cytokine IL-17 was induced by the addition of 0.1% FWBP. However, the influence of FWBP on the intestinal immune system in zebrafish is poorly investigated, thus, it is difficult to make any direct comparison. Therefore, this aspect requires further investigation. In addition, our findings proposed that the intestinal mucosal barrier-related genes (TJP1α, Occludin1, muc2.1, muc5.1, MMP9, and defensin1) exhibited higher gene expression in the FWBP-supplemented group than in the control groups, indicating that FWBP improved the zebrafish intestines' mechanical and chemical barrier function, by increasing the expression of related genes. This result may be due to the beneficial effects of FWBP on intestinal morphology. Similar results were also observed for other sources of natural polysaccharides in zebrafish diets. For instance, Li et al. reported that dietary supplementation with *Astragalus* polysaccharide up-regulated TJP1b, Occludin1 and IL-10 gene expression in the intestines of zebrafish [35].

The intestinal biological barrier is a mutually dependent and interrelated microecosystem, which is represented by the intestinal microbiota [33]. The intestinal microbiota plays important roles in immunity, the maintenance of homeostasis, and in digestion and nutrient absorption of the host [36]. A higher Shannon index value indicates a rich gut biodiversity. Our results demonstrate that supplementation with 0.05% FWBP can result in an enriched microbiota diversity in the zebrafish intestine. The possible explanation of the differences in results received from 0.05% FWBP and 0.1% FWBP could be that excess FWBP results in bacterial inhibition [37]. At the phylum level, the abundance of Proteobacteria was decreased, and the abundance of Firmicutes was increased in all FWBP groups. Proteobacteria is generally associated with dysbiosis, or an unstable gut microbial community; an increased abundance of Proteobacteria may bring potential risks to fish [38]. The Firmicutes are beneficial dominant bacteria in animals' guts, and many probiotics, such as *Lactobacillus*, *Enterococcus* and *Bacillus*, belong to the Firmicutes phylum [30]. In addition, dietary addition of 0.05% FWBP remarkably enhanced the abundance of Bacteroidetes and Actinobacteria. Studies have shown that *Bacteroides* have the ability to produce SCFAs by fermenting dietary fiber and other undigested food remnants, in order to regulate host immune homeostasis [39]. As probiotics, bacteria in the phylum Actinobacteria can produce abundant secondary metabolites, which can inhibit the growth of pathogenic intestinal bacteria, and enhance the host defense ability [2]. Our present study found that dietary 0.05% FWBP had a stronger effect on probiotic levels. This was consistent with our previous in vitro study, which found that, with the increase of FWBP supplementation, FWBP had the effect of first increasing, then decreasing, probiotic levels [37]. This result may be that the high concentration of FWBP has an inhibitory effect on bacterial growth [30]. Previous studies of the effects of polysaccharides on aquaculture were consistent with our results. Su et al. confirmed that dietary Yu-Ping-Feng polysaccharide supplementation decreased the intestinal Proteobacteria and Chlamydiae abundance, and increased the abundance of Bacteroidetes in *Litopenaeus vannamei* [40]. At the family and genus level, the abundances of the family Aeromonadaceae and genus *Aeromonas* were greatly decreased in the all FWBP groups. *Aeromonas*, belonging to Aeromonadaceae, is mostly considered a major opportunistic pathogen in fish, causing intestinal inflammation. The results indicated that FWBP positively changed the gut microbiota through decreasing the abundance of opportunistic pathogens, and increasing the abundance of beneficial bacterium.

5. Conclusions

In conclusion, FWBP can promote intestinal peristalsis in zebrafish larvae. FWBP can improve intestinal morphology, mitigate intestinal inflammation, improve the mechanical and chemical barrier, and positively modulate the intestinal microbiota of zebrafish, thus enhancing intestinal barrier function. These results indicate that FWBP could be developed as a functional food ingredient candidate, in order to promote intestinal health.

Author Contributions: Conceptualization, X.A.; methodology, Q.C.; software, N.L.; validation, Y.W.; formala anlysis, Q.C.; investigation, Y.Y.; resources, R.W.; data curation, N.Y.; writing—original draft preparation, Q.C.; writing—review and editing, J.M. and Y.W.; supervision, Y.Z.; project administration, J.Q.; funding acquisition, J.Q., X.A. and Y.W. All authors have read and agreed to the published version of the manuscript.

Funding: This research was funded by Major Science and Technology Program of Inner Mongolia Autonomous Region, grant number 2021ZD0023-3, Major Science and Technology Program of Inner Mongolia Autonomous Region, grant number 2021SZD0024-4, Major Science and Technology Program of Inner Mongolia Autonomous Region, grant number 2020ZD0004, Key Technology Project of Inner Mongolia Autonomous Region, grant number 2020GG0030.

Institutional Review Board Statement: Not applicable.

Informed Consent Statement: Not applicable.

Data Availability Statement: Not applicable.

Conflicts of Interest: The authors declare no conflict of interest.

References

1. Hao, Y.; Wang, X.; Yuan, S.; Wang, Y.; Liao, X.; Zhong, M.; He, Q.; Shen, H.; Liao, W.; Shen, J. Flammulina velutipes polysaccharide improves C57BL/6 mice gut health through regulation of intestine microbial metabolic activity. *Int. J. Biol. Macromol.* **2020**, *167*, 1308–1318. [CrossRef]
2. Shi, H.T.; Zhao, S.Z.; Wang, K.L.; Fan, M.X.; Han, Y.Q.; Wang, H.L. Effects of dietary Astragalus Membranaceus supplementation on growth performance, and intestinal morphology, microbiota and metabolism in common carp (*Cyprinus carpio*). *Aquac. Rep.* **2022**, *22*, 100955. [CrossRef]
3. Zhou, J.; Guo, S.-Y.; Zhang, Y.; Li, C.-Q. Human prokinetic drugs promote gastrointestinal motility in zebrafish. *Neurogastroenterol. Motil.* **2014**, *26*, 589–595. [CrossRef] [PubMed]
4. Wen, J.J.; Li, M.Z.; Gao, H.; Hu, J.L.; Nie, Q.X.; Chen, H.H.; Zhang, Y.L.; Xie, M.Y.; Nie, S.P. Polysaccharides from fermented Momordica charantia L. with Lactobacillus plantarum NCU116 ameliorate metabolic disorders and gut microbiota change in obese rats. *Food Funct.* **2021**, *12*, 2617–2630. [CrossRef]

5. Do, M.H.; Seo, Y.S.; Park, H.-Y. Polysaccharides: Bowel health and gut microbiota. *Crit. Rev. Food Sci. Nutr.* **2020**, *61*, 1212–1224. [CrossRef]

6. Jiang, H.; Dong, J.; Jiang, S.; Liang, Q.; Zhang, Y.; Liu, Z.; Ma, C.; Wang, J.; Kang, W. Effect of Durio zibethinus rind polysaccharide on functional constipation and intestinal microbiota in rats. *Food Res. Int.* **2020**, *136*, 109316. [CrossRef] [PubMed]

7. Shen, T.; Wang, G.; You, L.; Zhang, L.; Ren, H.; Hu, W.; Qiang, Q.; Wang, X.; Ji, L.; Gu, Z.; et al. Polysaccharide from wheat bran induces cytokine expression via the toll-like receptor 4-mediated p38 MAPK signaling pathway and prevents cyclophosphamide-induced immunosuppression in mice. *Food Nutr. Res.* **2017**, *61*, 1344523. [CrossRef] [PubMed]

8. Mendis, M.; Simsek, S. Arabinoxylans and human health. *Food Hydrocoll.* **2014**, *42*, 239–243. [CrossRef]

9. Hromádková, Z.; Paulsen, B.S.; Polovka, M.; Košťálová, Z.; Ebringerová, A. Structural features of two heteroxylan polysaccharide fractions from wheat bran with anti-complementary and antioxidant activities. *Carbohydr. Polym.* **2013**, *93*, 22–30. [CrossRef]

10. Lv, Q.Q.; Cao, J.J.; Liu, R.; Chen, H.Q. Structural characterization, α-amylase and α-glucosidase inhibitory activities of polysaccharides from wheat bran. *Food Chem.* **2021**, *341 Pt 1*, 128218. [CrossRef]

11. Chen, Q.; Wang, R.; Wang, Y.; An, X.; Liu, N.; Song, M.; Yang, Y.; Yin, N.; Qi, J. Characterization and antioxidant activity of wheat bran polysaccharides modified by Saccharomyces cerevisiae and Bacillus subtilis fermentation. *J. Cereal Sci.* **2020**, *97*, 103157. [CrossRef]

12. Shi, J.X. Research on the Microbial Fermentation of Weat Bran Polysaccharide and the Antioxidant Activity of Crude Polysaccharide. Master's Thesis, Inner Mongolia Agricultural University, Hohhot, China, 2017.

13. Chen, Q.; Wang, Y.; Yin, N.; Wang, R.; Zheng, Y.; Yang, Y.; An, X.; Qi, J. Polysaccharides from fermented wheat bran enhanced the growth performance of zebrafish (Danio rerio) through improving gut microflora and antioxidant status. *Aquac. Rep.* **2022**, *25*, 101188. [CrossRef]

14. Caro, M.; Iturria, I.; Martinez-Santos, M.; Pardo, M.A.; Rainieri, S.; Tueros, I.; Navarro, V. Zebrafish dives into food research: Effectiveness assessment of bioactive compounds. *Food Funct.* **2016**, *7*, 2615–2623. [CrossRef] [PubMed]

15. Zhang, C.N.; Zhang, J.L.; Ren, H.T.; Zhou, B.H.; Wu, Q.J.; Sun, P. Effect of tributyltin on antioxidant ability and immune responses of zebrafish (Danio rerio). *Ecotoxicol. Environ. Saf.* **2017**, *138*, 1–8. [CrossRef]

16. Hoseinifar, S.H.; Yousefi, S.; Capillo, G.; Paknejad, H.; Khalili, M.; Tabarraei, A.; Van Doan, H.; Spanò, N.; Faggio, C. Mucosal immune parameters, immune and antioxidant defence related genes expression and growth performance of zebrafish (Danio rerio) fed on Gracilaria gracilis powder. *Fish Shellfish. Immunol.* **2018**, *83*, 232–237. [CrossRef]

17. Jin, Y.; Xia, J.; Pan, Z.; Yang, J.; Wang, W.; Fu, Z. Polystyrene microplastics induce microbiota dysbiosis and inflammation in the gut of adult zebrafish. *Environ. Pollut.* **2018**, *235*, 322–329. [CrossRef]

18. Livak, K.J.; Schmittgen, T.D. Analysis of relative gene expression data using real-time quantitative PCR and the $2^{-\Delta\Delta CT}$ method. *Methods* **2001**, *25*, 402–408. [CrossRef] [PubMed]

19. Ikeda-Ohtsubo, W.; Nadal, A.L.; Zaccaria, E.; Iha, M.; Kitazawa, H.; Kleerebezem, M.; Brugman, S. Intestinal Microbiota and Immune Modulation in Zebrafish by Fucoidan From Okinawa Mozuku (*Cladosiphon okamuranus*). *Front. Nutr.* **2020**, *7*, 67. [CrossRef]

20. Song, X.J.; Feng, Z.F.; Tan, J.B.; Wang, Z.Y.; Zhu, W. Dietary administration of Pleurotus ostreatus polysaccharides (POPS) modulates the non-specific immune response and gut microbiota diversity of *Apostichopus japonicus*. *Aquac. Rep.* **2021**, *19*, 100578. [CrossRef]

21. Maes, C.; Delcour, J.A. Structural characterisation of water-extractable and water-unextractable arabinoxylans in wheat bran. *J. Cereal. Sci.* **2002**, *35*, 315–326. [CrossRef]

22. Hussain, A.; Bose, S.; Wang, J.-H.; Yadav, M.K.; Mahajan, G.B.; Kim, H. Fermentation, a feasible strategy for enhancing bioactivity of herbal medicines. *Food Res. Int.* **2016**, *81*, 1–16. [CrossRef]

23. Li, X.; Wang, Y.; Wang, R.; Chen, Q.; Guo, J.; Feng, X.; Zheng, X.; An, X.; Qi, J. Enzyme-assisted extraction of fermented wheat bran polysaccharides and it biological activity analysis. *Nat. Prod. Res. Dev.* **2019**, *31*, 155–162.

24. Xia, C.; Ai, Q.; Xiang, N.; Zhan, H.; Liu, X.; Fang, B. The Function of Fermented Wheat Bran and Its Application in Animal Feed. *Chin. J. Anim. Sci.* **2017**, *53*, 17–21.

25. Wang Yuan, S.J.; Yuanxiao, D.; Qianru, Z.; Ziqi, M.; Xuan, L.; Xiaoping, A.; Jingwei, Q. Optimization of Fermentation Conditions for Improved Production of Polysaccharides from Wheat Bran and Anti-Inflammatory Effects of the Extracted Polysaccharides. *Food Sci.* **2018**, *39*, 192–198.

26. de Alvarenga, K.A.F.; Sacramento, E.K.; Rosa, D.V.; Souza, B.R.; de Rezende, V.B.; Romano-Silva, M.A. Effects of antipsychotics on intestinal motility in zebrafish larvae. *Neurogastroenterol. Motil.* **2016**, *29*, e13006. [CrossRef]

27. Kim, D.; Koun, S.; Kim, S.Y.; Ha, Y.R.; Choe, J.W.; Jung, S.W.; Hyun, J.J.; Jung, Y.K.; Koo, J.S.; Yim, H.J.; et al. Prokinetic effects of diatrizoate meglumine (Gastrografin) in a zebrafish for opioid-induced constipation model. *Anim. Cells Syst.* **2021**, *25*, 264–271. [CrossRef]

28. Li, W.; Liu, B.; Liu, Z.; Yin, Y.; Xu, G.; Han, M.; Xie, L. Effect of dietary histamine on intestinal morphology, inflammatory status, and gut microbiota in yellow catfish (*Pelteobagrus fulvidraco*). *Fish Shellfish. Immunol.* **2021**, *117*, 95–103. [CrossRef] [PubMed]

29. Zhao, R.; Cheng, N.; Nakata, P.; Zhao, L.; Hu, Q. Consumption of polysaccharides from Auricularia auricular modulates the intestinal microbiota in mice. *Food Res. Int.* **2019**, *123*, 383–392. [CrossRef]

30. Liu, W.C.; Zhou, S.H.; Balasubramanian, B.; Zeng, F.Y.; Sun, C.B.; Pang, H.Y. Dietary seaweed (*Enteromorpha*) polysaccharides improves growth performance involved in regulation of immune responses, intestinal morphology and microbial community in banana shrimp *Fenneropenaeus merguiensis*. *Fish Shellfish. Immunol.* **2020**, *104*, 202–212. [CrossRef]

31. Tan, X.; Sun, Z.; Zhou, C.; Huang, Z.; Tan, L.; Xun, P.; Huang, Q.; Lin, H.; Ye, C.; Wang, A. Effects of dietary dandelion extract on intestinal morphology, antioxidant status, immune function and physical barrier function of juvenile golden pompano Trachinotus ovatus. *Fish Shellfish Immunol.* **2018**, *73*, 197–206. [CrossRef]

32. Zahran, E.; Risha, E.; AbdelHamid, F.; Mahgoub, H.A.; Ibrahim, T. Effects of dietary Astragalus polysaccharides (APS) on growth performance, immunological parameters, digestive enzymes, and intestinal morphology of Nile tilapia (*Oreochromis niloticus*). *Fish Shellfish. Immunol.* **2014**, *38*, 149–157. [CrossRef]

33. Wang, K.; Wu, L.-Y.; Dou, C.-Z.; Guan, X.; Wu, H.-G.; Liu, H.-R. Research Advance in Intestinal Mucosal Barrier and Pathogenesis of Crohn's Disease. *Gastroenterol. Res. Pr.* **2016**, *2016*, 9686238. [CrossRef] [PubMed]

34. Li, M.Y.; Zhu, X.M.; Niu, X.T.; Chen, X.M.; Tian, J.X.; Kong, Y.D.; Zhang, D.M.; Zhao, L.; Wang, G.Q. Effects of dietary Allium mongolicum Regel polysaccharide on growth, lipopolysaccharide-induced antioxidant responses and immune responses in *Channa argus*. *Mol. Biol. Rep.* **2019**, *46*, 2221–2230. [CrossRef] [PubMed]

35. Li, Y.; Ran, C.; Wei, K.; Xie, Y.; Xie, M.; Zhou, W.; Yang, Y.; Zhang, Z.; Lv, H.; Ma, X.; et al. The effect of Astragalus polysaccharide on growth, gut and liver health, and anti-viral immunity of zebrafish. *Aquaculture* **2021**, *540*, 736677. [CrossRef]

36. Chi, X.; Liu, Z.; Wang, H.; Wang, Y.; Wei, W.; Xu, B. Royal jelly enhanced the antioxidant activities and modulated the gut microbiota in healthy mice. *J. Food Biochem.* **2021**, *45*, e13701. [CrossRef] [PubMed]

37. Li, X. Extraction, Separation Purification and Biological Activity Analysis of Fermented Wheat Bran Polysaccharide. Master's Thesis, Inner Mongolia Agricultural University, Hohhot, China, 2019.

38. Li, Z.; Tran, N.T.; Ji, P.; Sun, Z.; Wen, X.; Li, S. Effects of prebiotic mixtures on growth performance, intestinal microbiota and immune response in juvenile chu's croaker, *Nibea coibor*. *Fish Shellfish. Immunol.* **2019**, *89*, 564–573. [CrossRef]

39. Su, C.; Liu, X.; Lu, Y.; Pan, L.; Zhang, M. Effect of dietary Xiao-Chaihu-Decoction on growth performance, immune response, detoxification and intestinal microbiota of pacific white shrimp (*Litopenaeus vannamei*). *Fish Shellfish. Immunol.* **2021**, *114*, 320–329. [CrossRef]

40. Su, C.; Fan, D.; Pan, L.; Lu, Y.; Wang, Y.; Zhang, M. Effects of Yu-Ping-Feng polysaccharides (YPS) on the immune response, intestinal microbiota, disease resistance and growth performance of *Litopenaeus vannamei*. *Fish Shellfish. Immunol.* **2020**, *105*, 104–116. [CrossRef]

fermentation

MDPI

Article

Novel Fermented Ice Cream Formulations with Improved Antiradical and Anti-Inflammatory Features

Andrea Polo [1,*], Ali Zein Alabiden Tlais [1], Pasquale Filannino [2], Alessio Da Ros [1], Kashika Arora [1], Vincenzo Cantatore [2], Olimpia Vincentini [3], Anja Nicolodi [4], Renzo Nicolodi [4], Marco Gobbetti [1] and Raffaella Di Cagno [1]

[1] Faculty of Sciences and Technology, Libera Università di Bolzano, 39100 Bolzano, Italy
[2] Department of Soil, Plant and Food Science, University of Bari Aldo Moro, 70126 Bari, Italy
[3] U.O Alimentazione, Nutrizione e Salute, Dipartimento Sicurezza Alimentare, Nutrizione e Sanità Pubblica Veterinaria, Istituto Superiore di Sanità, 00161 Roma, Italy
[4] Zuegg Com srl/GmbH, 39011 Lana, Italy
[*] Correspondence: andrea.polo@unibz.it

Abstract: Autochthonous strains belonging to *Lacticaseibacillus paracasei* and *Lacticaseibacillus casei* were screened based on pro-technological (kinetics of growth and acidification, exopolysaccharides biosynthesis), functional (cell viability during processing and storage, in vitro protein digestibility, and in vitro gastrointestinal batch digestion), and sensory properties to ferment milk intended to be included in the ice-cream formulation. The main discrimination among the strains resulted from the sensory evaluation by the panelists, who assigned the highest scores to the ice cream produced with milk fermented with *L. casei* F14. Antiradical and anti-inflammatory features of such fermented ice cream (with and without the addition of hydroxytyrosol) were shown in Caco-2 cell culture. The pretreatment of Caco-2 cells with fermented ice creams counteracted the accumulation of pro-inflammatory interleukin mediators IL-8 and IL-6 and reactive oxygen species (ROS) induced by detrimental stimuli, and preserved the Caco-2 cell monolayer integrity. The fermentation with *L. casei* F14 per se played a key role, whereas the inclusion of hydroxytyrosol only contributed to further enhancing the antiradical activity of ice cream. No protective effect on Caco-2 cells was brought about by the addition of dietary fiber.

Keywords: cow milk; fermentation; lactic acid bacteria; functional food

Citation: Polo, A.; Tlais, A.Z.A.; Filannino, P.; Da Ros, A.; Arora, K.; Cantatore, V.; Vincentini, O.; Nicolodi, A.; Nicolodi, R.; Gobbetti, M.; et al. Novel Fermented Ice Cream Formulations with Improved Antiradical and Anti-Inflammatory Features. *Fermentation* **2023**, *9*, 117. https://doi.org/10.3390/fermentation9020117

Academic Editor: Maria Aponte

Received: 30 December 2022
Revised: 20 January 2023
Accepted: 23 January 2023
Published: 26 January 2023

1. Introduction

During the last decades, the development of new functional foods with health-promoting features attracted the attention of the scientific community more and more [1]. The formulation of many functional products is based on the exploitation of microbial fermentations, in particular by lactic acid bacteria (LAB) [2]. In this case, the added value of fermented functional foods can be both direct (probiotic effect) and indirect due to the bioconversions realized by LAB and the ingestion of microbial metabolites synthesized during fermentation (biogenic effect). Overall, the positive effects due to LAB fermentations include the production and accumulation of bioactive compounds with health benefits (e.g., peptides, γ-amino butyric acid, and conjugated linoleic acid), the growth of functional microorganisms making the fermented food a carrier of them through the ingestion. The interaction of ingested functional microorganisms with human epithelial cells from the intestine can modulate several physiological functions [3]. More recently, LAB and fermentations were also exploited as a new and sustainable strategy for decreasing the phenomenon of food intolerance and allergy [4]. Finally, the binomial LAB and fermentation often also represent an effective biotechnology to modulate and enhance flavor and organoleptic and texture properties of foods [5].

Currently, fermented milk and derivatives are among the most diffuse functional products in the market, and the consumers' demand for alternative products with high acceptance and functionality is increasing more and more. Among them, milk ice cream can be used as a probiotic vehicle because of its composition and it has a potential for the development of new functional formulations. Ice cream is a complex multiphase system consisting of dispersed air cells, partially coalesced fat globules, ice crystals, and a continuous aqueous phase in which dissolved (lactose and mineral salts) and suspended (polysaccharides and proteins) substances are dispersed [6]. The first attempts to produce ice cream containing functional microorganisms were made in the 1990s, aiming to use frozen dairy desserts as carriers for probiotics [7]. In these first studies, the enrichment of LAB into the food matrix was obtained by blending ice cream mix with probiotics immediately before freezing, without a fermentation step [8]. This strategy was supposed to guarantee a better survival of the probiotics due to a higher final pH [9]. Based on this idea, several studies investigated the survival of microbial cells during the storage of different kinds of ice creams (e.g., produced with cow, goat, or sheep milk) at $-20\,^\circ$C and searched for the most effective incorporation methods of cells [10–13]. More recently, novel ice cream formulations have been developed with improved antioxidant and functional properties, nutritional profiles, and innovative flavors by using different technological methods [14,15]. However, with this background, the fermentation of cow milk as biotechnology to functionalize ice creams received lower attention.

It is known that lactic fermentation can represent an alternative and natural tool to develop new ice cream formulations with improved nutritional, textural, physical, and rheological properties [16]. Through fermentation, cow milk can be enriched by bioactive compounds and postbiotics, which, together with microorganisms, can contribute to the functional properties of the final product [3]. Moreover, fermentation by LAB can also modulate the sensory characteristics adding welcome flavors and increasing the palatability of products without the addition of additives [5]. The selection of starters is a crucial step due to the different capacities of strains to tolerate the relatively high sugar content that characterizes the ice cream composition, the freezing steps, and the presence of oxygen in the cream [17]. The addition of dietary fibers can also enhance the rheological and textural characteristics and melting resistance of ice creams preserving sensory properties and probiotics [18]. According to the National Academies of Science (Washington, DC, USA), the average daily consumption of dietary fiber is significantly lower than the adequate intakes of 38 and 25 g for men and women, respectively [19]. Therefore, dietary fiber could be included in new ice cream formulations to improve health effects (such as prebiotic activity, lowering cholesterol, and reducing blood glucose responses) and provide laxation effects. Besides, the synthesis of exopolysaccharides (EPS) by lactic acid bacteria has gained a special interest over the last decade due to the functional properties of these biopolymers [20]. Despite this, a limited number of studies dealing with the application of an in situ EPS production to functionalize ice cream are available [21,22]. Antioxidant activity is also a functional property that can be improved in new formulations through different methods. To this aim, hydroxytyrosol (HT) represents a good candidate to be tested as a functional ingredient in new ice cream formulations. It is a phenolic compound drawn as a by-product obtained from the manufacturing of olive oil, and it is considered one of the most powerful antioxidant compounds. Moreover, it is known for beneficial health effects such as anti-inflammatory, anticancer, and as a protector of skin and eyes [23].

This paper aimed to set up a new biotechnological protocol for manufacturing of functional cow milk ice creams. To this purpose, we hypothesize combining the lactic fermentation with autochthonous strains, the addition of dietary fibers and HT, and the EPS synthesis. The work included: (i) a selection of a tailored mix of autochthonous starters based on pro-technological properties; (ii) the definition of a suitable production process; and (iii) the characterization of microbiological, physical, nutritional, sensory, and especially, functional (antioxidant) properties of produced ice creams. To the best of our knowledge, this is the first time that such a holistic approach has been applied to the formulation of functional fermented cow milk ice creams. Overall, we believe that such an approach can represent a suitable model that can be applied and transferred to develop other new functional ice cream formulations.

2. Materials and Methods

2.1. Isolation of Lactic Acid Bacteria

Mesophilic lactic acid bacteria were isolated from cheese made with raw cow's milk produced in the South Tyrol region (north Italy). Ten grams of cheese were suspended into 90 mL of sterile physiological solution (NaCl, 0.9%, w/v) and homogenized with a Stomacher® 400 Circulator Lab Blender (Seward Ltd., Worthing, UK). Decimal serial dilutions were prepared in 9 mL of sterile physiological solution and plated on MRS agar medium (Oxoid Ltd., Hampshire, UK) containing 0.1% v/v cycloheximide, for both homogenized cheese and raw milk. After incubation at 30 °C for 48 h, colonies with different morphologies were isolated from the highest plate dilution: twice, catalase-negative, nonmotile rod isolates were cultivated in MRS broth (Oxoid Ltd., Hampshire, UK) at 30 °C for 24 h and restreaked into MRS. All the isolates considered for further analyses showed the capacity to acidify the culture medium. Microbial cultures were stored at -20 °C in 20% (v/v) glycerol.

2.2. Random Amplified Polymorphic DNA-Polymerase Chain Reaction (RAPD-PCR) Analysis

Genomic DNA from each strain was extracted from 2 mL samples of overnight cultures grown in MRS broth, as reported by [24]. RAPD-PCR was carried out as described by [25] for the biotyping of isolates using primers M13, P7, and P4 (Invitrogen Life Technologies). RAPD-PCR profiles were acquired by the MCE-202 MultiNA microchip electrophoresis system (Shimadzu s.r.l., Milan, Italy) and analyzed as described by [26]. Briefly, for each isolate, the presence or absence of fragments into profiles was recorded as 1 or 0, respectively, and imported into an Excel data matrix. Finally, the data matrix was analyzed to assess the similarity of the electrophoretic profiles by determining the Dice coefficients of similarity using the unweighted-pair group method with average linkages (UPGMA) algorithm.

2.3. Genotypic Identification of Isolates

LpigF and LpigR primer pair (Invitrogen Life Technologies, Carlsbad, CA, USA) was used to amplify the 16S rRNA gene fragment of presumptive lactic acid bacteria through polymerase chain reaction (PCR) (C1000 Touch Thermal Cycler, Bio-Rad, Hercules, CA, USA). The reaction mixture contained 0.2 mM of dNTP mix, 1 μM of each primer, 2 mM $MgCl_2$, 2 U Taq DNA polymerase (Promega, Madison, WI, USA), 1X PCR buffer, 25 ng DNA, and sterile double-distilled water to 25 μL. The PCR program comprised 30 cycles of denaturation for 45 sec at 94 °C, annealing for 1 min at 55 °C, and elongation for 1 min at 72 °C; the cycles were preceded by denaturation at 94 °C for 5 min and followed by elongation at 72 °C for 7 min. All amplicons were identified by sequencing (Eurofins, Luxembourg). The sequences were analyzed using the BLASTN software (https://blast.ncbi.nlm.nih.gov/Blast.cgi accessed on 10 June 2019).

2.4. Selection of Isolates Based on Kinetics of Growth and Acidification

Isolates were singly used as starters for the fermentation of whole ultra-high temperature processing (UHT) cow's milk. Cells were first cultivated in MRS broth (Oxoid Ltd., Hampshire, UK) at 30 °C until the late exponential growth phase was reached (ca. 12 h), harvested by centrifugation (7000 rpm, 10 min, 4 °C), washed twice in sterile physiological solution, finally re-suspended in physiological solution, and used to inoculate whole UHT cow's milk to reach a cell density of ca. 7.0 log CFU/mL (the inoculum was 4% v/v). Milk fermentations were performed in triplicate at 40 °C. The kinetics of growth and acidification during fermentation was determined: growth and pH were monitored at t0 and every 4 h for 48 h. Milk without bacterial inoculum was incubated under the same conditions and used as a control.

Viable cells were enumerated by plating in triplicate on MRS agar (Oxoid Ltd., Hampshire, UK) containing 0.1% v/v of cycloheximide, and viable cell counts were calculated as $\log_{10}$ colony forming units per mL (log CFU/mL). Dilutions with less than 30 or more than 300 colonies were discarded. The kinetics of growth were determined and modeled according to the Gompertz equation as modified by [27]: $y = k + A \exp\{-\exp[(\mu_{max} e/A)(\lambda_A - t) + 1]\}$, where k is the initial level of the dependent variable to be modeled (log CFU/mL), A is the difference in cell density between inoculation and the stationary phase, μ_{max} is the maximum growth rate (expressed as Δlog CFU/mL/h), λ_A is the length of the lag phase (expressed in hours), and t is the time.

The pH was measured in triplicate by using a pH meter (Sension$^+$ PH3, Hach-Lange, Lainate, Italy). Acidification data were modeled according to the Gompertz equation as modified by [27]: $y = k + \Delta pH \exp\{-\exp [(V_{max} e/\Delta pH)((\lambda_{pH} - t) + 1]\}$; where y is dpH/dt (units of pH/min); k is the initial level of the dependent variable to be modeled; ΔpH is the difference in pH (units) between the initial value and the value reached in the stationary phase of lactic acid fermentation; V_{max} is the maximum acidification rate as dpH/min; λ_{pH} is the length of the latency phase of acidification expressed in min; and t is the incubation time.

Experimental data were modeled by the non-linear regression procedure of the Statistica 8.0 software package (Statsoft, Hamburg, Germany).

2.5. Screening of Exopolysaccharides (EPS) Biosynthesis

EPS synthesis was screened for isolates by growing cell colonies on plate dishes containing MRS agar added with 292 mM sucrose, 146 mM glucose, and 146 mM fructose. The experiment was performed in triplicate. The mucous colony formation was assessed after incubation at 30 °C for 24–48 h. The strains which produced slimy colonies were recorded as capable of producing EPS [28].

2.6. Manufacture of the Ice Cream

A mixture of 50% whole and 50% skimmed UHT cow's milk was used for ice cream production at a pilot plant scale. The milk had the following nutritional composition for 100 mL: protein 3.5 g, fat 2.0 g, carbohydrates 5.1 g (4.9 g was lactose). The initial pH was 6.6. The procedure for manufacture is described in Figure 1. Table 1 shows all theses considered in this study, with corresponding codes, descriptions, and performed analyses. The codes are presented in the form XX.YY.(H), where XX indicates the thesis (I1, I2, I3, I4, and CO) and YY represents with (5% w/w) or without the addition of dietary commercial fibers (AF or WF, respectively), namely inulin, citrus fiber, flour of carob seed, and guar gum. H represents the addition of 50 ppm of HT (purity $\geq$ 99%. Nova Mentis, Dublin, Ireland); the latter option was tested only on the best performing of the previous thesis. In addition to the strains isolated in this study, *Streptococcus thermophilus* 446 (belonging to the Culture Collection of the Department of Soil, Plant and Food Science, University of Bari Aldo Moro) was also included as a starter because it is a recognized producer of EPS. I1 represents ice cream produced with milk fermented by *Lacticaseibacillus paracasei* F1, *Lacticaseibacillus casei* F14, and *S. thermophilus* 446; I2 represents ice cream produced with milk fermented by *L. paracasei* F1 and *L. casei* F14; I3 represents ice cream produced with milk fermented by *L. paracasei* F1; I4 represents ice cream produced with milk fermented by *L. casei* F14; CO represents ice cream produced with no fermented milk (control). Fermentations were started as already described for the kinetics of growth and acidification. *S. thermophilus* 446 was routinely propagated at 37 °C for 24 h in M17 medium (Oxoid Ltd., Hampshire, UK) before starting fermentations.

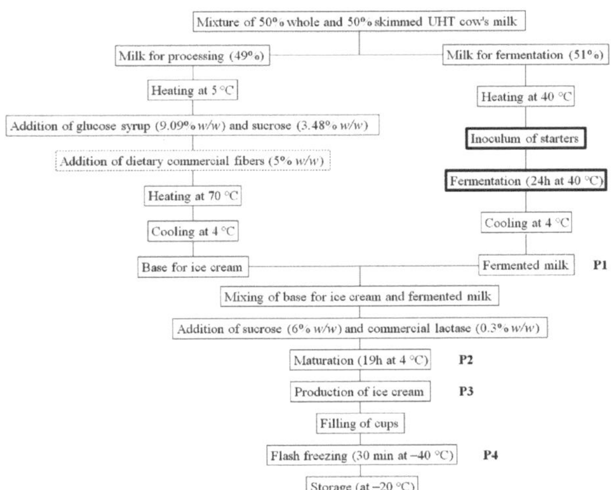

Figure 1. Manufacture of ice cream with fermented milk. The step in the dotted text box ("Addition of dietary commercial fibers (5%, w/w)") was made only for the theses with the addition of fibers, while this step was not made for the theses without the addition of fibers (in this case the total amount of fibers is 0.2%, w/w). The control's theses CO were made with the same protocol exception being the steps in bold rectangles that were not included: "Inoculum of starters" and "Fermentation (24 h at 40 °C)". P1–P4 indicates the sampling points during the ice cream preparation.

Table 1. Conditions considered in this study. The sample codes are presented in the form XX.YY. where XX indicates the thesis (I1. I2. I3. I4, and CO) and YY represents the addition of dietary commercial fibers (AF means with the addition of 5% w/w dietary commercial fibers. WF means without the addition of fibers). When dietary commercial fibers were not added, the total amount in ice cream was 0.2% w/w. ".H" suffix represents the addition of 50 ppm of HT.

Code (XX.YY.(H))	Thesis	Description
I1.WF	Ice cream produced with milk fermented by *Lacticaseibacillus paracasei* F1. *Lacticaseibacillus casei* F14. and *Streptococcus thermophilus* 446	Without addition of dietary commercial fibers.
I1.AF		Addition of 5% w/w dietary commercial fibers
I2.WF	Ice cream produced with milk fermented by *L. paracasei* F1 and *L. casei* F14	Without addition of dietary commercial fibers
I2.AF		Addition of 5% w/w dietary commercial fibers
I3.WF	Ice cream produced with milk fermented by *L. paracasei* F1	Without addition of dietary commercial fibers
I3.AF		Addition of 5% w/w dietary commercial fibers
I4.WF	Ice cream produced with milk fermented by *L. casei* F14	Without addition of dietary commercial fibers
I4.AF		Addition of 5% w/w dietary commercial fibers
I4.AF.H		Addition of 5% w/w dietary commercial fibers and 50 ppm of HT

Table 1. *Cont.*

Code (XX.YY.(H))	Thesis	Description
CO.WF	Ice cream produced with no fermented milk	Without addition of dietary commercial fibers
CO.AF		Addition of 5% w/w dietary commercial fibers

2.7. Microbiological Analysis and pH during Ice Cream Production and Storage

Cell density and pH were monitored at four different phases of ice cream production and after 0.5, 1, 2, and 3 months of storage at $-20\ ^\circ$C. During the ice cream production, samples were collected after the milk fermentation (P1), the maturation step (P2), the production of ice cream (P3), and the flash freezing step (P4). Figure 1 reports the sampling points during the ice cream preparation. The pH values were determined in triplicate using a pH meter (Sension$^+$ PH3, Hach-Lange, Lainate, Italy). Cell density was evaluated through a plate count on MRS (for all ices creams) and M17 (Oxoid Ltd., Hampshire, UK) agar media added with 0.1% v/v of cycloheximide after incubation for 48 h at 30 and 37 $^\circ$C, respectively. Plate count on M17 agar medium was carried out only for thesis I1, where *S. thermophilus* 446 was used as a starter, and for control samples. All plate counts were performed in triplicate. The matching between counted colonies and starters was confirmed after processing and storage by RAPD-PCR analyses and sequencing of 16s rRNA gene fragments with the same procedure already described.

2.8. In Situ Viability Staining and Confocal Laser Scanning Microscope (CLSM) Imaging

The cells' viability in ice creams after 3 months of storage at $-20\ ^\circ$C was also assessed by using the LIVE/DEAD BacLight viability kit (Invitrogen, Waltham, MA, USA) and observation by CLSM. This kit comprises two fluorescent nucleic acid stains: SYTO9 and propidium iodide. SYTO9 (excitation and emission maxima, 480 and 500 nm) penetrates both viable and nonviable bacteria, while propidium iodide (excitation and emission maxima, 490 and 635 nm) penetrates bacteria with damaged plasma membranes only, quenching the green SYTO9 fluorescence. Therefore, bacterial cells with compromised membranes emit red fluoresce and those with intact membranes emit green fluoresce [29]. A working solution containing both fluorescent stains was prepared according to the manufacturer's instructions. One hundred microliters of the stain was then mixed with an equal volume of ice cream sample. After incubation for 30 min at room temperature under dark conditions, microscope slides were prepared for each sample, covered with a coverslip and qualitatively visualized using a Leica SP8LIA CLSM (Leica Microsystems, Wetzlar, Germany) with a 488 laser. Fluorescence emission was observed between 525–575 nm (for SYTO9) and 535–665 nm (for propidium iodide). Images of twenty randomly selected view fields were captured from each slide with a 63X oil immersion objective by LAS X software (Leica Microsystems, Wetzlar, Germany).

2.9. Determination of Carbohydrates, Organic Acids, and Free Amino Acids

Sugars (lactose, glucose, and galactose) and organic acids (lactic acid) were determined by Ultimate 3000 high-performance liquid chromatography (Thermo Fisher Scientific, Massachusetts, MA, USA). Five grams of sample were added to 25 mL of 5 mM H_2SO_4 and homogenized for 10 min with a Stomacher$^\circledR$ 400 Circulator Lab Blender (Seward Ltd., Worthing, UK). The extract was subsequently centrifuged for 5 min at 5000 rpm and the supernatant was filtered through a 0.20 μm syringe membrane filter (VWR, Radnor, PA, USA). To analyze the concentrations of glucose, galactose, and lactose, HPLC was equipped with a Spherisorb column (80 Å, 4.6 mm $\times$ 250 mm, 5 μm) (Waters, Milford, CT, USA) and a 200a refractive index detector (Perkin Elmer, Waltham, MA, USA) [30]. Elution was at 32 $^\circ$C, with a flow rate of 1 mL min^{-1}, using acetonitrile 80% v/v as the mobile phase. The injection volume was 2 μL. For analysis of lactic acid, HPLC was equipped with an

Aminex HPX-87H column (300 mm × 7.8 mm, 9 μm) (ion exclusion, Bio-Rad, Hercules, CA, USA) and a UV detector operating at 210 nm [31]. Elution was at 70 °C, with a flow rate of 0.6 mL min^{-1}, using H_2SO_4 5 mM as the mobile phase. The injection volume was 20 μL. Calibration curves were prepared using pure standards (range 0.5–15 g L^{-1}). The calibration plots indicate a good correlation between peak areas and analyte concentrations, and regression coefficients were higher than 0.99 in all cases (Supplementary Table S1). Lactic acid and sugars used as standards were purchased from Sigma-Aldrich (St. Louis, MO, USA).

Total free amino acids concentration was determined through cadmium–ninhydrin spectrophotometric method [32], by using a UV-1800 spectrophotometer (Shimadzu s.r.l., Milan, Italy). Sixty grams of sample were diluted with 120 mL of distilled water, adjusted pH to 4.6 using HCl 50 mM, centrifuged (10,000 rpm, 10 min at 4 °C), and the supernatant was recovered and used for the assay. All analyses were performed in triplicate.

2.10. Total Titratable Acidity (TTA)

TTA was determined on 10 g of ice cream sample homogenized with 90 mL of distilled water and expressed as the amount (mL) of 0.1 M NaOH to get a pH of 8.3. The pH values were determined by the Sension$^+$ PH3 pH meter with a food penetration probe (Hach-Lange, Lainate, Italy). Analysis was performed in triplicate.

2.11. In Vitro Protein Digestibility of Ice-Cream

The in vitro digestibility was determined according to the method of [33], with slight modifications. A known amount of sample was incubated with 1.5 mg of pepsin, in 15 mL of 0.1 M HCl, at 37 °C for 3 h. After neutralization with 2 M NaOH and the addition of 4 mg of pancreatin in 7.5 mL of phosphate buffer (pH 8.0); 1 mL of toluene was added to prevent microbial growth, and the solution was incubated for 24 h at 37 °C. After 24 h, the enzyme was inactivated by the addition of 10 mL of trichloroacetic acid (20%, w/v), and the undigested protein was precipitated. The volume was made up to 100 mL with distilled water and centrifuged at 5000 rpm for 20 min. The concentration of protein in the supernatant was determined by the Bradford method [34]. The in vitro protein digestibility was expressed as the percentage of protein dissolved after enzyme hydrolysis. The analysis was performed in triplicate.

2.12. In Vitro Gastrointestinal Batch Digestion of Ice-Cream

An in vitro gastrointestinal batch digestion process was carried out as described by [35] with slight modifications. Samples of 100 ml were added to 50 mL of distilled water and mixed in a stomacher for 2 min. A control batch without a substrate was also used as a negative control. The solution was mixed with α-amylase (20 mg) in $CaCl_2$ (1 mM, 6.25 mL) and incubated at 37 °C for 30 min under stirring conditions (100 rpm). Pepsin (2.7 g) was dissolved in 25 mL 0.1 M HCl, and the mixed sample was added. The pH was adjusted to 2.0 using 6 M HCl and incubated at 37 °C for 2 h under stirring conditions (100 rpm). Pancreatin (560 mg) and bile (3.5 g) were dissolved in 125 mL of 0.1 M $NaHCO_3$, and the sample was added. The pH value was adjusted to 7.0 by using 6 M NaOH and incubated at 37 °C for 3 h under stirring conditions (100 rpm). Afterward, basal growth complements and fecal slurries were prepared as described by [36] and added to the batches in order to reproduce the colon phase of digestion. Oxygen was removed by nitrogen flushing and fecal batch were incubated at 37 °C in a shaking incubator for 48 h (0.15 mg/mL). Sample aliquots were collected from each batch at time points (0, 16, 24, and 48 h) for short-chain fatty acid (SCFA) analysis.

2.13. Short Chain Fatty Acid (SCFA) Analysis

Aliquots (1.0 mL) of samples were centrifuged (11,000 rpm, 10 min). The supernatant was acidified with 6 M HCl (3:1, v/v) and incubated at room temperature for 10 min. The mixture was re-centrifuged at 11,000 rpm for 5 min, filtered using a 0.2 μm PVDF filter

(Millipore, Burlington, VT, USA), and used to determine SCFA by HPLC [37]. An Äkta purifier system (GE Healthcare, Chicago, IL, USA) was equipped with an Aminex HPX-87H column (ion exclusion, particle size of 9 μm, column size of 250 × 4.0 mm; Bio-Rad, Hercules, CA, USA) and a UV detector operating at 210 nm. Elution was at 35 °C with a flow rate of 0.6 mL/min, and 10 mM H_2SO_4 was used as the mobile phase. Standard solutions (1000 mM) of lactic, acetic, propionic, butyric, and succinic acids were separately prepared with ultra-pure water and, after adequate dilution, were used for calibration. The method underwent validation for several parameters (Supplementary Table S2). The evaluation of the linearity range was through the determination coefficient (r^2). The calibration curves showed good linearity (r^2 value was from 0.981 to 0.994). Limits of quantification (LOQ) values ranged from 1.03 to 1.27 mM, being different for each compound. All chemicals were of analytical grade. Glucose, galactose, lactose, lactic acid, acetic acid, propionic acid, butyric acid, and succinic acid used as standards were obtained from Sigma-Aldrich (St. Louis, MO, USA).

2.14. Sensory Analysis

Sensory analysis was carried out by 10 non-trained panelists (between 25 and 50 years of age, with equal distribution between male and female). The method was adapted from the analysis reported by [18]. Before analysis, a selection of panelists was made among a larger number of people based on their capacity to distinguish and describe the sensory attributes adopted for the study. Then, the selected panelists were instructed to sniff each sample before tasting. Samples of ice creams were randomly coded and served in plastic cups (25 g) at −20 °C, together with non-salted table biscuits and still water. Samples were evaluated in three replicate sessions using three experimental replicates for each type of ice cream. Panelists were placed separately in rooms for unbiased evaluation of sensory attributes. Ice creams were scored based on continuous 0–10 scales, from 0 (lowest) to 10 (highest). The appearance was evaluated as white intensity. The texture was evaluated as smoothness. Olfactory attributes referred to odor intensity, and gustatory attributes were evaluated as flavor intensity and acidity.

2.15. Measurement of Transepithelial Electrical Resistance (TEER), Release of Inflammatory Cytokines Interleukin-6 (IL-6) and -8 (IL-8), and of Intracellular Reactive Oxygens Species (ROS) of Caco-2 Cells

The most promising thesis of fermented ice cream (both with and without the addition of fiber and with the addition of 50 ppm HT to the formulation) and the no fermented ice cream (as control) was selected for these tests. Human intestinal Caco-2 cells HTB-37 obtained from American Type Culture Collection (ATCC, Manassas, VA, USA) were routinely cultured in Dulbecco modified eagle's medium (DMEM) high glucose supplemented with 10% v/v heat-inactivated fetal bovine serum, 1% v/v non-essential amino acids, 1% v/v L-glutamine and 1% v/v penicillin and streptomycin as previously reported [38]. Caco-2 cells (passage 10–25) were sub-cultured once a week by dissociating with trypsin (Tryple L-select) and seeded at a density 30 × 10^3/cm^2 (split 1:6). All the reagents used for the cultures were purchased from Gibco (Gibco-Thermo Fisher Scientific, Waltham, MA, USA).

For TEER experiments, Caco-2 cells were seeded at a concentration of 250 × 10^3 on 1 μm pore size inserts (polyethylene terephthalate transwells®, Merck KGaA, Darmstadt, Germany), allocated in 12-well culture plates (Falcon) with 0.5 mL of complete culture medium in the Ap compartment, while the Bl compartment was filled with 1.5 mL, and let them differentiate in culture. The medium was changed every two days. After 21 days of culture to allow differentiation, cells were exposed for 24 h to 100 μg/mL of lyophilized ice creams alone and in combination with TNF-α (2 ng/mL) and IL-1β (25 ng/mL) (Peprotech-Thermofisher Waltham, MA, USA). DMEM alone was used as the negative control, and DMEM containing TNF-α (2 ng/mL) and IL-1β (25 ng/mL) was used as the positive control. After cell treatment, the integrity of the monolayer was monitored for 48 h by measuring TEER with the Millicell ERS device (Millipore-Sigma, San Luis, MO, USA).

The measurements were expressed in Ohms·cm^2, after subtracting the mean values of the resistance from cell-free inserts. TEER data were recorded at room temperature.

For IL-6 and IL-8 cytokines release detection, Caco-2 cells were plated in 12-well plates at a concentration of 200×10^3 cells/well, cultivated for 21 days, and exposed for 24 h at 37 °C to lyophilized ice creams at the concentration of 100 µg/mL alone or in combination with IL-1 beta (25 ng/mL). DMEM was used as the negative control, while IL-1 beta (25 ng/mL) was used as the positive control. Synthesis of the pro-inflammatory IL-6 and IL-8 was measured after 24 h using the enzyme-linked immunosorbent assay (ELISA) (Bender MedSystems, Wien, Austria). One hundred µL from each well was collected and transferred into a 96 well-plate (Perkin Elmer, Waltham, MA, USA) and analyzed in triplicate by a spectrophotometer (Victor3, Perkin-Elmer, Waltham, MA, USA) at 450 nm. Quantification was carried out using a reference standard curve as provided by the manufacturer.

The intracellular generation of ROS was assessed by measuring the oxidation of the probe 20,70-dichlorofluorescin diacetate (DCFH-DA) (Molecular Probes, Lifesciences, Eugene, OR, USA). Briefly, the confluent Caco-2 cells in the 24-well plates were pretreated for 6 h with freeze-dried ice cream powder (100 µg/mL). A set of the samples were also treated with TNF-α (2 ng/mL) and IL-1β (25 ng/mL) for 6 h before the end of the treatment to induce oxidative stress, and after 24 h 100 µL from each well were collected and transferred into a black microtiter 96 well-plate. The fluorescence intensity of DCFH-DA (relative fluorescence units, FU) was measured at an excitation and emission wavelength of 485 nm and 520 nm, respectively, using a Victor3 spectrofluorimeter (Perkin Elmer, Waltham, MA, USA).

For these tests, three biological replicates were considered, which were analyzed in triplicate.

2.16. Statistical Analyses

The mean values, standard deviation of the means, Student's *t*-test analysis, and variance analysis with one-way ANOVA of all replicates were calculated using Prism 4 (GraphPad, Boston, MA, USA). Differences were considered significant with *p*-values (p) < 0.05. Individual comparisons were made post hoc with the Tukey-Kramer test. Clustering analysis was performed with the default method available in R and based on Euclidean distance and McQuitty linkage.

3. Results

3.1. Isolation, Typing and Identification of Lactic Acid Bacteria

A total of 24 putative lactic acid bacteria were isolated from cheese. Gram-positive, catalase-negative, nonmotile isolates able to acidify the cultural medium were subjected to RAPD-PCR analysis. The isolates were grouped into five clusters at a similarity level of 70% based on RAPD patterns. For each cluster, one isolate was selected as representative and identified through the partial sequencing of the 16S gene. The isolates were identified as *L. paracasei* and *L. casei* (Table 2).

Table 2. Identification of the 16S gene sequences from selected bacterial isolates.

Isolation Source	Isolates Code	Closest Relative Strain	Accession Number	Similarity (%)
Cheese	F1	*Lacticaseibacillus paracasei*	KR006310	99
Cheese	F2	*L. paracasei*	KU315106	99
Cheese	F5	*L. paracasei*	KU315108	100
Cheese	F13	*L. paracasei*	MK614505	99
Cheese	F14	*Lacticaseibacillus casei*	KJ764639	99

3.2. Pro-Technological Properties of Isolated Strains

Five strains representative of each cluster (*L. paracasei* F1, *L. paracasei* F2, *L. paracasei* F5, *L. paracasei* F13, *L. casei* F14) were singly used as a starter to ferment (40 °C, for 48 h) whole UHT cow's milk and the pro-technological properties were assessed based on the extent of pH decrease (ΔpH), rate of pH reduction (Vmax), length of lag phase (λ_{pH}) (about acidification), increase of cell density (A), maximum growth rate (μ_{max}), and length of the lag phase (λ_A) (about growth). The ranges of the variables for all strains tested are given in Table 3.

Table 3. Parameters of the kinetics of growth and acidification of representative bacterial strains during fermentation of whole UHT cow's milk for 48 h at 40 °C: increase of cell density (A), maximum growth rate (μ_{max}), length of the lag phase (λ_A), pH decrease (ΔpH), rate of pH reduction (Vmax) and length of lag phase (λ_{pH}). Each value was expressed as the mean of three analytical replicates $\pm$ standard deviations. One-way ANOVA and individual post hoc comparisons with the Tukey–Kramer was performed separately for each column: values with different superscript letters differ significantly ($p < 0.05$).

Strains	Growth			Acidification		
	A (Δlog CFU/mL)	μ_{max} (Δlog CFU/mL/h)	λ_A (h)	ΔpH (pH units)	Vmax (ΔpH/min)	λ_{pH} (min)
Lacticaseibacillus paracasei F1	2.00 ± 0.03 [ab]	0.15 ± 0.06 [ab]	13.50 ± 0.20 [a]	2.37 ± 0.03 [a]	0.16 ± 0.02 [a]	13.15 ± 0.20 [ab]
L. paracasei F2	2.02 ± 0.01 [a]	0.31 ± 0.01 [c]	18.72 ± 0.20 [b]	2.40 ± 0.01 [a]	0.17 ± 0.01 [a]	14.08 ± 0.20 [d]
L. paracasei F5	1.63 ± 0.06 [c]	0.13 ± 0.03 [a]	12.49 ± 0.30 [c]	2.23 ± 0.02 [b]	0.16 ± 0.01 [a]	13.86 ± 0.30 [cd]
L. paracasei F13	1.82 ± 0.05 [d]	0.24 ± 0.02 [b]	15.32 ± 0.20 [d]	2.40 ± 0.01 [a]	0.16 ± 0.01 [a]	13.04 ± 0.20 [a]
Lacticaseibacillus casei F14	1.92 ± 0.04 [ad]	0.17 ± 0.05 [ab]	14.23 ± 0.20 [e]	2.41 ± 0.01 [a]	0.17 ± 0.01 [a]	13.46 ± 0.20 [bc]

Starting from an initial cell density of ca. 7 log CFU/mL, after 48 h of fermentation, the cell density of tested strains increased to 8.74 ± 0.11–9.10 ± 0.12 log CFU/mL. *L. paracasei* F1, *L. paracasei* F2 and *L. casei* F14 had increases in cell density significantly higher ($p < 0.05$) than other strains (2.00 ± 0.03, 2.02 ± 0.01, and 1.92 ± 0.04 Δlog CFU/mL, respectively). *L. paracasei* F2 and *L. paracasei* F13 displayed a significantly ($p < 0.05$) higher value of μ_{max} (0.31 ± 0.01 and 0.24 ± 0.02 Δlog CFU/mL/h, respectively), while *L. paracasei* F1, *L. paracasei* F5, and *L. casei* F14 resulted in a length of the growth lag phase significantly ($p < 0.05$) lower than other strains (13.50 ± 0.20, 12.49 ± 0.30, and 14.23 ± 0.20 h, respectively). The initial pH value of whole UHT cow's milk was 6.76 ± 0.01. After 48 h of fermentation, *L. paracasei* F1, *L. paracasei* F2, *L. paracasei* F13, and *L. casei* F14 showed a significantly ($p < 0.05$) higher decrease of pH (2.37 ± 0.03, 2.40 ± 0.01, 2.40 ± 0.01, and 2.41 ± 0.01 pH units, respectively). The rate of pH reduction had no significant ($p > 0.05$) variation among strains. The mean values ranged between 0.16 and 0.17 ΔpH/min. However, *L. paracasei* F1, *L. paracasei* F13, and *L. casei* F14 showed a significantly ($p < 0.05$) lower length of the pH lag phase (13.15 ± 0.20, 13.04 ± 0.20, and 13.46 ± 0.20 min, respectively).

None of the strains isolated in this study was found to be EPS producers at the colony level.

3.3. Microbiological Analysis and pH during the Ice Cream Manufacture and Storage

The initial pH value of whole UHT cow's milk (P0) was 6.65 ± 0.05. After fermentation (production step P1), pH values significantly ($p < 0.05$) decreased to values between 3.78 ± 0.02–4.36 ± 0.03. After the maturation step (P2), pH values slightly but significantly ($p < 0.05$) increased in almost all fermented ice creams to 4.32 ± 0.04–4.64 ± 0.03 and remained mostly stable until the end of processing. In the negative controls produced with no fermented milk (CO.AF and CO.WF), the pH values were between 6.26 ± 0.03 and 6.68 ± 0.02 all along the process.

Plate counts confirmed that the cell density after the inoculum (P0) was ca 7.0 ± 0.0 log CFU/mL for all theses, with the exception being CO.AF and CO.WF where no inoculum was made (cell density < 2 log CFU/mL). After fermentation (P1), in all theses with fermented milk, the cell densities significantly ($p < 0.05$) increased by 2–3 logarithmic cycles compared to the inoculum (P0). During the following production steps, the cell density of lactobacilli decreased in all theses; nevertheless, the cell densities were still consistent at the end of the process. The final cell density of lactobacilli in the thesis without the addition of dietary fibers was 8.0 ± 0.0, 9.2 ± 0.1, 8.8 ± 0.4, and 8.5 ± 0.2 log CFU/g for I1.WF, I2.WF, I3.WF, and I4.WF, respectively. In the thesis with the addition of dietary fibers, cell densities were 7.6 ± 0.0, 8.8 ± 0.1, 8.9 ± 0.1, and 9.1 ± 0.1 log CFU/g for I1.AF, I2.AF, I3.AF and I4.AF, respectively. The final level of streptococci was 8.7 ± 0.1 and 8.9 ± 0.6 log CFU/mL for I1.WF and I1.AF, respectively. When the ice cream was produced with no fermented milk, the cell density ranged between <2 and 2.6 ± 0.0 log CFU/mL and between 3.5 ± 0.0 and 5.7 ± 0.0 log CFU/mL for lactobacilli and streptococci, respectively, with no significant ($p > 0.05$) variation during the process.

During storage, the pH varied slightly but not significantly ($p > 0.05$) (data not shown).

After three months of conservation at $-20\ ^\circ$C, the cell density of lactobacilli in the theses without the addition of dietary fibers decreased to 7.8 ± 0.2, 9.0 ± 0.0, 8.6 ± 0.2, and 8.6 ± 0.1 log CFU/g for I1.WF, I2.WF, I3.WF, and I4.WF, respectively. In the theses with the addition of dietary fibers, cell densities decreased to 7.4 ± 0.1, 8.3 ± 0.0, 8.5 ± 0.1, and 8.7 ± 0.1 log CFU/g for I1.AF, I2.AF, I3.AF, and I4.AF, respectively. With the exception of I3.AF thesis, these changes were not statistically significant. In the theses with no fermented milk (control), cell density ranged between 3.0 ± 0.1 and 3.3 ± 0.0 log CFU/g for samples without the addition of dietary fibers (CO.WF) and between <2 and 2.3 ± 0.1 log CFU/g for samples with the addition of dietary fibers (CO.AF).

In the thesis I1, cell densities of streptococci decreased to 8.5 ± 0.1 log CFU/g for samples without the addition of dietary fibers (I1.WF) and to 8.2 ± 0.1 log CFU/g for samples with the addition of dietary fibers (I1.AF), but only for I1.WF was it statistically significant. The control thesis without the addition of dietary fibers (CO.WF) showed a statistically significant decrease in cell density from 5.6 ± 0.0 to 5.2 ± 0.2 log CFU/g, while the control with the addition of dietary fibers (CO.AF) ranged between 3.0 ± 0.1 and 3.6 ± 0.1 log CFU/g and changes were not statistically significant.

The persistence of the selected starters throughout the processing and storage period was confirmed using the RAPD-PCR profile of bacterial starters and 16s rRNA sequences (data not shown).

Qualitative observations of samples after 3 months of storage by CLSM confirmed that a high density of living cells was present in all theses with fermented milk, both with and without the addition of dietary fibers (green fluorescence), while dead cells (red fluorescence) are negligible (Figure 2). The control samples (CO.WF and CO.AF) showed lower cell density compared to samples with fermented milk.

3.4. TTA, Carbohydrate, Organic acid, Free Amino Acid Analyses, and In Vitro Protein Digestibility

In samples with fermented milk, TTA ranged between 5.1 ± 0.5 and 6.5 ± 0.5 mL of 0.1 M NaOH (measured in I3.AF and I2.AF, respectively), while the values in control samples (CO.AF and CO.WF) were lower ($p < 0.05$) and between 0.7 ± 0.1 and 0.9 ± 0.2 mL of 0.1 M NaOH. Theses with fermented milk did not differ significantly among them, while they have significantly higher values compared to no fermented ice creams (CO.WF and CO.AF). Changes after 3 months of storage were not statistically significant, with the exception being I2.AF and I3.AF, which had final values statistically different compared to t0. The concentration of lactose, glucose, galactose, and lactic acid free amino acids were measured in fresh ice creams and after 3 months of storage. In fresh ice creams produced with fermented milk, lactic acid concentrations ranged between 49.6 ± 0.1 and 69.2 ± 0.3 mM, and it had a significant ($p < 0.05$) decrease after 3 months of storage in all theses, while it was not found in ice creams produced with no fermented milk. Glucose concentration in fresh ice creams ranged between 185.6 ± 0.6 and 221.3 ± 6.4 mM, and after storage, it displayed a significant decrease ($p < 0.05$) only in I3.AF and I4.WF. Galactose concentrations were between ca. 9.3 ± 0.1 and 14.2 ± 0.5 mM at t0, and significantly decreased ($p < 0.05$) in all theses except CO.AF. Lactose concentrations were under the detection limit (< 0.6 mM) for all samples both in fresh ice cream and after storage. Free amino acids concentration in fresh ice cream ranged between 60.6 ± 0.3 and 105.7 ± 0.3 mg/kg. Overall, it was higher in fermented ice creams without the addition of fiber compared to fermented ice creams with fiber addition. However, after storage, it increased in all ice creams produced with the addition of dietary fibers, and such variation was significant ($p < 0.05$) for all theses with fermented milk. On the contrary, the concentration significantly decreased ($p < 0.05$) in all theses without the addition of dietary fiber, except for I4.FW, in which the decrease resulted in no significant ($p > 0.05$) (Figure 3). Protein digestibility ranged from 93.31 to 94.25%, with no significant ($p > 0.05$) differences between theses and during storage.

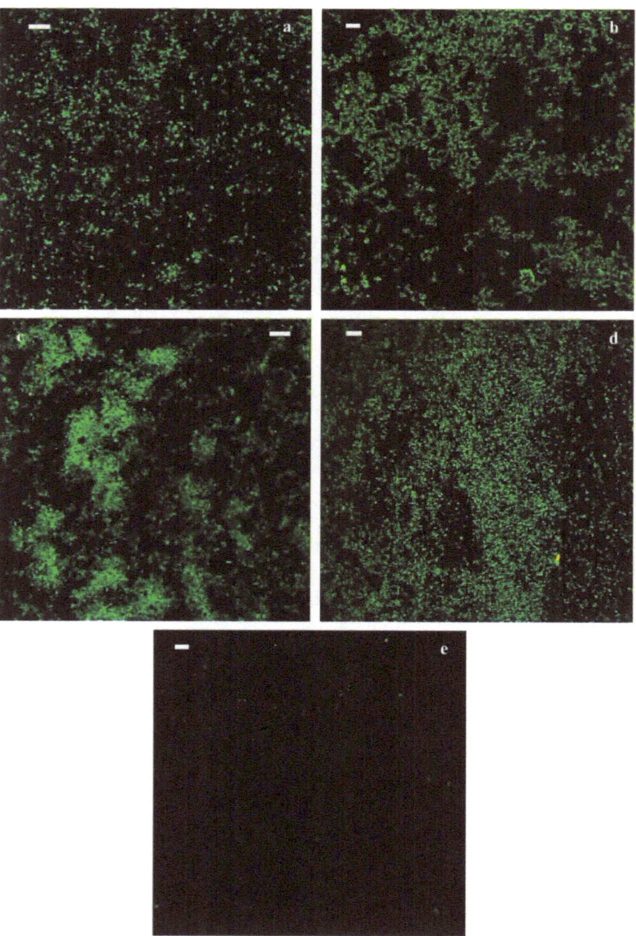

Figure 2. Confocal laser scanning microscope images of ice creams stained with LIVE/DEAD BacLight viability stain. (**a**) I1.WF; (**b**) I2.WF; (**c**) I3.WF; (**d**) I4.WF; and (**e**) CO.WF. Green fluorescence indicates live bacterial cells, red fluorescence indicates dead bacteria cells. All panels refer to theses without the addition of dietary fibers. Bar = 10 μm.

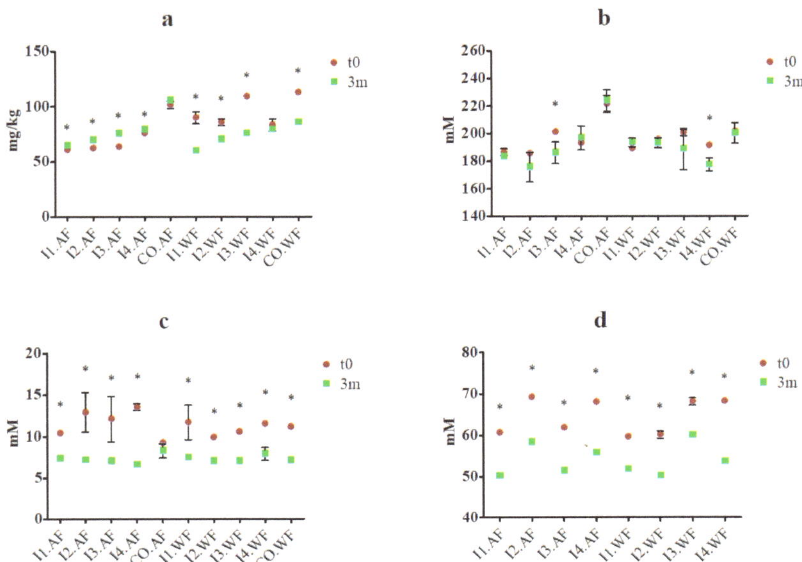

Figure 3. Free amino acids (**a**), glucose (**b**), galactose (**c**), and lactic acid (**d**) measured in fresh ice creams (t0) and after 3 months of storage (3 m). The thesis codes are as reported in Table 1: "I1."–"I4." prefix represents the thesis "CO." prefix represents the negative control (without fermentation), "WF" and "AF" suffix means without and with the addition of dietary commercial fibers (5%, w/w), respectively. Lactic acid was not found in negative control without fermentation. Data are the mean of three analytical replicates. Student's t-test analysis was performed separately for each thesis. * is reported when values differ significantly ($p < 0.05$).

3.5. In Vitro Gastrointestinal Batch Digestion of Ice-Cream and Short Chain Fatty Acid (SCFA) Analysis

HPLC analysis was used to identify and quantify the main SCFA (succinic, acetic, butyric, lactic, and propionic acids) in fecal batch cultures. Analysis was carried out at 0, 16, 24, and 48 h of incubation (Figure 4). Overall, the fatty acid trends determined by fresh ice cream and ice cream stored for 1 month at −20 °C were mostly similar (Figure 4 and Supplementary Figure S1, respectively). At the beginning of incubation, fermented ice cream determined higher ($p < 0.05$) lactic and succinic acid values than CO.AF and CO.WF. During incubation, the level of succinic acid followed a fluctuating trend for all thesis, first increasing and then decreasing, or toward stable levels. A similar trend was found for lactic acid. Fiber fortification resulted in the highest ($p < 0.05$) levels of lactic and butyric acid. For the latter, the highest values were found for thesis I3.AF, I4.AF, and I4.AF.H after 48 h of fecal batch incubation. The addition of the fiber also changed the acetic and propionic acid profiles (Figures 4 and S1).

3.6. Sensory Analysis

Sensory analysis of milk ice creams (CO.WF and CO.AF) was compared to that of ice creams fermented with selected strains. The pseudo-heat map (Figure 5) summarizes the result highlighting differences among different theses. Three main clusters can be identified corresponding to control theses (CO.AF and CO.WF), I4 theses (both with and without fiber addition), and all other fermented theses (I1, I2 and I3). The starters of fermentations appear as the main driver of clustering, while the fiber addition did not affect the overall perception of panelists. In terms of color appearance (white intensity), thesis I4.AF and I4.WF obtained the highest scores. In terms of smoothness, CO.AF and I4.AF supplied the highest scores. Panelists perceived higher odor intensity and acidity in fermented ice creams compared to no fermented ice creams. About the flavor intensity, the I4 theses resulted in the most appreciated, with I4.AF showing the highest score.

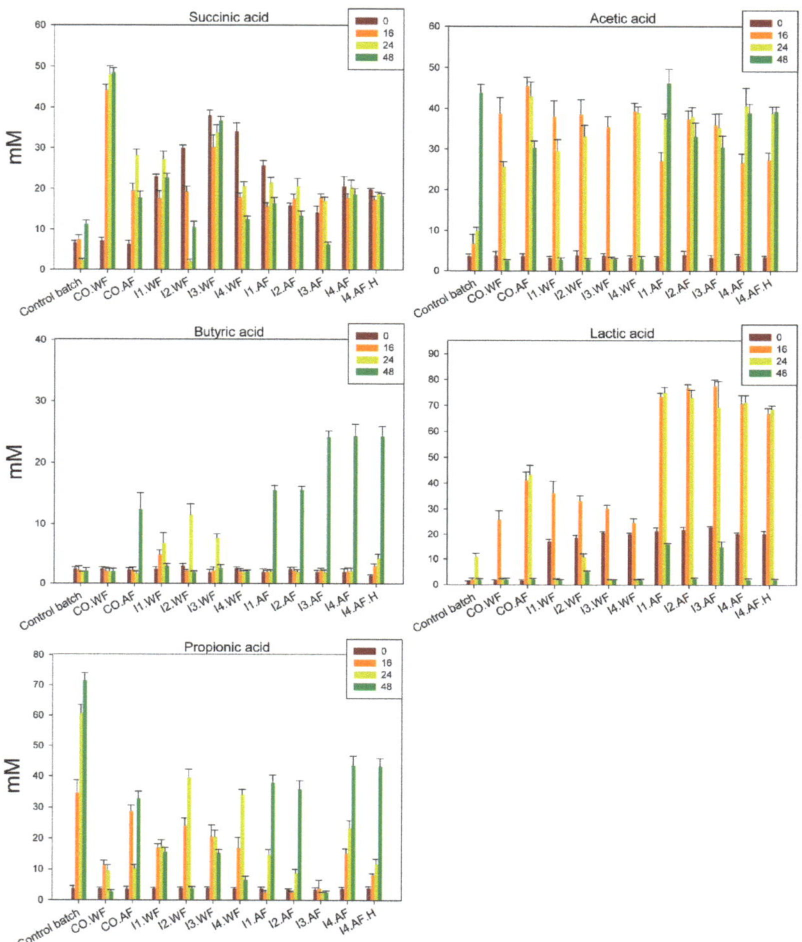

Figure 4. Concentrations (mM) of main organic acids in fecal batch culture over 48 h of incubation. Fecal batches were supplemented with ice creams, and a batch culture without supplementation was used as a negative control.

3.7. HT Addition

The addition of 50 ppm HT to the recipe of I4.AF ice cream, the most promising thesis based on sensory analysis, was also investigated. Lactobacilli were detected at a cell density of 9.2 ± 0.1 log CFU/g in fresh ice-cream, and 9.0 ± 0.1 log CFU/g after one month of storage at $-20\ °C$. The value of pH was 4.4 ± 0.0, without significant ($p > 0.05$) changes during storage.

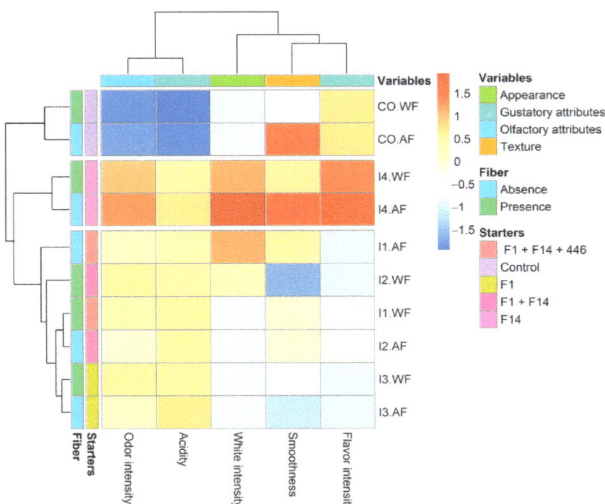

Figure 5. Pseudo-heat map showing the sensory attributes of the panelist's evaluation. The appealing appearance was evaluated as color (white) intensity; texture characters were evaluated as smoothness; olfactory attributes referred to odor and aroma pleasantness; and gustatory attributes were evaluated as overall flavor intensity and acidity. "I1."–"I4." prefix represents the theses produced with milk fermented with different starters (*Lacticaseibacillus paracasei* (F1), *Lacticaseibacillus casei* (F14), and *Streptococcus thermophilus* (446)), "CO." prefix represents the negative control (without fermentation). "WF" and "AF" suffix means without and with the addition of dietary commercial fibers (5%, *w*/*w*), respectively. Rows are clustered using Euclidean distance and McQuitty linkage (red line). The color scale shows the differences between the standardized data.

3.8. Antioxidant and Anti-Inflammatory Features of Fermented Ice Cream in Caco-2 Cell Culture

The protection of the most promising thesis (I4) of fermented ice cream towards oxidative stress was investigated on human intestinal Caco-2 cells. TEER is a widely accepted method to quantitatively measure the dynamics of epithelial integrity through an in vitro model. Compared to the negative control (DMEM, 486.7 ± 19.5 ohms $\times$ cm^2 at 48 h), the fresh fermented ice cream (theses I4.WF, I4.AF, and I4.AF.H) induced a significant ($p < 0.05$) increase of TEER in Caco-2 cells both at 24 and 48 h of incubation (values at 48 h between 580.0 ± 17.1 and 622.0 ± 11.1 ohms x cm^2), while no fermented ice creams (CO.WF and CO.AF) induced any significant variation (values at 48 ranged between 448.3 ± 56.5 and 455.3 ± 11.9 ohms x cm^2) (Figure 6a). The Figure 6b reports the TEER at 48 h of incubation, when ice cream (both fresh and after one month of storage at -20 °C) was used to treat Caco-2 cells, both alone and in combination with TNF-α (2 ng/mL) and IL-1β (25 ng/mL). In Caco-2 not subjected to the inflammatory stimulus and treated with ice creams after one month of storage, the results confirmed the significant increase of TEER induced by the fermented ice cream (values between 594.0 ± 26.1 and 595.7 ± 43.8 ohms $\times$ cm^2) compared to the TEER measured in the negative control (DMEM, 486.7 ± 19.5 ohms $\times$ cm^2) and in Caco-2 cells treated with no fermented ice cream (values between 428.7 ± 11.7 and 426.0 ± 28.1 ohms $\times$ cm^2), with the exception being I4.WF (577.7 ± 33.1 ohms $\times$ cm^2), which resulted not significantly dissimilar compared to DMEM. Instead, if also subjected to the inflammatory stimulus, the TEER of positive control (DMEM containing TNF-α and IL-1β) and of Caco-2 cells treated with no fermented ice cream was lower and ranged between 170.7 ± 24.6 (CO.AF after 1 month of storage) and 199.0 ± 19.1 (fresh CO.WF) ohms $\times$ cm^2. However, the values measured in cell lines treated with fermented ice creams (both fresh and after 1 month of storage) were near to those of negative control (DMEM) and resulted significantly higher

($p < 0.05$) than those displayed by the positive control and Caco-2 treated with no fermented ice cream.

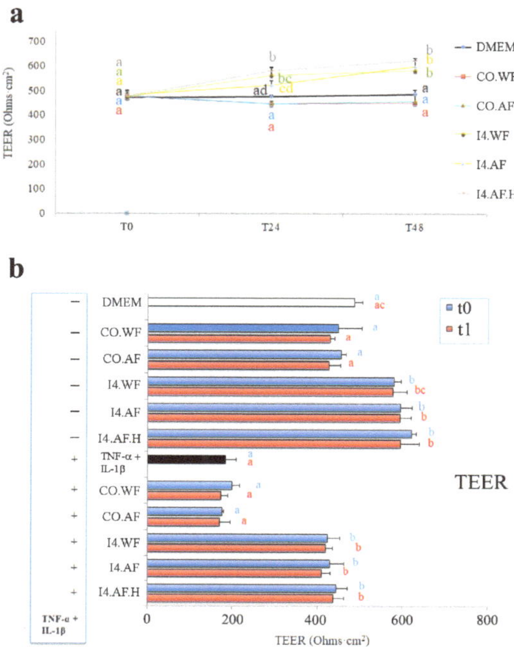

Figure 6. TEER (Ohms·cm^2) of Caco-2 cells. (**a**) Caco-2 cells were treated with lyophilized ice creams (100 µg/mL) and TEER was measured at time 0 (T0) and after 24 and 48 h (T24 and T48, respectively). (**b**) Caco-2 cells were treated with lyophilized ice creams (100 µg/mL), both fresh (t0) and after one month of storage at -20 °C (t1), alone or in combination with TNF-α (2 ng/mL) and IL-1β (25 ng/mL), and TEER was measured after 24 h. CO.AF and CO.WF represent Caco-2 cells treated with no fermented ice creams, respectively, with and without the addition of dietary fibers. I4.AF and I4.WF are Caco-2 cells treated with fermented ice creams, with and without the addition of dietary fibers, respectively. I4.AF.H represents the Caco-2 cells treated with fermented ice cream with the addition of both dietary fibers and HT. DMEM alone was used as the negative control, while DMEM containing TNF-α (2 ng/mL) and IL-1β (25 ng/mL) was used as a positive control. Data are the means ($\pm$SD) of three biological replicates analyzed in triplicate. One-way ANOVA and individual post hoc comparisons with the Tukey–Kramer was performed separately for each thesis (panel (**a**)) and for t0 and t1 (panel (**b**)). Each analysis displays superscript letters with a different color. For every analysis, measurements with different superscript letters differ significantly ($p < 0.05$).

In Caco-2 cells not subjected to the inflammatory stimulus, ELISA analysis revealed no significant changes in IL-6 and IL-8 levels, with the only exception of cells treated with fresh CO.WF and with CO.AF and I4.AF after 1 month of storage. However, in this case, differences were small even if significant and, overall, the absolute values ranged between 15.5 ± 3.1 (DMEM) and 30.7 ± 4.2 (I4.AF after 1 month of storage) pg/mL for IL-6, and between 46.0 ± 11.1 (I4.WF after 1 month of storage) and 145.3 ± 39.1 (CO.AF) pg/mL for IL-8. In Caco-2 cells stimulated with IL-1β, significant decreases in levels of IL-6 and IL-8 were obtained when a treatment with fermented ice creams was applied compared to positive controls (DMEM treated with IL-1β) and cells treated with no fermented ice creams. The decreases were significant both if fresh ice creams and if ice creams after one month of storage were used (Figure 7a,b, respectively). In particular, the concentrations of IL-6 were 114.7 ± 12.3 pg/mL in the positive control, between 124.0 ± 18.2 and 125.7 ± 15.0 pg/mL when cells were treated with no fermented ice creams (CO.AF and CO.WF) and between 24.7 ± 1.5 and 36.0 ± 6.6 pg/mL if treated with fermented ice creams, while the concentration of IL-8 was 673.7 ± 91.7 pg/mL in the positive control, and it ranged between 614.0 ± 65.0 and 651.7 ± 73.5 pg/mL and between 104.0 ± 13.0 and 244.3 ± 21.0 pg/mL when the treatment was performed with no fermented and with fermented ice creams, respectively.

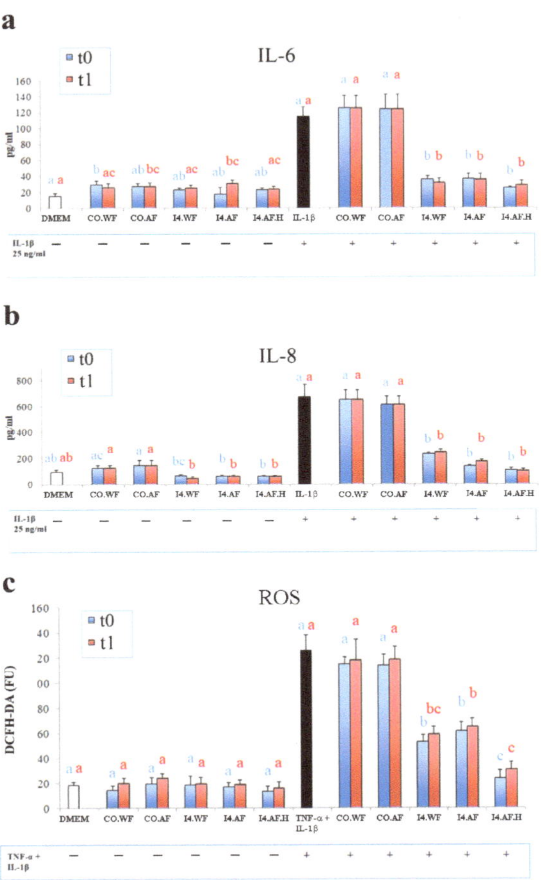

Figure 7. IL-6 (**a**), IL-8 (**b**), and intracellular ROS (**c**) measured at 24 h in Caco-2 cells after exposure to lyophilized ice creams (100 mg/mL) both fresh (t0) and after one month of storage at −20 °C (t1). +, Caco-2 cells were also stimulated with IL-1β (25 ng/mL) for IL-6 and IL-8 release and with TNF-α (2 ng/mL) and IL-1β (25 ng/mL) for ROS induction. CO.AF and CO.WF represent Caco-2 cells treated with no fermented ice creams, respectively, with and without the addition of dietary fibers. I4.AF and I4.WF are Caco-2 cells treated with fermented ice creams, with and without the addition of dietary fibers, respectively. I4.AF.H represents the Caco-2 cells treated with fermented ice cream with the addition of both dietary fibers and HT. DMEM alone was used as the negative control. DMEM containing the corresponding stimulus was used as the positive control. Data are the means (±SD) of three biological replicates analyzed in triplicate. One-way ANOVA and individual post hoc comparisons with the Tukey–Kramer was performed separately for t0 and t1. Different analyses display superscript letters with different colors. For every analysis, values with different superscript letters differ significantly ($p < 0.05$).

No significant differences in intracellular ROS levels were detected in human intestinal cells treated with ice cream only (both fermented and unfermented) compared to the negative control, both with fresh and with 1-month old ice cream (Figure 7c). In this case, the ROS level was 18.7 ± 2.5 DFCH-DA (FU) for negative control, while it ranged between 13.7 ± 4.0 and 24.0 ± 3.6 DFCH-DA (FU) if cells were treated with ice creams. If subjected to the oxidative stress with TNF-α and IL-1β, a significant decrease of the intracellular level of ROS resulted in human intestinal cells treated with fermented ice cream compared to the positive control (125.7 ± 12.0 CFCH-DA FU) and to those subjected to a treatment with no fermented ice creams (values between 113.7 ± 8.7 and 118.3 ± 10.5 CFCH-DA FU). In this case, the treatment with fermented ice cream, both fresh and after 1 month of storage, displayed lower final levels of ROS compared to the unfermented ice cream. The lowest ($p < 0.05$) values were detected with the formulation added of HT (I4.AF.H): 23.7 ± 6.8 and 31.0 ± 6.2 CFCH-DA FU for fresh and 1-month old ice cream, respectively.

4. Discussion

The intrinsic chemical, structural and physical features of ice cream make this matrix suitable for delivering functional microorganisms. The enrichment of functional microorganisms into ice cream can be obtained by blending it with probiotics immediately before freezing [8] or through the fermentation of milk [21]. In this paper, we explored the second case as, with this approach, the ice cream can also be enriched with positive postbiotics which, together with microorganisms, enhance the functional properties [39]. The screening of starters was based on strains belonging to *Lacticaseibacillus* genus that naturally inhabits the raw milk to have a better adaptation and to guarantee high fermentation performance. *L. paracasei* F1 and *L. casei* F14 were selected as they resulted in both fast-growing and fast-acidifying strains. Both species were very often isolated in milk and are largely recognized for their multiple pro-biotic activities [40,41]. As EPS produced by LAB are known for their functional properties and the Generally Recognized as Safe (GRAS) status [21], the EPS producer *S. thermophilus* 446 strain was also considered for starters formulations considering that none of the selected autochthonous lactobacilli showed EPS production at the colony level. According to previous studies, to ensure high adaptability and fast acidification of milk, starters were inoculated at a cell density of ca. 7 log CFU/mL [42]. Ice creams manufactured with fermented milk differed for the used starters. Despite a slight decreasing trend of cell densities being observed, starter strains remained viable at high densities both along the production process, and during three months of storage. The decrease observed at the maturation step can be explained with the dilution effect due to the mixing of fermented milk with skimmed UHT milk. *L. paracasei* F1 and *L. casei* F14 cell viability were more affected by the co-presence of *S. thermophilus* 446 (thesis I1). This is in agreement with previous studies, which hypothesized an inhibition effect due to the metabolites (like organic acids and antimicrobial agents) secreted by *S. thermophilus* and the competition effect for the nutrients present in the milk [43]. Apart from the control produced with no fermented milk, all ice creams had pH values < 5 at the end of the production process and after storage. The slight increase in pH at the maturation step is due to the addition of the base for ice cream (containing no fermented milk, sugars, stabilizer, fibers, and emulsifier) to the fermented milk (Figure 1), which is characterized by higher pH (ca. 6), increasing the overall pH of the mixture. The addition of fibers to the recipe did not affect pH, TTA, and cell viability in the mixtures. The commercial lactase enzyme added to the mixture just before the maturation step (0.3% w/w) contributed to lactose transformation as demonstrated by the low level of such sugar and the presence of galactose in no fermented ice creams samples (controls), where lactic acid is not present.

Regarding the content of glucose, the ice creams studied in this work represent complex systems. The raw milk contained 130 mM of lactose as only sugar, but the glucose values measured in the ice creams were higher. This is due to the ice cream recipe: in fact, additional sugars were added both to the no fermented milk fraction (9.1% w/w of glucose syrup (Roquette, Lestrem, France. Dextrose equivalent 59–62, glucose 21–31% w/w, and disaccharides 35% *w/w*, as for manufacturer specifications) and 3.5% *w/w* of sucrose) and to the base for ice cream mixture after milk fermentation (6% *w/w* of sucrose) (Figure 1). Overall, the amount of glucose added to the mixture after the fermentation step was 5 time higher compared to the quantity originally present before the fermentation step. Therefore, final glucose values resulted from the combination of two opposite factors: from one side the glucose consumption by bacterial starters during the fermentation, from the other side the addition to the mixture of further and higher amount of glucose and sucrose after the fermentation step. As matter of fact, in general glucose values were higher in no fermented ice creams (controls) compared to fermented counterpart. Fluctuation of galactose concentrations were found after storage as observed also in previous studies [44]. Even galactose concentrations resulted from the combination of two opposite factors occurring during the production process: first it is consumed by bacterial starters, then it is added due to the hydrolysis of lactose.

As microbiological and chemical analyses did not highlight differences among fermented ice creams, a sensory analysis was used to select the most promising microbial starters. White intensity, smoothness, acidity, odor, and flavor intensity of ice cream samples were evaluated by no trained panelists, who assigned the highest scores to the theses I4.WF and I4.AF. Therefore, these were chosen for further experiments. HT is a well-known functional ingredient used to develop new functional products. It is claimed to exert antioxidant activity thanks to its ability to break down peroxidation chain reactions, scavenging of free radicals, and the prevention of metal ions [45]. In this paper, its addition to the recipe of the most promising thesis was also considered (thesis I4.AF,H) since it did not affect cell viability and fermentation. Such findings agree with [44] that investigated the addition of phenolic compounds for the production of functional milk (yogurt-like) beverages, and neither interference with the fermentation by LAB nor impact on volatile compounds was observed.

We used in vitro digestion and fecal batches to investigate possible links between human consumption of ice cream and implications at the gut level. In particular, HPLC analysis was used to identify and quantify the main SCFA produced in fecal batch cultures, though we are aware that synthesis of SCFA in the human colon is a highly dynamic process that follows complex enzymatic pathways and involves an extensive number of bacterial species belonging to different phyla. In the initial stages of fecal batches incubation, fermented ice cream provided substantial amounts of lactic and succinic acids. These two organic acids are generated by the metabolism of *L. casei* and *L. paracasei* during milk fermentation and represent important SCFA precursors at a gut level, as well as playing a role in the modulation of intestinal inflammation [46]. Trends in lactic and succinic acid levels during fecal batches incubation were in agreement with those of butyric acid, whose production by the gut microbiota prevents the accumulation of organic acids and stabilizes the intestinal environment [46]. An interesting result was represented by the high levels of butyric acid in batches containing fermented ice cream, since butyric acid is the preferred energy source for colonocytes, affect peripheral organs indirectly by activation of hormonal and nervous systems, and act as a potent anti-inflammatory agent [46].

As intestinal cells are extremely susceptible to oxidative injury when exposed to lipid peroxidation products and luminal oxidants derived by food ingestion, the antiradical and anti-inflammatory properties of I4.AF, I4.WF and I4.AF.H ice cream was further investigated on human intestinal Caco-2 cells, which represent one of the most widely used in vitro models to mimic the intestinal mucosa, by measuring the TEER, ROS, cytokines IL-6 and IL-8 and comparing with those of control theses (CO.AF and CO.WF) [47]. Under the conditions of our study, the treatment with fermented ice cream halted the dramatic increase of IL-8 and IL-6 levels induced by pro-inflammatory stimulus in Caco-2 cells. Previous studies have highlighted the key role of pro-inflammatory mediators, such as IL-8 and IL-6, in the acute phase inflammatory reaction, which represents one of the main targets for the treatment of inflammatory disorders of the gastrointestinal tract [48,49]. Fermented ice cream significantly counteracted also the TNF-α and IL-1β-induced intracellular accumulation of ROS. Such activity is useful when antioxidant defenses are overwhelmed by the accumulation of ROS, resulting in oxidative stress that leads to damage of cellular structures and biomacromolecules in the human body [50,51].

As, under investigation conditions, cell lines develop functional and morphological traits of enterocytes that include their intercellular tight junctions, we assessed the TEER, which measures the dynamics of epithelial integrity [52]. During 48 h of incubation in DMEM, TEER was stable. However, the pretreatment of cells with fermented ice cream induced, after 48 h, an increase of TEER between ca. 19 and 28%. For I4.AF and I4.AF.H, such effect was still evident even after one month of storage at $-20°$ (an increase of ca. 22%). Treatment of human colon cells with TNF-α and IL-1β induced a cellular inflammatory response by stimulating the secretion of primary histopathologic indicators of inflammation like IL-8 and IL-6 [53] that caused a TEER decrease of 62%. However, when cells were preliminary treated (apical compartment) with fermented ice creams and subsequently stimulated (basolateral compartment) with TNF-α and IL-1β, the negative effect of cytokine mix on the TEER was attenuated by ca. 79–86%. Such beneficial effects are also displayed when dietary fibers and/or HT are added to the ice cream recipe (I4.AF and I4.AF.H), and it persisted even if ice cream is stored at -20 °C for one month before the assay (attenuation between 75–84%).

The beneficial functional properties of fermented ice creams are likely attributable to the presence of bioactive peptides, which can be released during milk fermentation from caseins and serum proteins of milk. Bioactive peptides may exert multiple actions, including antioxidant, anti-inflammatory, and immunomodulatory roles [54–57]. Furthermore, fermentation can affect the activity of other functional components that normally occur in fermented milk, such as amino acids, uric acid, vitamins, enzymes (superoxide dismutase, catalase, glutathione peroxidase), fatty acids, and coenzyme Q_{10}. Finally, we cannot exclude a direct role of *L. casei* F14, since it was detected at high cell densities during all processing and storage stages of fermented ice cream. Many strains of *L. casei* were previously reported as valuable probiotics with anti-inflammatory activity [58,59].

The antiradical and anti-inflammatory properties of fermented ice creams proposed in the present study are of high importance not only because they were not reported before in ice cream products, but also as they were obtained without additives, simply through the fermentation of milk with the selected starter during the production process. The milk fermentation by *L. casei* F14 per se played a key role, whereas the addition of HT only contributed to further enhance the antiradical activity of ice cream, while no contribution was detected for anti-inflammatory activity, and no protective effect on Caco-2 cells was brought about by the addition of fibers.

In conclusion, the biotechnological protocol we proposed for manufacturing a new ice cream formulation resulted in promising antioxidant properties acquired through the fermentation of milk with a selected starter. The milk fermentation by *L. casei* F14 conferred to the resulting ice cream the ability to counteract the accumulation of pro-inflammatory mediators (IL-8 and IL-6) and ROS, and to preserve the Caco-2 cell monolayer integrity. Since it was reported that the human antioxidant defense system benefits from the intake of foods allowing a concentration of antioxidant compounds in gastrointestinal environments [50], it can reasonably be hypothesized that the fermented ice cream developed in this work could help to counteract the oxidative stress and inflammatory process at the intestinal level and to support the maintenance of barrier integrity also in vivo subjects [38]. Of course, further studies focused on the gastrointestinal tract by using simulators and/or through in vivo trials should confirm the potential and clarify the individual compounds responsible for antioxidant and anti-inflammatory effects. However, the results of the present study provide solid evidence to support one more time the fermentation by selected starters as a tool to improve the functional values and to propose novel products to benefit the food industry and consumers.

Supplementary Materials: The following supporting information can be downloaded at: https://www.mdpi.com/article/10.3390/fermentation9020117/s1, Figure S1: Concentrations (mM) of main organic acids in fecal batch culture over 48 h of incubation; Table S1: Calibration data used for the HPLC quantification of sugars and lactic acid; Table S2: Calibration data used for the HPLC quantification of short chain fatty acids.

Author Contributions: Conceptualization, R.D.C.; methodology, A.P. and R.D.C.; validation, A.P. and R.D.C.; formal analysis, A.P., A.Z.A.T., P.F., A.D.R., K.A., V.C. and O.V.; investigation, A.P.; resources, A.N. and R.N.; data curation, A.P.; writing—original draft preparation, A.P.; writing—review and editing, A.P., P.F.; visualization, A.N., R.N.; supervision, R.D.C.; project administration, R.D.C.; funding acquisition, R.D.C., M.G. and R.N. All authors have read and agreed to the published version of the manuscript.

Funding: This research received no external funding. The APC was funded by the Open Access Publishing Fund of the Free University of Bozen-Bolzano.

Data Availability Statement: Data sharing is not applicable to this article.

Acknowledgments: The project SYSTEMIC "an integrated approach to the challenge of sustainable food systems: adaptive and mitigatory strategies to address climate change and malnutrition", a Knowledge hub on Nutrition and Food Security that has received funding from national research funding parties in Belgium (FWO), France (INRA), Germany (BLE), Italy (MIPAAF), Latvia (IZM), Norway (RCN), Portugal (FCT), and Spain (AEI) in a joint action of JPI HDHL, JPI-OCEANS and FACCE-JPI launched in 2019 under the ERA-NET ERA-HDHL (n° 696295). The authors want to thank EFRE FESR Südtirol Alto Adige—Micro4Food.

Conflicts of Interest: The authors declare no conflict of interest.

References

1. Gobbetti, M.; Di Cagno, R.; de Angelis, M. Functional Microorganisms for Functional Food Quality. *Crit. Rev. Food Sci. Nutr.* **2010**, *50*, 716–727. [CrossRef] [PubMed]
2. Filannino, P.; Di Cagno, R.; Gobbetti, M. Metabolic and Functional Paths of Lactic Acid Bacteria in Plant Foods: Get out of the Labyrinth. *Curr. Opin. Biotechnol.* **2018**, *49*, 64–72. [CrossRef] [PubMed]
3. Mathur, H.; Beresford, T.P.; Cotter, P.D. Health Benefits of Lactic Acid Bacteria (Lab) Fermentates. *Nutrients* **2020**, *12*, 1679. [CrossRef] [PubMed]
4. Rizzello, C.G.; De Angelis, M.; Coda, R.; Gobbetti, M. Use of Selected Sourdough Lactic Acid Bacteria to Hydrolyze Wheat and Rye Proteins Responsible for Cereal Allergy. *Eur. Food Res. Technol.* **2006**, *223*, 405–411. [CrossRef]
5. Peyer, L.C.; Zannini, E.; Arendt, E.K. Lactic Acid Bacteria as Sensory Biomodulators for Fermented Cereal-Based Beverages. *Trends Food Sci. Technol.* **2016**, *54*, 17–25. [CrossRef]
6. Marshall, R.T.; Goff, H.D.; Hartel, R.W. *Ice Cream*, 6th ed.; Kluwer/Plenum Publishing: New York, NY, USA, 2003; ISBN 9788578110796.
7. Başyiğit, G.; Kuleaşan, H.; Karahan, A.G. Viability of Human-Derived Probiotic Lactobacilli in Ice Cream Produced with Sucrose and Aspartame. *J. Ind. Microbiol. Biotechnol.* **2006**, *33*, 796–800. [CrossRef] [PubMed]
8. Tamime, A.Y.; Thomas, L.V. *Probiotic Dairy Products*, 2nd ed.; Tamime, A.Y., Thomas, L.V., Eds.; John Wiley and Sons Ltd: Hoboken, NJ, USA, 2018; ISBN 978-1-119-21410-6.
9. Akin, N.; Öztürk, H.I. The effects of probiotic cultures on quality characteristics of ice cream. In *Microbial Cultures and Enzymes in Dairy Technology*; IGI Global: Hershey, PA, USA, 2018; pp. 297–315. ISBN 9781522553649.
10. Arslan, A.A.; Gocer, E.M.C.; Demir, M.; Atamer, Z.; Hinrichs, J.; Kücükcetin, A. Viability of Lactobacillus Acidophilus ATCC 4356 Incorporated into Ice Cream Using Three Different Methods. *Dairy Sci. Technol.* **2016**, *96*, 477–487. [CrossRef]
11. Pankiewicz, U.; Góral, M.; Kozłowicz, K.; Góral, D. Application of Pulsed Electric Field in Production of Ice Cream Enriched with Probiotic Bacteria (L. Rhamnosus B 442) Containing Intracellular Calcium Ions. *J. Food Eng.* **2020**, *275*. [CrossRef]
12. Balthazar, C.F.; Silva, H.L.A.; Esmerino, E.A.; Rocha, R.S.; Moraes, J.; Carmo, M.A.V.; Azevedo, L.; Camps, I.; Abud, Y.K.D.; Sant'Anna, C.; et al. The Addition of Inulin and Lactobacillus Casei 01 in Sheep Milk Ice Cream. *Food Chem.* **2018**, *246*, 464–472. [CrossRef] [PubMed]
13. da Silva, P.D.L.; Bezerra, M.d.F.; dos Santos, K.M.O.; Correia, R.T.P. Potentially Probiotic Ice Cream from Goat's Milk: Characterization and Cell Viability during Processing, Storage and Simulated Gastrointestinal Conditions. *LWT-Food Sci. Technol.* **2015**, *62*, 452–457. [CrossRef]
14. Sacchi, R.; Caporaso, N.; Squadrilli, G.A.; Paduano, A.; Ambrosino, M.L.; Cavella, S.; Genovese, A. Sensory Profile, Biophenolic and Volatile Compounds of an Artisanal Ice Cream ('Gelato') Functionalised Using Extra Virgin Olive Oil. *Int. J. Gastron. Food Sci.* **2019**, *18*, 100173. [CrossRef]

15. Gremski, L.A.; Coelho, A.L.K.; Santos, J.S.; Daguer, H.; Molognoni, L.; do Prado-Silva, L.; Sant'Ana, A.S.; da Silva Rocha, R.; da Silva, M.C.; Cruz, A.G.; et al. Antioxidants-Rich Ice Cream Containing Herbal Extracts and Fructooligossaccharides: Manufacture, Functional and Sensory Properties. *Food Chem.* **2019**, *298*, 125098. [CrossRef] [PubMed]

16. Aboulfazli, F.; Baba, A.S.; Misran, M. Effects of Fermentation by Bifidobacterium Bifidum on the Rheology and Physical Properties of Ice Cream Mixes Made with Cow and Vegetable Milks. *Int. J. Food Sci. Technol.* **2015**, *50*, 942–949. [CrossRef]

17. Alamprese, C.; Foschino, R.; Rossi, M.; Pompei, C.; Savani, L. Survival of Lactobacillus Johnsonii La1 and Influence of Its Addition in Retail-Manufactured Ice Cream Produced with Different Sugar and Fat Concentrations. *Int. Dairy J.* **2002**, *12*, 201–208. [CrossRef]

18. Akalın, A.S.; Kesenkas, H.; Dinkci, N.; Unal, G.; Ozer, E.; Kınık, O. Enrichment of Probiotic Ice Cream with Different Dietary Fibers: Structural Characteristics and Culture Viability. *J. Dairy Sci.* **2018**, *101*, 37–46. [CrossRef]

19. Sah, B.N.P.; Vasiljevic, T.; McKechnie, S.; Donkor, O.N. Physicochemical, Textural and Rheological Properties of Probiotic Yogurt Fortified with Fibre-Rich Pineapple Peel Powder during Refrigerated Storage. *LWT-Food Sci. Technol.* **2016**, *65*, 978–986. [CrossRef]

20. Duboc, P.; Mollet, B. Applications of Exopolysaccharides in the Dairy Industry. *Int. Dairy J.* **2001**, *11*, 759–768. [CrossRef]

21. Dertli, E.; Toker, O.S.; Durak, M.Z.; Yilmaz, M.T.; Tatlisu, N.B.; Sagdic, O.; Cankurt, H. Development of a Fermented Ice-Cream as Influenced by in Situ Exopolysaccharide Production: Rheological, Molecular, Microstructural and Sensory Characterization. *Carbohydr. Polym.* **2016**, *136*, 427–440. [CrossRef]

22. Zhang, J.; Zhao, W.; Guo, X.; Guo, T.; Zheng, Y.; Wang, Y.; Hao, Y.; Yang, Z. Survival and Effect of Exopolysaccharide-Producing Lactobacillus Plantarum YW11 on the Physicochemical Properties of Ice Cream. *Pol. J. Food Nutr. Sci.* **2017**, *67*, 191–200. [CrossRef]

23. Martínez, L.; Ros, G.; Nieto, G. Hydroxytyrosol: Health Benefits and Use as Functional Ingredient in Meat. *Medicines* **2018**, *5*, 13. [CrossRef] [PubMed]

24. De los Reyes-Gavilan, C.G.; Limsowtin, G.K.Y.; Tailliez, P.; Sechaud, L.; Accolas, J.P. A Lactobacillus Helveticus-Specific DNA Probe Detects Restriction Fragment Length Polymorphisms in This Species. *Appl. Environ. Microbiol.* **1992**, *58*, 3429–3432. [CrossRef]

25. Di Cagno, R.; Minervini, G.; Rizzello, C.G.; De Angelis, M.; Gobbetti, M. Effect of Lactic Acid Fermentation on Antioxidant, Texture, Color and Sensory Properties of Red and Green Smoothies. *Food Microbiol.* **2011**, *28*, 1062–1071. [CrossRef]

26. Pontonio, E.; Di Cagno, R.; Tarraf, W.; Filannino, P.; De Mastro, G.; Gobbetti, M. Dynamic and Assembly of Epiphyte and Endophyte Lactic Acid Bacteria during the Life Cycle of *Origanum Vulgare* L. *Front. Microbiol.* **2018**, *9*, 1–16. [CrossRef]

27. Zwietering, M.H.; Jongenburger, I.; Rombouts, F.M.; Van 'T Riet, K. Modeling of the Bacterial Growth Curve. *Appl. Environ. Microbiol.* **1990**, *56*, 1875–1881. [CrossRef] [PubMed]

28. Bounaix, M.S.; Gabriel, V.; Morel, S.; Robert, H.; Rabier, P.; Remaud-Siméon, M.; Gabriel, B.; Fontagné-Faucher, C. Biodiversity of Exopolysaccharides Produced from Sucrose by Sourdough Lactic Acid Bacteria. *J. Agric. Food Chem.* **2009**, *57*, 10889–10897. [CrossRef]

29. Auty, M.A.E.; Gardiner, G.E.; McBrearty, S.J.; O'Sullivan, E.O.; Mulvihill, D.M.; Collins, J.K.; Fitzgerald, G.F.; Stanton, C.; Ross, R.P. Direct in Situ Viability Assessment of Bacteria in Probiotic Dairy Products Using Viability Staining in Conjunction with Confocal Scanning Laser Microscopy. *Appl. Environ. Microbiol.* **2001**, *67*, 420–425. [CrossRef]

30. Rizzello, C.G.; Nionelli, L.; Coda, R.; De Angelis, M.; Gobbetti, M. Effect of Sourdough Fermentation on Stabilisation, and Chemical and Nutritional Characteristics of Wheat Germ. *Food Chem.* **2010**, *119*, 1079–1089. [CrossRef]

31. Zeppa, G.; Conterno, L.; Gerbi, V. Determination of Organic Acids, Sugars, Diacetyl, and Acetoin in Cheese by High-Performance Liquid Chromatography. *J. Agric. Food Chem.* **2001**, *49*, 2722–2726. [CrossRef]

32. Doi, E.; Shibata, D.; Matoba, T. Modified Colorimetric Ninhydrin Methods for Peptidase Assay. *Anal. Biochem.* **1981**, *118*, 173–184. [CrossRef] [PubMed]

33. Akeson, W.R.; Stahmann, M.A. A Pepsin Pancreatin Digest Index of Protein Quality Evaluation. *J. Nutr.* **1964**, *83*, 257–261. [CrossRef] [PubMed]

34. Bradford, M.M. A Rapid and Sensitive Method for the Quantitation of Microgram Quantities of Protein Utilizing the Principle of Protein-Dye Binding. *Anal. Bio* **1976**, *72*, 248–254. [CrossRef]

35. Wu, T.; Grootaert, C.; Voorspoels, S.; Jacobs, G.; Pitart, J.; Kamiloglu, S.; Possemiers, S.; Heinonen, M.; Kardum, N.; Glibetic, M.; et al. Aronia (Aronia Melanocarpa) Phenolics Bioavailability in a Combined in Vitro Digestion/Caco-2 Cell Model Is Structure and Colon Region Dependent. *J. Funct. Foods* **2017**, *38*, 128–139. [CrossRef]

36. Mandalari, G.; Faulks, R.M.; Bisignano, C.; Waldron, K.W.; Narbad, A.; Wickham, M.S.J. In Vitro Evaluation of the Prebiotic Properties of Almond Skins (*Amygdalus Communis* L.). *FEMS Microbiol. Lett.* **2010**, *304*, 116–122. [CrossRef] [PubMed]

37. Vernazza, C.L.; Gibson, G.R.; Rastall, R.A. In Vitro Fermentation of Chitosan Derivatives by Mixed Cultures of Human Faecal Bacteria. *Carbohydr. Polym.* **2005**, *60*, 539–545. [CrossRef]

38. Di Cagno, R.; Filannino, P.; Vincentini, O.; Cantatore, V.; Cavoski, I.; Gobbetti, M. Fermented Portulaca Oleracea L. Juice: A Novel Functional Beverage with Potential Ameliorating Effects on the Intestinal Inflammation and Epithelial Injury. *Nutrients* **2019**, *11*, 248. [CrossRef] [PubMed]

39. Borges, D.O.; Matsuo, M.M.; Bogsan, C.S.B.; da Silva, T.F.; Casarotti, S.N.; Penna, A.L.B. Leuconostoc Mesenteroides Subsp. Mesenteroides SJRP55 Reduces Listeria Monocytogenes Growth and Impacts on Fatty Acids Profile and Conjugated Linoleic Acid Content in Fermented Cream. *Lwt* **2019**, *107*, 264–271. [CrossRef]

40. Chen, Y.P.; Hsu, C.A.; Hung, W.T.; Chen, M.J. Effects of Lactobacillus Paracasei 01 Fermented Milk Beverage on Protection of Intestinal Epithelial Cell in Vitro. *J. Sci. Food Agric.* **2016**, *96*, 2154–2160. [CrossRef]
41. Kato-Kataoka, A.; Nishida, K.; Takada, M.; Suda, K.; Kawai, M.; Shimizu, K.; Kushiro, A.; Hoshi, R.; Watanabe, O.; Igarashi, T.; et al. Fermented Milk Containing Lactobacillus Casei Strain Shirota Prevents the Onset of Physical Symptoms in Medical Students under Academic Examination Stress. *Benef. Microbes* **2016**, *7*, 153–156. [CrossRef]
42. Levante, A.; Bancalari, E.; Tambassi, M.; Lazzi, C.; Neviani, E.; Gatti, M. Phenotypic Diversity of Lactobacillus Casei Group Isolates as a Selection Criterion for Use as Secondary Adjunct Starters. *Microorganisms* **2020**, *8*, 128. [CrossRef]
43. Ma, C.; Ma, A.; Gong, G.; Liu, Z.; Wu, Z.; Guo, B.; Chen, Z. Cracking Streptococcus Thermophilus to Stimulate the Growth of the Probiotic Lactobacillus Casei in Co-Culture. *Int. J. Food Microbiol.* **2015**, *210*, 42–46. [CrossRef]
44. Servili, M.; Rizzello, C.G.; Taticchi, A.; Esposto, S.; Urbani, S.; Mazzacane, F.; Di Maio, I.; Selvaggini, R.; Gobbetti, M.; Di Cagno, R. Functional Milk Beverage Fortified with Phenolic Compounds Extracted from Olive Vegetation Water, and Fermented with Functional Lactic Acid Bacteria. *Int. J. Food Microbiol.* **2011**, *147*, 45–52. [CrossRef] [PubMed]
45. Silva, A.F.R.; Resende, D.; Monteiro, M.; Coimbra, M.A.; Silva, A.M.S.; Cardoso, S.M. Application of Hydroxytyrosol in the Functional Foods Field: From Ingredient to Dietary Supplements. *Antioxidants* **2020**, *9*, 1246. [CrossRef]
46. Koh, A.; De Vadder, F.; Kovatcheva-Datchary, P.; Bäckhed, F. From Dietary Fiber to Host Physiology: Short-Chain Fatty Acids as Key Bacterial Metabolites. *Cell* **2016**, *165*, 1332–1345. [CrossRef]
47. Sottero, B.; Rossin, D.; Poli, G.; Biasi, F. Lipid Oxidation Products in the Pathogenesis of Inflammation-Related Gut Diseases. *Curr. Med. Chem.* **2018**, *25*, 1311–1326. [CrossRef] [PubMed]
48. Wang, W.; Xia, T.; Yu, X. Wogonin Suppresses Inflammatory Response and Maintains Intestinal Barrier Function via TLR4-MyD88-TAK1-Mediated NF-KB Pathway in Vitro. *Inflamm. Res.* **2015**, *64*, 423–431. [CrossRef]
49. Sinagra, E.; Pompei, G.; Tomasello, G.; Cappello, F.; Morreale, G.C.; Amvrosiadis, G.; Rossi, F.; Lo Monte, A.I.; Rizzo, A.G.; Raimondo, D. Inflammation in Irritable Bowel Syndrome: Myth or New Treatment Target? *World J. Gastroenterol.* **2016**, *22*, 2242–2255. [CrossRef] [PubMed]
50. Bouayed, J.; Bohn, T. Exogenous Antioxidants—Double-Edged Swords in Cellular Redox State: Health Beneficial Effects at Physiologic Doses versus Deleterious Effects at High Doses. *Oxid. Med. Cell Longev.* **2010**, *3*, 228–237. [CrossRef] [PubMed]
51. Halliwell, B.; Zhao, K.; Whiteman, M. The Gastrointestinal Tract: A Major Site of Antioxidant Action? *Free Radic. Res.* **2000**, *33*, 819–830. [CrossRef]
52. Tlais, A.Z.A.; Da Ros, A.; Filannino, P.; Vincentini, O.; Gobbetti, M.; Di Cagno, R. Biotechnological Re-Cycling of Apple by-Products: A Reservoir Model to Produce a Dietary Supplement Fortified with Biogenic Phenolic Compounds. *Food Chem.* **2021**, *336*, 127616. [CrossRef] [PubMed]
53. Kucharzik, T.; Hudson, J.T.; Lügering, A.; Abbas, J.A.; Bettini, M.; Lake, J.G.; Evans, M.E.; Ziegler, T.R.; Merlin, D.; Madara, J.L.; et al. Acute Induction of Human IL-8 Production by Intestinal Epithelium Triggers Neutrophil Infiltration without Mucosal Injury. *Gut* **2005**, *54*, 1565–1572. [CrossRef]
54. Tonolo, F.; Sandre, M.; Ferro, S.; Folda, A.; Scalcon, V.; Scutari, G.; Feller, E.; Marin, O.; Bindoli, A.; Rigobello, M.P. Milk-Derived Bioactive Peptides Protect against Oxidative Stress in a Caco-2 Cell Model. *Food Funct.* **2018**, *9*, 1245–1253. [CrossRef] [PubMed]
55. Tonolo, F.; Fiorese, F.; Moretto, L.; Folda, A.; Scalcon, V.; Grinzato, A.; Ferro, S.; Arrigoni, G.; Bindoli, A.; Feller, E.; et al. Identification of New Peptides from Fermented Milk Showing Antioxidant Properties: Mechanism of Action. *Antioxidants* **2020**, *9*, 117. [CrossRef] [PubMed]
56. Reyes-Díaz, A.; González-Córdova, A.F.; Hernández-Mendoza, A.; Reyes-Díaz, R.; Vallejo-Cordoba, B. Immunomodulation by Hydrolysates and Peptides Derived from Milk Proteins. *Int. J. Dairy Technol.* **2018**, *71*, 1–9. [CrossRef]
57. Reyes-Díaz, A.; Mata-Haro, V.; Hernández, J.; González-Córdova, A.F.; Hernández-Mendoza, A.; Reyes-Díaz, R.; Torres-Llanez, M.J.; Beltrán-Barrientos, L.M.; Vallejo-Cordoba, B. Milk Fermented by Specific Lactobacillus Strains Regulates the Serum Levels of IL-6, TNF-α and IL-10 Cytokines in a LPS-Stimulated Murine Model. *Nutrients* **2018**, *10*, 691. [CrossRef] [PubMed]
58. Fardet, A.; Rock, E. In Vitro and in Vivo Antioxidant Potential of Milks, Yoghurts, Fermented Milks and Cheeses: A Narrative Review of Evidence. *Nutr. Res. Rev.* **2018**, *31*, 52–70. [CrossRef] [PubMed]
59. Vaisberg, M.; Paixão, V.; Almeida, E.B.; Santos, J.M.B.; Foster, R.; Rossi, M.; Pithon-Curi, T.C.; Gorjão, R.; Momesso, C.M.; Andrade, M.S.; et al. Daily Intake of Fermented Milk Containing Lactobacillus Casei Shirota (Lcs) Modulates Systemic and Upper Airways Immune/Inflammatory Responses in Marathon Runners. *Nutrients* **2019**, *11*, 1678. [CrossRef]

Metabolomics Analysis Reveals the Effect of Fermentation to Secondary Metabolites of *Chenopodium album* L. Based on UHPLC-QQQ-MS

Na Liu [1,2], Xiaoping An [1,2], Yuan Wang [1,2] and Jingwei Qi [1,2,*]

1 College of Animal Science, Inner Mongolia Agricultural University, Hohhot 010018, China
2 Inner Mongolia Herbivorous Livestock Feed Engineering Technology Research Center, Hohhot 010018, China
* Correspondence: qijingwei@imau.edu.cn

Abstract: *Chenopodium album* L. (CAL) is an excellent vegetable crop that is rich in nutrients and possesses potential pharmaceutical value. However, the research on the secondary metabolites and the processing utilization of CAL has been rarely reported. In this study, the polyphenol content, microstructure and secondary metabolite composition of aerial parts of CAL (AC), including stems, leaves, inflorescence and grain, before and after fermentation were investigated. The results showed that the polyphenol content of fermented AC (FAC) was significantly higher than that of AC (increased by 38.62%). The AC had a compact surface, while FAC had a loose and cracked surface with large holes. A total of 545 secondary metabolites, including 89 alkaloids, 179 flavonoids, 25 lignans and coumarins, 163 phenolic acids, 35 terpenoids, 9 quinones, 6 tannins and 39 others, were identified in the AC and FAC by UHPLC-QQQ-MS metabolomics. Differential metabolites analysis reviewed 285 differential metabolites (117 upregulated and 168 downregulated) between AC and FAC. The decrease in parts of toxic alkaloids accompanied with the increase in some biologically active substances with small molecules, such as quercetin, kaempferol, p-coumaric acid and protocatechuic acid, indicated that fermentation is beneficial to enhance the bioavailability of AC. This study provides a reference value for the identification of secondary metabolites from AC and the application of fermentation in the deepness development of AC.

Keywords: *Chenopodium album* L.; fermentation; secondary metabolites; metabolomics

Citation: Liu, N.; An, X.; Wang, Y.; Qi, J. Metabolomics Analysis Reveals the Effect of Fermentation to Secondary Metabolites of *Chenopodium album* L. Based on UHPLC-QQQ-MS. *Fermentation* **2023**, *9*, 100. https://doi.org/10.3390/fermentation9020100

Academic Editors: Michela Verni and Carlo Giuseppe Rizzello

Received: 20 November 2022
Revised: 13 January 2023
Accepted: 20 January 2023
Published: 22 January 2023

1. Introduction

Chenopodium is a species considered to be an excellent vegetable crop [1] because it can provide proteins, dietary fiber, minerals, vitamins and essential fatty acids [2]. *Chenopodium album* L. (CAL), also known as lamb's quarter, is a natural plant which withstands harsh soil and climatic conditions, distributed around the world, including in semi-arid and light-saline environments of China [1,3]. The potential pharmaceutical value of this plant, such as antiscorbutic, diuretic, anthelmintic and cardiotonic, has been demonstrated [4]. It can also be used for the treatment of peptic ulcer, flatulence, hepatic disorders, spleen enlargement and burns [5]. CAL contains many secondary metabolites, including polyphenols, flavonoids and alkaloids, which exhibit strong antioxidant, antibacterial, anticancer and anti-inflammatory activity [1,2]. Laghari et al. (2011) [6] confirmed that the extract of CAL leaves had a high level of total phenolic compounds and antioxidant activity and revealed that the extract of CAL leaves has great potential as a source for natural health products.

A scanning of the literature has revealed that the CAL also contain some antinutritional factors such as saponins, tannins and phytic acid, which limits the digestion and utilization of proteins and carbohydrates by animals [7]. Fermentation is a traditional technique for food quality modification, which transforms complex organic substances into a simpler form, increases the nutritional value of raw material(s) and enhances specific biological functionalities due to the created secondary metabolites [8]. It has been proven that fermentation decreases the phytic acid level of wheat bran [9], and total tannin, oxalic acid, phytic acid and saponin of paper mulberry [10]. Therefore, it is worth trying to use fermentation technology for the depth development of CAL. The previous study in our lab established the fermentation conditions for aerial parts of CAL (AC) and found that fermented AC (FAC) had positive effects on growth, nutrient digestibility, immunity, carcass characteristics and meat quality of broilers [3]. However, the investigation on secondary metabolites analysis of CAL was limited, and no work about the effects of fermentation on secondary metabolites of CAL has been reported.

In recent years, widely targeted metabolomics analysis, mostly based on ultrahigh performance liquid chromatography coupled to triple quadrupole mass spectrometry (UHPLC-QQQ-MS), has been increasingly applied in the field of analysis and identification of secondary metabolites in plants, due to its advantages of high throughput, fast separation, high sensitivity and wide coverage [11–13]. It integrates the advantages of nontargeted and targeted metabolite detection technologies and provides an effective qualitative and quantitative method to identify secondary metabolites [14]. Therefore, the polyphenol content and microstructure of aerial parts of CAL (including stems, leaves, inflorescence and grain) before and after fermentation were compared. Then, the UHPLC-QQQ-MS metabolomics approach was employed to analyze the types and relative contents of secondary metabolites. These results enrich our understanding of the chemical components of AC and provide a reference for the application of fermentation in the deepness development of AC.

2. Materials and Methods

2.1. Materials

The CAL used in this work was collected from April to November in Hohhot (Inner Mongolia, Hohhot, China) and identified by Professor Zhaozhe Li of Inner Mongolia Agricultural University. The aerial parts of CAL, including stems, leaves, inflorescence and grain, were obtained, dried at 25 °C in the shade, ground to a fine powder and then combined for further fermentation. *Bacillus subtilis* (CGMCC 1.0892), *Lactobacillus plantarum* (CGMCC No.1.12934) and *Saccharomyces cerevisiae* (CGMCC No. 2.1190) were purchased from the China General Microbiological Culture Collection Centre (Beijing, China). Pectinase was obtained commercially (Beijing Solarbio Technology Co., Ltd., Beijing, China). Maize meal and ground cinnamon were obtained from a local market. All other chemicals were of analytical grade.

2.2. Fermentation of AC and Sample Preparation

The fermentation of AC was conducted based on previous study [3]. Briefly, AC was mixed with corn meal, ground cinnamon and pectinase at a ratio of 16.5:3.0:0.1:0.4 (4 kg in total), blended with 50% distilled water (w/v), and inoculated with 0.1% compound probiotics. The compound probiotics were prepared by mixing *S. cerevisiae*, *B. subtilis* and *L. plantarum* at a ratio of 1:1:1. Fermentation was conducted with multi-layer polythene bags (5 kg capacity) equipped with a gas pressure opening valve in an incubator for 24 h at 30 °C. The substrate before and after fermentation, the AC and FAC, were collected, allowed to dry (45 °C, 24 h) and used for further analysis.

2.3. Determination of Polyphenol Content

Extractions of AC and FAC were performed using the hot water extraction method. The sample was extracted at 85 °C with distilled water at a ratio of 1:40 (g/mL) for 90 min. The extracts were lyophilized and kept at 4 °C. The polyphenol content of the AC and FAC extracts was determined by the Folin–Ciocalteu method according to Liu et al. (2022) [15], and gallic acid was used as a standard. An appropriate concentration of the sample solution (0.5 mL) was mixed with 10% Folin–Ciocalteu reagent (2.5 mL) and shaken. After 5 min, 2.0 mL of the 7.5% Na_2CO_3 solution was added to the mixture. The reaction mixture was kept in the dark for 60 min, after which the absorbance was measured at 756 nm. All the determinations were carried out in triplicate.

2.4. Scanning Electron Microscopy Analysis

Microscopic images of the AC and FAC were recorded using a JSM-6390 LV scanning electron microscope. The AC and FAC samples were sputter-coated with gold under vacuum, and their images were viewed at an accelerating voltage of 20 kV.

2.5. Secondary Metabolites Analysis

2.5.1. Sample Preparation and Extraction Process

The AC and FAC samples are freeze-dried by vacuum freeze-dryer (Scientz-100F). The freeze-dried sample was crushed using a mixer mill (MM 400, Retsch) with a zirconia bead for 1.5 min at 30 Hz. Dissolve 100 mg of lyophilized powder with 1.2 mL 70% methanol solution, vortex 30 s every 30 min for 6 times in total, place the sample in a refrigerator at 4 °C overnight. Following centrifugation at 12,000 rpm for 10 min, the extracts were filtrated (SCAA-104, 0.22μm pore size; ANPEL, Shanghai, China) before UPLC-MS/MS analysis.

2.5.2. UPLC Conditions

The sample extracts were analyzed using an UPLC-ESI-MS/MS system (UPLC, SHIMADZU Nexera X2 system, Kyoto, Japan; MS, Applied Biosystems 4500 Q TRAP, Foster City, CA, USA). The analytical conditions were as follows. UPLC: column, Agilent SB-C18 (1.8 μm, 2.1 mm × 100 mm). The mobile phase consisted of solvent A, pure water with 0.1% formic acid, and solvent B, acetonitrile with 0.1% formic acid. Sample measurements were performed with a gradient program that employed the starting conditions of 95% A, 5% B. Within 9 min, a linear gradient to 5% A, 95% B was programmed, and a composition of 5% A, 95% B was kept for 1 min. Subsequently, a composition of 95% A, 5.0% B was adjusted within 1.1 min and kept for 2.9 min. The flow velocity was set as 0.35 mL per minute. The column oven was set to 40 °C. The injection volume was 4 μL. The effluent was alternatively connected to an ESI triple quadrupole linear ion trap (QTRAP)-MS.

2.5.3. ESI-Q TRAP-MS/MS

Linear ion trap (LIT) and triple quadrupole (QQQ) scans were acquired on a triple quadrupole linear ion trap mass spectrometer (QTRAP, Applied AB4500 QTRAP UPLC/MS/MS System, Foster City, CA, USA), equipped with an ESI Turbo Ion Spray interface, operating in positive and negative ion mode and controlled by Analyst 1.6.3 software (AB Sciex, Framingham, MA, USA). The ESI source operation parameters were as follows: ion source, turbo spray; source temperature 550 °C; ion spray voltage (IS) 5500 V (positive ion mode)/−4500 V (negative ion mode); ion source gas I (GSI), gas II(GSII), curtain gas (CUR) was set at 50, 60 and 25.0 psi, respectively; the collision-activated dissociation (CAD) was set to high. Instrument tuning and mass calibration were performed with 10 and 100 μmoL/L polypropylene glycol solutions in QQQ and LIT modes, respectively. QQQ scans were acquired as multiple reaction monitoring (MRM) experiments with collision gas (nitrogen) set to medium. DP and CE for individual MRM transitions were carried out with further DP and CE optimization. A specific set of MRM transitions were monitored for each period according to the metabolites eluted within this period [16].

2.6. Statistical Analysis

The polyphenol content data for the AC and FAC were sorted in Excel 2020 for Windows and analyzed in triplicate, and the results were expressed as the mean ± standard deviation (SD) using SAS 9.3 for Windows. One-way analysis of variance (ANOVA) and Duncan's multiple range tests were used to determine the significance of the differences among samples, with a significance level of 0.05.

Metabolite data were $\log_2$-transformed for statistical analysis to improve normality and were normalized. Hierarchical clustering analysis (HCA), principal component analysis

(PCA), and orthogonal partial least squares discriminant analysis (OPLS-DA) have been used to analyze the multivariate and differences of metabolites by R software. Based on OPLS-DA analysis, the differential metabolites were screened by the following criteria: (1) if the difference of metabolites content between the control group and the experimental group is more than 2 times or less than 0.5, the difference is considered to be significant; (2) on the basis of the above, the metabolites with VIP ≥ 1 are selected. The Kyoto Encyclopedia of Genes and Genomes (KEGG) database was used to annotate the differential metabolites and analyze metabolic pathways. All data were graphed using GraphPad Prism v6.01 (GraphPad Software Inc., La Jolla, CA, USA).

3. Results and Discussion

3.1. The Polyphenol Content

Polyphenols constitute a large group of plant secondary metabolites widely distributed throughout the plant-derived foods [17]. They exhibit several biological activities, for instance, functioning as antioxidant, antiaging, anti-inflammation properties, and have positive effects on human microbiota composition and functionality [18]. The levels of polyphenols of the AC and FAC are shown in Figure 1. The polyphenol content of the FAC increased to 18.59 ± 1.49 mg/g, which was a 38.62% increase compared to that of the AC (13.41 ± 1.40 mg/g). The result clearly indicated that the fermentation significantly improved the polyphenol content of AC ($p < 0.05$). Gan et al. (2016) [19] investigated the influences of fermentation on the polyphenols of eight common edible legumes, they found that both natural and LAB-mediated fermentation significantly enhanced polyphenol in the soluble fraction of all selected edible legumes compared to nonfermented. It coincides with the experimental results. An improvement in polyphenol content may be caused via secondary metabolic pathways in the fermentation process or released from the substrate by enzymes produced by the microorganisms [20].

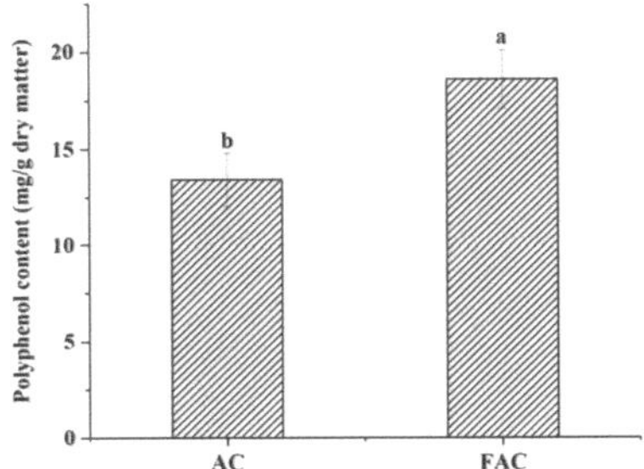

Figure 1. The polyphenol content of AC before and after fermentation. [a,b] $p < 0.05$ compared between AC and FAC.

3.2. SEM Analysis

SEM is a qualitative method for characterizing plant surface microstructure. Figure 2 shows the surface images of the AC and FAC at a magnification factor of 1000-fold. The two observations showed that both the AC and FAC had many discontinuous fragments. However, the AC had a compact surface structure and did not show obvious holes on the surface. In contrast, FAC had loose and cracked surface structure with large holes. It is reported that during the fermentation process, the micro-structure of substrate materials is altered, which further influences the retention and release of the bioactive substances in the substrate material [8]. Chen et al. (2019) compared SEM images of rice bran before and after fermentation, they found fermentation changed the microstructure of rice bran from being compact to loose and porous, and increased the extractability of bound phenolics [21]. It is suggested that fermentation treatment broke down the cell wall structure of rice bran and made the alkali ions more accessible to the covalent bonds between phenolics and fiber.

Similarly, in this experiment, the polyphenol content of the AC was increased by 38.63% by fermentation, which confirmed that fermentation promoted the release of the polyphenols through changing the structure of substrate.

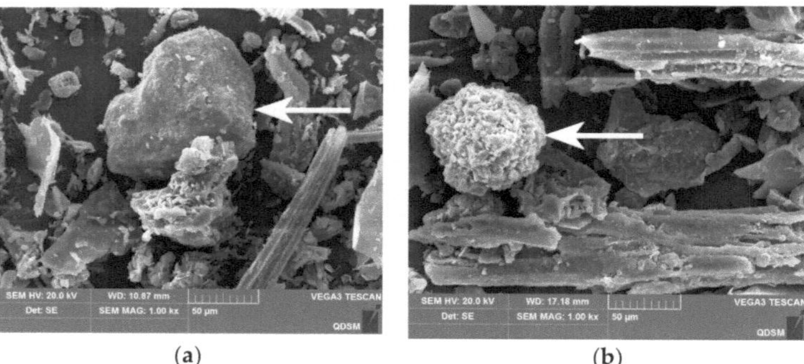

Figure 2. SEM images of (**a**) AC and (**b**) FAC.

3.3. Data Quality Assessment and Metabolites Identified

In order to have a clearer understanding of the changes in the secondary metabolites of AC before and after fermentation, the UPLC-MS platform was adopted. The total ion current (TIC) map of the mixed sample quality control (QC) is shown in Figure S1a and shows the summed intensity of all ions in the mass spectrum at different time points [16]. Based on the local metabolic database, the metabolites of the samples were qualitatively and quantitatively analyzed by mass spectrometry. The multipeak detection plot (XIC) of the metabolites in multiple reaction monitoring (MRM) mode is illustrated in Figure S1b, and the figure shows the substances that can be detected in the sample, and each mass peak of a different color represents a detected metabolite [22]. Through the overlap display analysis of the TIC map of different quality control QC samples (Figure S1c), the results showed the high overlap ratio of total ion current (TIC) curves of the QC samples, indicating that the signal stability was good and the test results were reliable.

A total number of 545 secondary metabolites were target identified in AC and FAC, including 89 alkaloids, 179 flavonoids, 25 lignans and coumarins, 163 phenolic acids, 35 terpenoids, 9 quinones, 6 tannins, and 39 others (Figure 3a). The accumulation pattern of secondary metabolites among AC and FAC could be visualized through a heatmap hierarchical cluster analysis (Figure 3b). The heatmap showed that the three biological replicates of each group were clustered together, indicating the good homogeneity between replicates and the high reliability of the data. Principal component analysis (PCA) is a multidimensional data statistical analysis method of unsupervised pattern recognition [22]. The PCA result showed that there are significant differences among AC and FAC groups, and no significant difference within groups (Figure 3c).

3.4. Identification of Differential Metabolites

Orthogonal partial least squares discriminant analysis (OPLS-DA) is a multivariate statistical analysis method with supervised pattern recognition which can maximize group differentiation between groups and help to find differential metabolites [13]. The score plots between AC and FAC in OPLS-DA are shown in Figure 4a. In the OPLS-DA model, R^2X and R^2Y were used to represent the interpretation rate to the X and Y matrices, respectively, and Q2 represented the prediction ability. A Q^2 value greater than 0.9 indicates that the model is excellent. According to the results ($R^2Y = 1$, $Q^2 = 0.986$) in Figure 4b, the models are stable and reliable and could be applied to further screen for differential metabolites. In addition, the OPLS-DA mode was verified through 200 alignment experiments, and the result confirmed that the model was meaningful (Figure 4b).

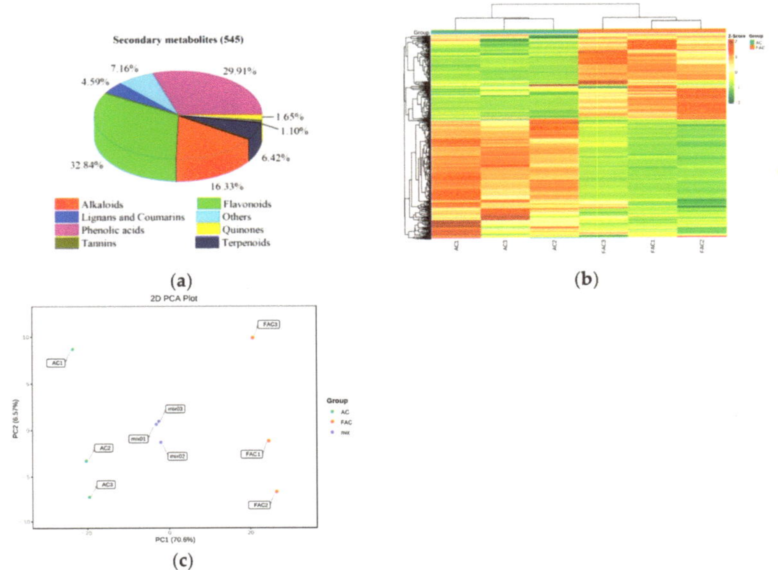

Figure 3. Secondary metabolites identified. (**a**) Classification of the 545 secondary metabolites of AC and FAC; (**b**) hierarchical cluster analysis (HCA); (**c**) principal component analysis (PCA).

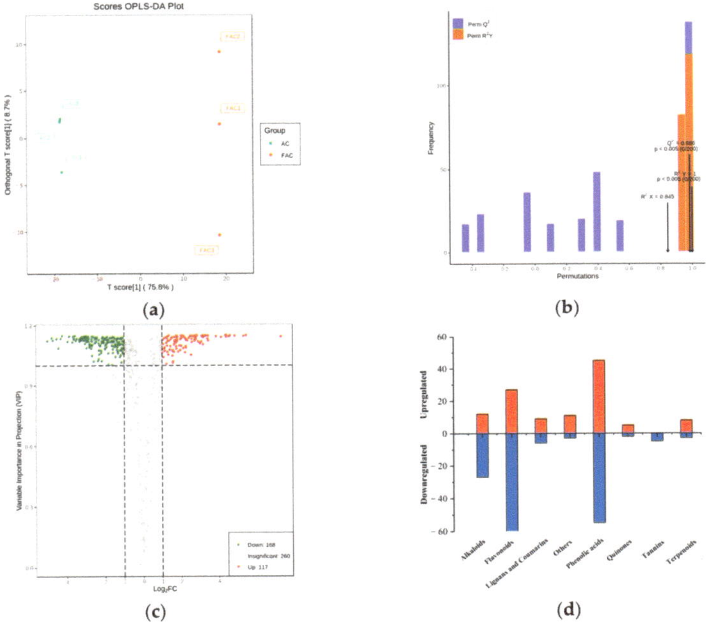

Figure 4. *Cont.*

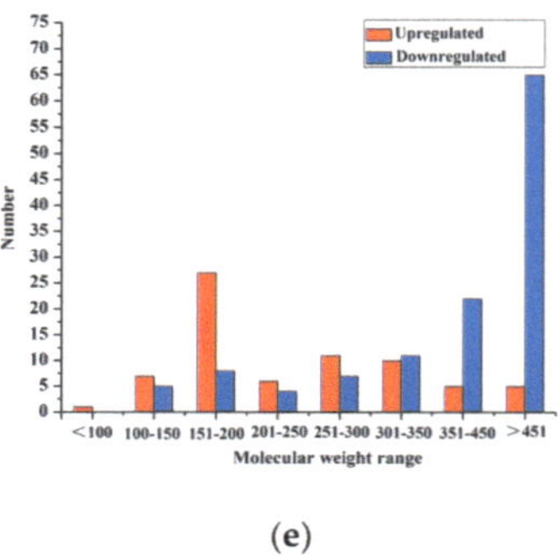

(**e**)

Figure 4. Differentially accumulated metabolites between AC and FAC. (**a**) Score scatter plot of the OPLS-DA model; (**b**) permutation test of the OPLS-DA model; (**c**) volcano plot of the differential metabolites; (**d**) classification of differentially expressed metabolites; (**e**) molecular weight range of differential phenolic acids and flavonoids.

The differential metabolites between two sample groups could be visualized through a volcano plot (Figure 4c). The criteria for significant differences included a fold change of ≥ 2 or ≤ 0.5 and a VIP score of ≥ 1. A total of 285 differential metabolites were identified (Table S1) between AC and FAC. Compared with AC, 117 metabolites in FAC were upregulated, while 168 metabolites were downregulated. The differential metabolites produced during fermentation were further classified and compared. Among all these differential metabolites, there were 39 alkaloids, 94 flavonoids, 100 phenolic acids, 15 lignans and coumarins, 7 quinones, 5 tannins, 11 terpenoids and 14 others (Table S1). The major categories were alkaloids, phenolic acids and flavonoids, accounting for more than 80% of the total detected differential substances (Figure 4d).

Alkaloids is one of the largest groups of naturally occurring plant defense compounds that have been used by humans as poisons, stimulants, sedatives and medicinal substances for thousands of years [23]. Although possessing significant benefits to humans and pharmaceutical industries, some of the plant alkaloids are categorized as main plant toxins due to their enormous structural diversity [24]. Natural toxins in the plants are often not popular due to their potential hazard for human health related to their presence as contaminants in food [25]. In this study, parts of toxic alkaloid content such as serotonin, cadaverine, spermine were significantly downregulated (Table S1) after fermentation. It was demonstrated that fermentation successfully degraded the parts of the toxic compound in the AC. Similar results were observed by Deus et al. (2021) who performed fermentation of cocoa [26]. Six amines were detected during cocoa on-farm fermentation, and the total levels of most amines decreased, in which the levels of serotonin decreased continuously and could not be detected [26]. We speculated that the reason why parts of toxic alkaloids declined in this study was the degradation caused by enzymes produced by microorganisms during the fermentation process.

Flavonoids are an important class of natural products and belong to phenolic compounds. Flavonoids are synthesized by plants via primary or secondary metabolisms. Flavonoids are also an integral part of diets for humans because they cannot be synthesized by humans [27,28]. Flavonoids are associated with a broad spectrum of health-promoting effects, such as anti-oxidative, anti-inflammatory, anti-mutagenic and anti-carcinogenic properties coupled with their capacity to modulate key cellular enzyme function [29]. Phenolic acids, as another important class of phenolic compounds, have one carboxylic acid group and are acknowledged as strong natural antioxidants. Additionally, phenolic acids exhibit a wide range of biological and pharmacological properties including anti-inflammatory, anticancer, antimicrobial, antiallergic, antiviral, antithrombotic, hepatoprotective and act as signaling molecules [30]. In comparison of differential phenolic acids and flavonoids before and after fermentation in this study, we found that the relative contents of compounds with high molecular weight were dramatically downregulated, while active substances with small molecular weight were significantly increased (Figure 4e). For instance, the relative contents of quercetin, kaempferol, p-Coumaric acid, protocatechuic acid, protocatechualdehyde and 4-Hydroxybenzoic acid of FAC were significantly higher than that of AC (Table S1). Quercetin and kaempferol are the major representative flavanols of flavonoids. Quercetin has been recognized for its multiple biological activities such as anti-oxidation, anti-inflammatory and anticancer [31]. Kenan Kinaci et al. (2012) found that quercetin increased GSH levels and decreased the eNOS and NF-κB expression levels in renal ischemia/reperfusion (I/R) injury in rats [32]. In a study performed by Yuan et al. (2020), the therapeutic mechanism of quercetin for rheumatoid arthritis (RA) was by inhibiting neutrophil activities [33]. Yousuf and colleagues (2020) have investigated the anticancer activity of quercetin by establishing Cyclin-dependent kinase 6 (CDK6) as a target. They found that quercetin could decrease the expression of CDK6 and inhibit the viability and colony formation potential of selected cancer cells [34]. Kaempferol displays several pharmacological properties, such as anti-inflammatory, antioxidant, anticancer and anti-tumor activities [35]. Rho et al. (2011) isolated kaempferol from kenaf (Hibiscus cannabinus L.) leaves and evaluated its anti-inflammatory activity. They found that kaempferol showed potent NO inhibitory activity by suppressing the expression of iNOS mRNA in a dose-dependent manner without cytotoxicity [36]. Wang et al. (2021) explored the underlying molecular mechanism of kaempferol in suppressing pancreatic cancer inducing ROS-dependent apoptosis via tissue transglutaminase (TGM2)-mediated Akt/mTOR signaling [37]. P-coumaric acid is one of the most important phenolic acids, shows antimicrobial, antioxidant and anticancer activity and plays an important role in human health [38]. One study found that p-coumaric acid decreased basal oxidative stress more effectively than vitamin E according to DNA damages in rat colonic mucosa [39]. Lou et al. (2012) specified that p-coumaric acid exhibits an antimicrobial effect against *Salmonella typhimurium*, *Shigella dysenteriae* and *Escherichia coli* [40]. A study performed by Sharma et al. (2017) revealed that p-coumaric acid induces a significant dose-dependent reduction of polyp incidence in the colon of rats exposed to the procarcinogen DMH and suppresses the formation of preneoplastic lesions [41]. Protocatechuic acid is a main anthocyanin metabolite and is reported to have antioxidant activities by decreasing lipid peroxidation and increasing the scavenging of free radical scavenging such as hydrogen peroxide (H_2O_2) and diphenylpicrylhydrazyl (DPPH) [42]. Protocatechuic aldehyde is the primary metabolites of proanthocyanidins and is demonstrated to have antiadipogenic, anti-proliferative and anti-inflammatory properties both in vivo and in vitro [43]. Wei et al. (2013) investigated the anti-inflammatory effect of protocatechuic aldehyde on myocardial ischemia/reperfusion (MI/R) injury. It was suggested that the levels of tumor necrosis factor-α, interleukin-6, intracellular adhesion molecule-1, phosphorylated IκB-α and the nuclear translocation of nuclear factor-kappa B (NF-κB) were all evidently decreased by protocatechuic aldehyde both in vivo and in vitro. Furthermore, protocatechuic aldehyde could exert great protective effects against MI/R injury in rats and ischemia/reperfusion (SI/R) injury in cultured neonatal rat cardiomyocytes [44]. 4-Hydroxybenzoic acid is a valuable intermediate for the synthesis of several bioproducts with potential applications in food, cosmetics, pharmacy, fungicides, etc. [45]. In this study, fermentation processes increased the relative contents of these active substances with small molecular weight mentioned above. The increase in active substances with small molecular weight in FAC is partially attributed to the liberation of polyphenols induced by enzymes degradation during fermentation. It is confirmed by the looser surface structure of FAC, which makes enzymes more easily able to access the interior structure of AC [21]. The results of differential metabolites of AC and FAC indicated that fermentation is beneficial to the partial degradation of toxic alkaloids and increases in biologically active substances with small molecules in aerial parts of CAL.

3.5. KEGG Annotation and Enrichment Analysis of Differential Metabolites

According to the KEGG annotation and enrichment results, the major pathways are presented in bubble plots in Figure 5. On the basis of metabolic pathway analysis, the results indicated that a total of 134 differential metabolites were annotated to metabolic pathways; the top five metabolic pathways, ranked in terms of the *p*-value, were "biosynthesis of secondary metabolites", "metabolic pathways", "ubiquinone and other terpenoid-quinone biosynthesis", "biosynthesis of various secondary metabolites—part 2" and "phenylpropanoid biosynthesis" in AC versus FAC. The pathway of the biosynthesis of secondary metabolites changed dramatically before and after fermentation, which explained that the synthesis and accumulation of secondary metabolites was influenced by fermentation.

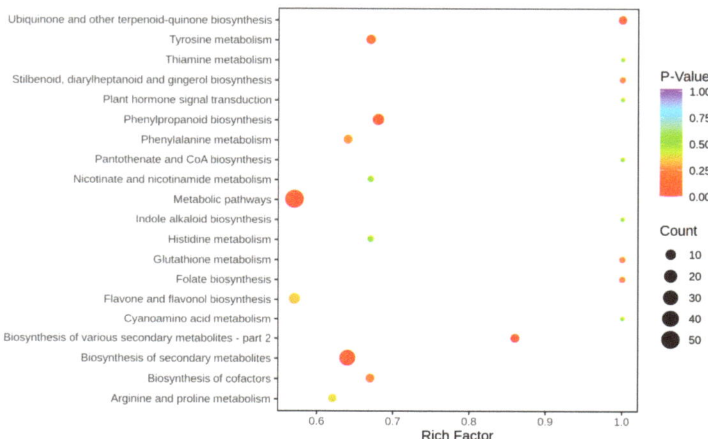

Figure 5. Top 20 enriched KEGG pathways for differential metabolites of AC and FACL.

We focused on the biosynthesis of secondary metabolites to uncover metabolite changes. According to the biosynthesis of the secondary metabolites pathway, the relative contents of many valuable compounds were obviously increased after fermentation, for example, coumarin, scopoletin and cinnamic acid. Coumarins are a broad family of secondary metabolites found in plants [46]. Due to their physiological, bacteriostatic and anti-tumor activity, they have attracted much attention for further backbone derivatization and screening as novel therapeutic agents [47]. The coumarin nucleus has proved to be easily decorated, giving the possibility of designing new coumarin-based compounds and investigating their potential in the treatment of various diseases [46]. Scopoletin is a derivative of coumarin; one of the most widespread coumarins in nature, it has been reported to exhibit pharmacological activity, including anti-hypertensive, anti-bacterial, anti-inflammatory and anti-diabetic properties [48]. Armenia et al. (2019) investigated the blood pressure lowering effect of scopoletin in oxidative stress-associated hypertensive rats, and found that scopoletin significantly decreased the systolic blood pressure (SBP), diastolic blood pressure (DBP) and mean arterial blood pressure (MAP) of the rats [49]. Cinnamic acid is a natural aromatic carboxylic acid found in plants, and has been studied in antioxidant, anticancer, anti-inflammatory and anti-diabetic properties [50]. A study conducted by Balli et al. (2020) showed that fermented millet had an approximate 30% higher cinnamic acid content and played higher antioxidant protection effects on human erythrocytes [51]. In sum, the enhancement of valuable compounds proved that fermentation improved the bioavailability of aerial parts of CAL.

4. Conclusions

In summary, the surface structure, polyphenol content and secondary metabolites of the aerial parts of CAL were considerably altered by fermentation. A total of 545 secondary metabolites, including 89 alkaloids, 179 flavonoids, 25 lignans and coumarins, 163 phenolic acids, 35 terpenoids, 9 quinones, 6 tannins and 39 others, were identified by the liquid chromatography tandem mass spectrometry-based metabolomics approach, which is used to given us a better understanding of secondary metabolites in the aerial parts of CAL. There are 285 differential metabolites that were identified (117 upregulated and 168 downregulated), mainly concentrating on alkaloids, phenolic acids and flavonoids. After fermentation, the relative amount of biologically active substance such as quercetin, kaempferol, protocatechuic acid, protocatechualdehyde and coumarin were significantly increased, while parts of toxic alkaloids such as serotonin, cadaverine and spermine were significantly decreased. The present work lays the foundation for the application of fermentation technology for deep exploitation of the aerial parts of CAL.

Supplementary Materials: The following supporting information can be downloaded at: https://www.mdpi.com/article/10.3390/fermentation9020100/s1, Figure S1: (**a**) Total ion current of one quality control sample by mass spectrometry; (**b**) Multi-peak detection plot of metabolites in the multiple reaction monitoring mode; (**c**) Total ion current overlaps of the quality control sample by mass spectrometry detection. Table S1: Differential metabolites of AC and FAC.

Author Contributions: Conceptualization, X.A.; methodology, N.L.; software, N.L.; validation, Y.W.; formal analysis, N.L.; investigation, Y.W.; resources, J.Q; data curation, X.A.; writing—original draft preparation, N.L.; writing—review and editing, N.L.; supervision, X.A.; project administration, J.Q.; funding acquisition, J.Q. and X.A. All authors have read and agreed to the published version of the manuscript.

Funding: This research was funded by Major Science and Technology Program of Inner Mongolia Autonomous Region (grant number 2020ZD0004); Major Science and Technology Program of Inner Mongolia Autonomous Region (grant number 2021ZD0023-3) and Symbolic Achievements Special Fund Project of College of Animal Science of Inner Mongolia Agricultural University(grant number BZCG202110).

Institutional Review Board Statement: Not applicable.

Informed Consent Statement: Not applicable.

Data Availability Statement: Not applicable.

Conflicts of Interest: The authors declare no conflict of interest.

References

1. Choudhary, S.P.; Sharma, D.K. Bioactive constituents, phytochemical and pharmacological properties of *Chenopodium album*: A miracle weed. *Int. J. Pharmacogn.* **2014**, *1*, 545–552.
2. Poonia, A.; Upadhayay, A. *Chenopodium album* Linn: Review of nutritive value and biological properties. *J. Food Sci. Technol. Mysore* **2015**, *52*, 3977–3985. [CrossRef]
3. Xie, M.; Wang, R.; Wang, Y.; Liu, N.; Qi, J. Effects of dietary supplementation with fermented *Chenopodium album* L. on growth, nutrient digestibility, immunity, carcase characteristics and meat quality of broilers. *Ital. J. Anim. Sci.* **2021**, *20*, 2063–2074. [CrossRef]
4. Hussain, S.; Asrar, M.; Rasul, A.; Sultana, S.; Saleem, U. *Chenopodium album* extract ameliorates carbon tetrachloride induced hepatotoxicity in rat model. *Saudi J. Biol. Sci.* **2022**, *29*, 3408–3413. [CrossRef] [PubMed]
5. Suleman, M.; Faiz, A.U.H.; Abbas, F.I. Antibacterial, antiparasitic and phytochemical activities of *Chenopodium album* (Bathua) plant extract. *Bangladesh J. Bot.* **2021**, *50*, 417–421. [CrossRef]
6. Laghari, A.H.; Memon, S.; Nelofar, A.; Yasmin, K.M. Determination of free phenolic acids and anti-oxidant activity of methanolic extracts obtained from fruits and leaves of *Chenopodium album*. *Food Chem.* **2011**, *126*, 1850–1855. [CrossRef] [PubMed]
7. Jan, R.; Saxena, D.C.; Singh, S. Analyzing the effect of optimization conditions of germination on the antioxidant activity, total phenolics, and antinutritional factors of Chenopodium (*Chenopodium album*). *J. Food Meas. Charact.* **2017**, *11*, 256–264. [CrossRef]
8. Xiang, H.; Sun-Waterhouse, D.; Waterhouse, G.I.; Cui, C.; Ruan, Z. Fermentation-enabled wellness foods: A fresh perspective. *Food Sci. Hum. Wellness* **2019**, *8*, 203–243. [CrossRef]

9. Tanasković, S.J.; Šekuljica, N.; Jovanović, J.; Gazikalović, I.; Grbavčić, S.; Đorđević, N.; Luković, N.; Hao, J.; Vukašinović Sekulić, M.; Knežević-Jugović, Z. Upgrading of valuable food component contents and anti-nutritional factors depletion by solid-state fermentation: A way to valorize wheat bran for nutrition. *J. Cereal Sci.* **2021**, *99*, 103159. [CrossRef]

10. Wang, N.; Xiong, Y.; Wang, X.; Guo, L.; Lin, Y.; Ni, K.; Yang, F. Effects of *Lacto bacillus plantarum* on Fermentation Quality and Anti-Nutritional Factors of Paper Mulberry Silage. *Fermentation* **2022**, *8*, 144. [CrossRef]

11. Li, W.; Wen, L.; Chen, Z.; Zhang, Z.; Pang, X.; Deng, Z.; Liu, T.; Guo, Y. Study on metabolic variation in whole grains of four proso millet varieties reveals metabolites important for antioxidant properties and quality traits. *Food Chem.* **2021**, *357*, 129791. [CrossRef] [PubMed]

12. Wang, F.; Chen, L.; Chen, H.; Chen, S.; Liu, Y. Analysis of flavonoid metabolites in citrus peels (*Citrus reticulata* "Dahongpao") using UPLC-ESI-MS/MS. *Molecules* **2019**, *24*, 2680. [CrossRef] [PubMed]

13. Xiao, J.; Gu, C.; He, S.; Zhu, D.; Huang, Y.; Zhou, Q. Widely targeted metabolomics analysis reveals new biomarkers and mechanistic insights on chestnut (*Castanea mollissima* Bl.) calcification process. *Food Res. Int.* **2021**, *141*, 110128. [CrossRef] [PubMed]

14. Tang, Y.C.; Liu, Y.J.; He, G.R.; Cao, Y.W.; Bi, M.M.; Song, M.; Yang, P.P.; Xu, L.F.; Ming, J. Comprehensive analysis of secondary metabolites in the extracts from different lily bulbs and their antioxidant ability. *Antioxidants* **2021**, *10*, 1634. [CrossRef]

15. Liu, N.; Wang, Y.; An, X.; Qi, J. Study on the Enhancement of Antioxidant Properties of Rice Bran Using Mixed-Bacteria Solid-State Fermentation. *Fermentation* **2022**, *8*, 212. [CrossRef]

16. Wang, F.; Huang, Y.; Wu, W.; Zhu, C.; Zhang, R.; Chen, J.; Zeng, J. Metabolomics analysis of the peels of different colored citrus fruits (*Citrus reticulata* cv. 'Shatangju') during the maturation period based on UHPLC-QQQ-MS. *Molecules* **2020**, *25*, 396. [CrossRef]

17. Condezo-Hoyos, L.; Gazi, C.; Pérez-Jiménez, J. Design of polyphenol-rich diets in clinical trials: A systematic review. *Food Res. Int.* **2021**, *149*, 110655. [CrossRef] [PubMed]

18. Gutiérrez-Escobar, R.; Aliaño-González, M.J.; Cantos-Villar, E. Wine polyphenol content and its influence on wine quality and properties: A review. *Molecules* **2021**, *26*, 718. [CrossRef]

19. Gan, R.Y.; Shah, N.P.; Wang, M.F.; Lui, W.Y.; Corke, H. Fermentation alters antioxidant capacity and polyphenol distribution in selected edible legumes. *Int. J. Food Sci. Technol.* **2016**, *51*, 875–884. [CrossRef]

20. Handa, C.L.; de Lima, F.S.; Guelfi, M.F.G.; da Silva Fernandes, M.; Georgetti, S.R.; Ida, E.I. Parameters of the fermentation of soybean flour by *Monascus purpureus* or *Aspergillus oryzae* on the production of bioactive compounds and antioxidant activity. *Food Chem.* **2019**, *271*, 274–283. [CrossRef]

21. Chen, Y.; Ma, Y.; Dong, L.; Jia, X.; Liu, L.; Huang, F.; Chi, J.; Xiao, J.; Zhang, M.; Zhang, R. Extrusion and fungal fermentation change the profile and antioxidant activity of free and bound phenolics in rice bran together with the phenolic bioaccessibility. *LWT* **2019**, *115*, 108461. [CrossRef]

22. Gao, J.; Ren, R.; Wei, Y.; Jin, J.; Ahmad, S.; Lu, C.; Wu, J.; Zheng, C.; Yang, F.; Zhu, G. Comparative metabolomic analysis reveals distinct flavonoid biosynthesis regulation for leaf color development of *Cymbidium sinense* 'Red Sun'. *Int. J. Mol. Sci.* **2020**, *21*, 1869. [CrossRef] [PubMed]

23. Gershenzon, J. *Alkaloids: Biochemistry, Ecology, and Medicinal Applications*; Crop Science Society of America: New York, NY, USA, 1999; p. 486.

24. Adibah, K.Z.M.; Azzreena, M.A. Plant toxins: Alkaloids and their toxicities. *GSC Biol. Pharm. Sci.* **2019**, *6*, 21–29.

25. Qie, M.; Li, S.; Guo, C.; Yang, S.; Zhao, Y. Study of the occurrence of toxic alkaloids in forage grass by liquid chromatography tandem mass spectrometry. *J. Chromatogr. A* **2021**, *1654*, 462463. [CrossRef]

26. Deus, V.L.; Bispo, E.S.; Franca, A.S.; Gloria, M.B.A. Understanding amino acids and bioactive amines changes during on-farm cocoa fermentation. *J Food Compos. Anal.* **2021**, *97*, 103776. [CrossRef]

27. Addi, M.; Elbouzidi, A.; Abid, M.; Tungmunnithum, D.; Elamrani, A.; Hano, C. An overview of bioactive flavonoids from Citrus fruits. *Appl. Sci.* **2022**, *12*, 29. [CrossRef]

28. Liu, N.; Song, M.; Wang, N.; Wang, Y.; Wang, R.; An, X.; Qi, J. The effects of solid-state fermentation on the content, composition and in vitro antioxidant activity of flavonoids from dandelion. *PLoS ONE* **2020**, *15*, e0239076. [CrossRef]

29. Panche, A.N.; Diwan, A.D.; Chandra, S.R. Flavonoids: An overview. *J. Nutr. Sci.* **2016**, *5*, e47. [CrossRef]

30. Kumar, N.; Goel, N. Phenolic acids: Natural versatile molecules with promising therapeutic applications. *Biotechnol. Rep.* **2019**, *24*, e00370. [CrossRef]

31. Zou, H.; Ye, H.; Kamaraj, R.; Zhang, T.; Zhang, J.; Pavek, P. A review on pharmacological activities and synergistic effect of quercetin with small molecule agents. *Phytomedicine* **2021**, *92*, 153736. [CrossRef]

32. Kenan Kinaci, M.; Erkasap, N.; Kucuk, A.; Koken, T.; Tosun, M. Effects of quercetin on apoptosis, NF-κB and NOS gene expression in renal ischemia/reperfusion injury. *Exp. Ther. Med.* **2012**, *3*, 249–254. [CrossRef] [PubMed]

33. Yuan, K.; Zhu, Q.; Lu, Q.; Jiang, H.; Zhu, M.; Li, X.; Huang, G.; Xu, A. Quercetin alleviates rheumatoid arthritis by inhibiting neutrophil inflammatory activities. *J. Nutr. Biochem.* **2020**, *84*, 108454. [CrossRef] [PubMed]

34. Yousuf, M.; Khan, P.; Shamsi, A.; Shahbaaz, M.; Hasan, G.M.; Haque, Q.M.R.; Christoffels, A.; Islam, A.; Hassan, M.I. Inhibiting CDK6 activity by quercetin is an attractive strategy for cancer therapy. *ACS Omega* **2020**, *5*, 27480–27491. [CrossRef] [PubMed]

35. Imran, M.; Salehi, B.; Sharifi-Rad, J.; Aslam Gondal, T.; Saeed, F.; Imran, A.; Shahbaz, M.; Fokou, P.; Arshad, M.; Khan, H.; et al. Kaempferol: A key emphasis to its anticancer potential. *Molecules* **2019**, *24*, 2277. [CrossRef]

36. Rho, H.S.; Ghimeray, A.K.; Yoo, D.S.; Ahn, S.M.; Kwon, S.S.; Lee, K.H.; Cho, D.H.; Cho, J.Y. Kaempferol and kaempferol rhamnosides with depigmenting and anti-inflammatory properties. *Molecules* **2011**, *16*, 3338–3344. [CrossRef]
37. Wang, F.; Wang, L.; Qu, C.; Chen, L.; Geng, Y.; Cheng, C.; Yu, S.; Wang, D.; Yang, L.; Meng, Z.; et al. Kaempferol induces ROS-dependent apoptosis in pancreatic cancer cells via TGM2-mediated Akt/mTOR signaling. *BMC Cancer* **2021**, *21*, 396. [CrossRef] [PubMed]
38. Boz, H. p-Coumaric acid in cereals: Presence, antioxidant and antimicrobial effects. *Int. J. Food Sci. Technol.* **2015**, *50*, 2323–2328. [CrossRef]
39. Guglielmi, F.; Luceri, C.; Giovannelli, L.; Dolara, P.; Lodovici, M. Effect of 4-coumaric and 3,4-dihydroxybenzoic acid on oxidative DNA damage in rat colonic mucosa. *Br. J. Nutr.* **2003**, *89*, 581–587. [CrossRef]
40. Lou, Z.; Wang, H.; Rao, S.; Sun, J.; Ma, C.; Li, J. p-Coumaric acid kills bacteria through dual damage mechanisms. *Food Control* **2012**, *25*, 550–554. [CrossRef]
41. Sharma, S.H.; Chellappan, D.R.; Chinnaswamy, P.; Nagarajan, S. Protective effect of p-coumaric acid against 1, 2 dimethylhydrazine induced colonic preneoplastic lesions in experimental rats. *Biomed. Pharmacother.* **2017**, *94*, 577–588. [CrossRef]
42. Sroka, Z.; Cisowski, W. Hydrogen peroxide scavenging, antioxidant and anti-radical activity of some phenolic acids. *Food Chem. Toxicol.* **2003**, *41*, 753–758. [CrossRef] [PubMed]
43. Zhang, S.; Gai, Z.; Gui, T.; Chen, J.; Chen, Q.; Li, Y. Antioxidant Effects of Protocatechuic Acid and Protocatechuic Aldehyde: Old Wine in a New Bottle. *Evid.-Based Complement. Altern.* **2021**, *2021*, 6139308. [CrossRef] [PubMed]
44. Wei, G.; Guan, Y.; Yin, Y.; Duan, J.; Zhou, D.; Zhu, Y.; Quan, W.; Xi, M.; Wen, A. Anti-inflammatory effect of protocatechuic aldehyde on myocardial ischemia/reperfusion injury in vivo and in vitro. *Inflammation* **2013**, *36*, 592–602. [CrossRef] [PubMed]
45. Wang, S.; Bilal, M.; Hu, H.; Wang, W.; Zhang, X. 4-Hydroxybenzoic acid—A versatile platform intermediate for value-added compounds. *Appl. Microbiol. Biotechnol.* **2018**, *102*, 3561–3571. [CrossRef] [PubMed]
46. Annunziata, F.; Pinna, C.; Dallavalle, S.; Tamborini, L.; Pinto, A. An overview of coumarin as a versatile and readily accessible scaffold with broad-ranging biological activities. *Int. J. Mol. Sci.* **2020**, *21*, 4618. [CrossRef] [PubMed]
47. Jain, P.K.; Joshi, H. Coumarin: Chemical and pharmacological profile. *J. Appl. Pharm. Sci.* **2012**, *6*, 236–240.
48. He, B.T.; Liu, Z.H.; Li, B.Z.; Yuan, Y.J. Advances in biosynthesis of scopoletin. *Microb. Cell Factories* **2022**, *21*, 152. [CrossRef]
49. Armenia, A.; Hidayat, R.; Meiliani, M.; Yuliandra, Y. Blood pressure lowering effect of scopoletin on oxidative stress-associated hypertensive rats. *J. Res. Pharm.* **2019**, *23*, 249–258. [CrossRef]
50. Ruwizhi, N.; Aderibigbe, B.A. Cinnamic acid derivatives and their biological efficacy. *Int. J. Mol. Sci.* **2020**, *21*, 5712. [CrossRef]
51. Balli, D.; Bellumori, M.; Pucci, L.; Gabriele, M.; Longo, V.; Paoli, P.; Melani, F.; Mulinacci, N.; Innocenti, M. Does fermentation really increase the phenolic content in cereals? A study on millet. *Foods* **2020**, *9*, 303. [CrossRef]

Article

Optimization of Solid-Phase Lactobacillus Fermentation Conditions to Increase γ-Aminobutyric Acid (GABA) Content in Selected Substrates

Hee-yeon Kwon [1], Ji-soo Choi [1], Soo-jin Kim [1], Eun-min Kim [1], Ji-hyun Uhm [1], Bo-kyung Kim [1], Jae-yeon Lee [1], Yong-deok Kim [1] and Kwon-tack Hwang [2,*]

[1] R&D Center, NSTBIO Co., Ltd., Incheon 21984, Republic of Korea
[2] Department of Food and Nutrition, Nambu University, Gwangju 62271, Republic of Korea
* Correspondence: hwangskt@nambu.ac.kr; Tel.: +82-62-970-0174

Abstract: The purpose of this study was to optimize conditions of solid-phase fermentation of lactic acid bacteria to enhance GABA contents in grains. Optimal solid-phase fermentation conditions that could enhance the GABA content after fermenting *Oryza sativa* (brown rice) were investigated by changing the Lactobacillus strain, fermentation temperature, fermentation time, and inoculated bacteria number. *Avena sativa*, *Cicer arietinum*, and red and brown *Lens culinaris* were then fermented using the optimal solid-phase fermentation conditions to measure changes in GABA content and antioxidant activity. As a result of the experiment, the optimal solid-phase fermentation conditions to enhance the GABA contents in grains were: fermentation time, 48 h; amounts of bacteria, inoculating 5% of 1×10^7 CFU/mL of lactic acid bacteria; and fermentation temperature, 36 °C. When fermented under this condition, the GABA content increased from 4.64 mg/g to 6.93 mg/g (49.0%) compared to unfermented raw material. The results of the DPPH and ABTS radical scavenging activity assays confirmed that both the GABA content and radical scavenging activity were increased after fermentation. Such solid fermentation conditions developed in this study can be used to support the development of health functional food materials with enhanced GABA content and antioxidant activity.

Keywords: fermentation; GABA; antioxidant; Lactobacillus; grain

Citation: Kwon, H.-y.; Choi, J.-s.; Kim, S.-j.; Kim, E.-m.; Uhm, J.-h.; Kim, B.-k.; Lee, J.-y.; Kim, Y.-d.; Hwang, K.-t. Optimization of Solid-Phase Lactobacillus Fermentation Conditions to Increase γ-Aminobutyric Acid (GABA) Content in Selected Substrates. *Fermentation* **2023**, *9*, 22. https://doi.org/10.3390/fermentation9010022

Academic Editors: Michela Verni and Carlo Giuseppe Rizzello

Received: 29 November 2022
Revised: 21 December 2022
Accepted: 21 December 2022
Published: 26 December 2022

1. Introduction

In recent years, due to rapid industrialization, the living environment has changed. The incidences of emotional diseases such as mental stress, depression, bipolar disorder, and panic disorder are also increasing [1,2]. In addition, with advances in medical science, the incidences of geriatric diseases are increasing as we enter the age of aging. Among geriatric diseases, Alzheimer's disease is the neurological disease that has the highest percentage. It is expected to increase to 130 million by 2050 [3,4]. γ-Aminobutyric acid (GABA) is a four-carbon free amino acid that constitutes living organisms. It is widely present in bacteria, fungi, plants, and animals. GABA is produced together with CO_2 by the irreversible decarboxylation of glutamic acid by glutamate decarboxylase (GAD) [1,5]. GABA is a neurotransmitter in the nervous system that can improve brain blood circulation and promote brain cell metabolism. It is known to be an excellent substance for improving neurological diseases such as dementia prevention, memory enhancement, and alleviation of depression. In addition, it has various physiologically active functions such as lowering blood pressure, diuretic action, improving liver function, and promoting alcohol metabolism [6–10].

GABA is involved in nitrogen and carbon metabolism in plants. GABA in plants plays the role of messengers of enzyme action, regulation of gene expression, and intra- and intercellular transport of intermediates. It is widely distributed in various fruits and vegetables and miscellaneous grains such as rice and beans [11]. However, GABA contents in grains are very low: at 1–4 mg/100 g in regular rice, 4–8 mg/100 g in brown rice, and 10–100 mg/100 g in germinated brown rice. Such low contents cause it to be difficult to show its significant physiological activity in the body because it can be lost due to heat generated during the cooking and processing of food [12]. For this reason, various studies are being conducted to enhance GABA contents in natural products. Ding et al. (2016) have conducted a study on changes in GABA content due to germination under different oxygen conditions using two rice varieties [13]. Han et al. (2014) have found that the optimal production conditions to enhance GABA content are by varying conditions for soaking and germination of rice [14]. Zushi and Matsuzoe (2007) have reported that GABA contents in tomatoes could be improved by using salt-stress. Nejad-Alimoradi et al. (2019) have confirmed that GABA content can be enhanced in pumpkins by using salt-stress [15,16]. In a recent study using *Saccharomyces cerevisiae*, a method for manufacturing a functional fermented apple beverage with an increased GABA content by fermenting apple juice was reported [17].

Lactobacilli are Gram-positive, acid-resistant, non-spore-forming bacteria in the form of cocci or bacilli that share common physiological and metabolic properties. They can anaerobically utilize carbohydrates to produce lactic acid. They are widely used as a starter for traditional and industrial food fermentation [18]. Among them, lactic acid bacteria are known to be capable of effectively producing GABA from glutamic acid by possessing excellent GAD activity [19,20]. Park and Oh (2006) have presented a method to increase GABA content by manufacturing fermented soymilk yogurt using the *L. brevis* strain. Han et al. (2019) have used lactic acid bacteria fermentation to increase the GABA content in soymilk [21,22]. Lee et al. (2010) have confirmed the conversion of glutamic acid to GABA from fermented kelp broth with lactic acid bacteria isolated from kimchi and salted fish, which are traditional fermented foods [23]. Siragusa et al. (2007) have isolated strains from 22 types of cheese, identified lactic acid bacteria with excellent GABA synthesis ability, and confirmed the possibility of developing health-promoting foods using them [24].

Studies to increase GABA contents in natural products conducted so far have only been actively conducted by changing the properties of the raw material itself through germination or chemical treatment or through submerged fermentation (SMF). Studies that focus on solid-state fermentation (SSF) are insufficient. SSF refers to a fermentation process in which microorganisms use solid substances as nutrients to produce secondary metabolites. Compared to SMF, when producing new substances such as enzymes, fragrances, and pigments, SSF is receiving attention from many researchers because of its advantages such as high yield, high quality characteristics, and production cost reduction. SSF is economically advantageous because less solvent is used for product extraction and separation compared to SMF and the production cost is lowered by reducing waste treatment costs and sterilization costs. In addition, SSF has better industrial value because it is easier to mass-produce and build facilities due to the higher amount of substrate per unit volume of fermentation tank than SMF. Since SSF is carried out under a low moisture content, the possibility of bacterial contamination is low, causing it to be easier to manage production compared to SMF [25]. The advantage of using lactic acid bacteria for fermentation through a solid-state culture is that they can physiologically enhance active substances without significantly changing the form, taste, or aroma of natural products. SSF is continuously being studied in the biotechnology industry and is attracting attention as an alternative to SMF because it can potentially be applied to the production of secondary metabolites in various fields such as feed, fuel, food, and medicine [26].

Miscellaneous grains refer to grains such as millet, soybean, and buckwheat, excluding white rice among cereals. In the past, miscellaneous grains were considered inferior crops, but as their excellent nutrition and various physiological functions such as anti-cancer, immunity increasing, and antioxidant effects have been revealed, their use value for new well-being food has recently been increased [27–29]. Miscellaneous grains contain from two to three times as many vitamins, minerals, and dietary fiber that rice lacks and they are nutritionally excellent and contain a large amount of various physiologically active substances such as GABA [30–32]. As the health functional aspects of miscellaneous grains have emerged, the consumer preference for processed foods using them is increasing [33]. According to a recent report examining the consumption of processed foods using grains, as for the consumption patterns of cereals, alcohol manufacturing, rice cake manufacturing, and instant rice showed the largest market size. Among them, consumption in the instant rice market is steadily increasing, while consumption of alcoholic beverages and rice cakes is gradually decreasing [34]. Recently, as single-person households and women's participation in society have increased, the home meal replacement (HMR) market has expanded, which is rapidly increasing the demand for processed rice such as aseptically processed rice, retort rice, and frozen rice [35,36].

Maintaining the raw form of miscellaneous grains and strengthening the nutritional components is considered to be very important industrially in the growing processed food market using grains. [37]. The development of a technology that can increase the GABA content in miscellaneous grains by using the SFF technology that can preserve the shape of the raw material is expected to have great applicability in the processed food market and is also expected to be used as a health functional food material. Therefore, the objective of this study was to establish optimal fermentation conditions for enhancing GABA contents while maintaining the original form of grains through SSF of lactic acid bacteria.

2. Materials and Methods

2.1. Materials and Reagents

Oryza sativa (brown rice) used in this study was cultivated in Naju, Jeollanam-do, Korea, in 2021. Other grains (*Avena sativa, Cicer arietinum*, red and brown *Lens culinaris*) were purchased from Hyundainongsan CO, LTD in Namyangju, Gyeonggi-do, Republic of Korea, in 2021. The Lactobacillus strains used in this study (*Lactiplantibacillus plantarum* P1, *Lacticaseibacillus casei* C1, *Limosilactobacillus fermentum* F1, and *Lacticaseibacillus rhamnosus* R1) were purchased from Lactomason Co., Ltd. in Jinju, Gyeongsangnam-do, Republic of Korea. The potassium persulfate, 2,2-diphenyl-1-picrylhydrazyl (DPPH), 2,2'-azino-bis(3-ethylbenzothiazoline-6-sulfonic acid) (ABTS), and 2-hydroxy-1-naphthaldehyde (HN) were purchased from Sigma-Aldrich Co., Ltd. (St. Louis, MO, USA). The borate buffer (pH 10) was purchased from Agilent Technologies (Santa Clara, CA, USA).

2.2. Lactic Acid Bacteria Fermentation Conditions

To select the optimal fermentation strain, four types of lactic acid bacteria (*L. plantarum, L. casei, L. fermentum*, and *L. rhamnosus*) in a powder state were diluted with sterile distilled water to 1×10^9 CFU/mL and activated at 36 °C for 1 h. Then, 200 g of *Oryza sativa* was precisely weighed and mixed with 10 mL of each lactic acid bacteria culture. The fermentation was then performed in an incubator at 36 °C for 48 h.

For the optimization of the fermentation temperature, two excellent lactic acid strains (*L. plantarum, L. casei*) selected above were diluted with sterile distilled water to 1×10^9 CFU/mL, respectively, and activated at 36 °C for 1 h. Then, 200 g of *Oryza sativa* was precisely weighed and mixed with 10 mL of activated lactic acid bacteria solution. The fermentation was performed in an incubator at 24, 28, 32, 36, and 40 °C for 48 h.

To determine the optimal bacterial inoculum amount, *L. plantarum* and *L. casei* were diluted with sterile distilled water to a concentration of 1×10^3, 1×10^5, 1×10^7, 1×10^9, and 1×10^{11} CFU/mL, respectively, and activated at 36 °C for 1 h. Then, 200 g of *Oryza sativa* was precisely weighed and mixed with 10 mL of activated lactic acid bacteria solution. The fermentation was then performed at 36 °C for 48 h.

To obtain the optimal fermentation time, *L. plantarum* and *L. casei* were diluted with sterile distilled water to a concentration of 1×10^7 CFU/mL, respectively, and activated at 36 °C for 1 h. Then, 200 g of brown rice was precisely weighed and mixed with 10 mL of activated lactic acid bacteria solution. The fermentation was then performed at 36 °C for 2, 4, 8, 12, 24, 48, and 72 h, respectively.

For the production of four types of fermented grains (*Avena sativa*, *Cicer arietinum*, and red and brown *Lens culinaris*), *L. plantarum* and *L. casei* were diluted with sterile distilled water to a concentration of 1×10^7 CFU/mL, respectively, and activated at 36 °C for 1 h. Then, 200 g of each grain was precisely weighed, mixed with 10 mL of activated lactic acid bacteria solution, and fermented at 36 °C for 48 h.

After each fermentation, all the samples were freeze-dried to remove moisture in order to comparatively analyze the GABA content, DPPH, and ABTS radical scavenging activity. After 20 g of the dried sample was precisely weighed, 200 mL of 50% ethanol was added and ultrasonic extraction was performed for 3 h [6]. The extract was concentrated under reduced pressure to a certain concentration and used as a sample after freeze-drying.

2.3. Analysis of Glutamic Acid and GABA Content

The HPLC analysis of glutamic acid and GABA was performed by modifying the GABA-HN derivatization analysis method reported by Panrod (2016) and Hayat (2014) et al. [38,39]. For the test solution, 100 mg of a freeze-dried fermented grain extract powder sample was precisely weighed and added to a 10 mL volumetric flask. The distilled water was then added up to the mark to reach a concentration of 10 mg/mL. The HN solution for HN derivatization was prepared by precisely weighing 0.15 g of HN, then it was added into a 50 mL volumetric flask to create 0.3% solution with methanol. For derivatization, 0.5 mL of the test solution, 0.3 mL of borate buffer (pH 10), and 0.5 mL of 0.3% HN solution were mixed in a 1.5 mL tube and reacted at 75 °C in a constant temperature water bath for 10 min. After that, it was cooled in a dark room for 10 min and purified with methanol in a 10 mL volumetric flask. All the samples were filtered through a 0.22 μm filter before being injected into HPLC. For the analysis, Shimadzu HPLC system and Shimadzu SPD-M20A Photodiode Array Detector (Shimadzu Corp., Kyoto, Japan) were used. The analysis conditions are shown in Table 1. The symmetry C18 (4.6 × 250 mm, 5 μm) was used as the analysis column.

Table 1. HPLC condition for analysis of glutamic acid and GABA.

Instrument	Conditions
Column	Symmetry C18 (4.6 × 250 mm, 5 um), Waters Corporation
Column temp.	35 °C
Mobile phase (isocratic)	100 mM acetate buffer (in water): Methanol = 60: 40, *v/v*
Detector	UV detector (235 nm)
Flow rate	0.8 mL/min
Injection volume	10 μL
Run time	50 min

2.4. DPPH Readical Scavenging Acitivity

The DPPH radical scavenging activity was measured with reference to the method of Jang et al. (2021) [40]. After precisely weighing 50 mg of each sample, it was adjusted with distilled water in a 10 mL volumetric flask and was prepared at 5 mg/mL and used in the experiment. Briefly, for 0.2 mL of each sample, 0.4 mM DPPH solution was added. After reacting at room temperature for 10 min, the absorbance was measured at 517 nm using a microplate reader (Spectramax ABS plus, Molecular Devices, Sunnyvale, CA, USA). The DPPH radical scavenging activity was determined with the following formula. The ascorbic acid was used as a positive control.

$$\text{DPPH radical scavenging activity (\%)} = \{1 - (A_{\text{experiment}}/A_{\text{control}})\} \times 100 \qquad (1)$$

2.5. ABTS Readical Scavenging Acitivity

The ABTS radical scavenging activity was measured by modifying the method of Re et al. (1999) [41]. To generate ABTS radicals, 7 mM ABTS and 2.45 mM potassium persulfate solution were mixed at a 2:1 ratio and reacted in a dark place for 16 h. After mixing 50 μL of sample and 950 μL of ABTS solution, it was reacted in a dark place for 10 min. Then, the absorbance was measured at 734 nm with a microplate reader. The ABTS radical scavenging activity was calculated according to the formula. The ascorbic acid was used as a positive control.

$$\text{ABTS radical scavenging activity (\%)} = \{1 - (A_{\text{experiment}}/A_{\text{control}})\} \times 100 \tag{2}$$

2.6. Statistical Analysis

The results of glutamic acid, GABA contents, and antioxidant activity are expressed as mean $\pm$ SD (standard deviation) of the samples in triplicate. The values were analyzed with a Student *t*-test on Microsoft 365 Excel computerized software or using a one-way ANOVA followed by Duncan's multiple range tests using IBM SPSS Statistics version 28.0 (IBM, Armonk, NY, USA). The differences were considered statistically significant at $p < 0.05$.

3. Results and Discussions

3.1. Identification and Qualification of Glutamic Acid and GABA

The HPLC analysis was performed at 235 nm for the analysis of glutamic acid and GABA content; the established chromatogram is shown in Figure 1. Glutamic acid and GABA derivatized with HN were detected and were analyzed in high concentrations in *Oryzae sativa* extracts. The retention times of glutamic acid and GABA were 5.10 and 12.590 min, respectively.

(A)

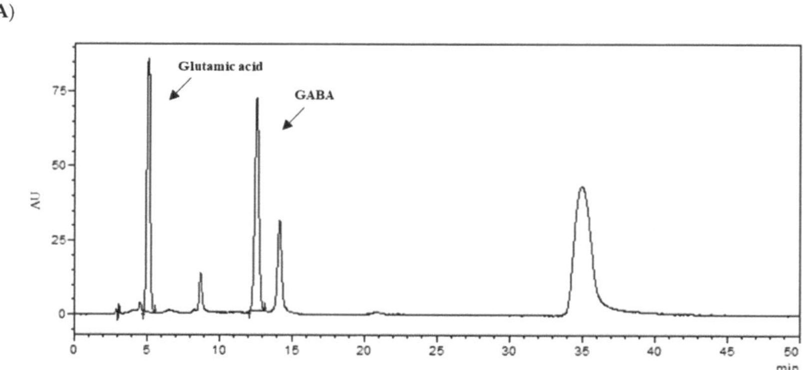

Figure 1. *Cont.*

(B)

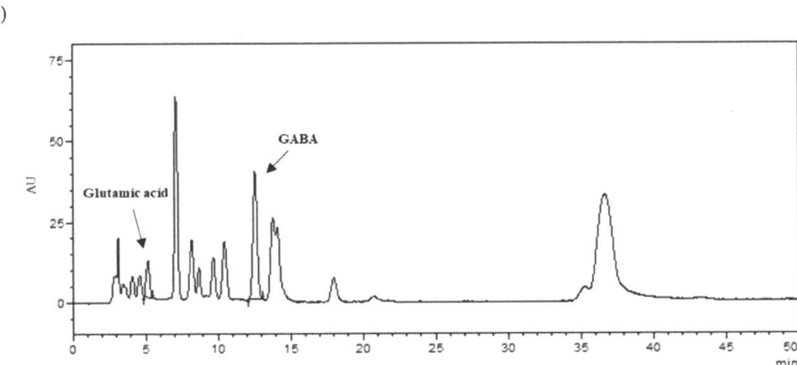

Figure 1. HPLC chromatograms of glutamic acid and GABA standard solution (**A**) and *Oryzae sativa* extracts (**B**).

3.2. Changes in GABA Content According to Fermentation Conditions

3.2.1. Effect of Fermentation Strains on GABA Content

To set the optimal conditions for lactic acid bacteria fermentation, the GABA content was comparatively analyzed by fermenting *Oryza sativa* while changing the major factors involved in fermentation (fermentation strain, fermentation temperature, inoculated bacteria number, and fermentation time). It has been reported that the lactic acid bacteria have an excellent ability to produce GABA. Their ability varies greatly between species and strains [42]. Among them, *L. plantarum*, *L. casei*, *L. fermentum*, and *L. rhamnosus* are reported to have excellent GABA bioconversion ability and various studies are being conducted [43–46]. Therefore, first, the GABA content was analyzed after fermentation using four types of lactic acid bacteria (*L. plantarum*, *L. casei*, *L. fermentum*, and *L. rhamnosus*) in *Oryza sativa* to select a strain with a high GABA production ability (Figure 2). In the unfermented *Oryza sativa* extract, the GABA content was measured to be 5.87 ± 0.24 mg/dry weight g. However, the GABA contents in the experimental group fermented in the solid phase using *L. plantarum*, *L. casei*, *L. fermentum*, and *L. rhamnosus* were 6.81 ± 0.16 mg/dry weight g, 6.69 ± 0.22 mg/g, 6.30 ± 0.17 mg/g, and 6.49 ± 0.16 mg/g, respectively. The GABA content was significantly increased in all the experimental groups compared to that in the control group. Among them, the *L. plantarum* P1 and *L. casei* C1 enhanced GABA content the most. The content of glutamic acid in the unfermented *Oryza sativa* extract was measured to be 5.33 ± 0.18 mg/dry weight g. The *Oryza sativa* extract fermented with *L. casei* had the highest decrease at 3.97 ± 0.09 mg/dry weight g and the extract fermented with *L. plantarum* decreased the second most with 4.28 ± 0.03 mg/dry weight g. This means that the two strains had the best ability to convert glutamic acid to GABA. When referring to the glutamic acid content and GABA production, the strain with the best GABA conversion ability was *L. casei* while *L. plantarum* had the second highest. Since this study focused on finding the optimal fermentation conditions for converting glutamic acid to GABA, *L. casei* and *L. plantarum* strains were used to set up the optimal extraction conditions.

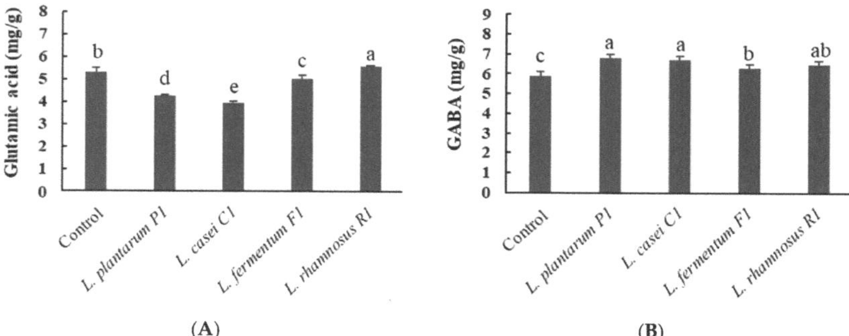

(A)　　　　　　　　　　　　　　　　(B)

Figure 2. Glutamic acid (**A**) and GABA (**B**) contents of *Oryza sativa* extracts depending on the Lactobacillus strains. Each bar is the mean ± standard deviation of the results from three different analyses (n = 3). The bars with different letters (a–e) indicate significant differences at $p < 0.05$ using Duncan's multiple range test.

3.2.2. Effect of Fermentation Temperature on GABA Content

The fermentation was carried out at 24 °C, 28 °C, 32 °C, 36 °C, and 40 °C to measure the change in the GABA production ability of lactic acid bacteria according to the fermentation temperature (Figure 3). As a result of the experiment, the glutamic acid content decreased significantly up to 36 °C. The GABA content showed a tendency to increase significantly in inverse proportion to this. Both strains showed the highest GABA content at 36 °C. The *L. casei* strain fermented at 40 °C showed a tendency to partially decrease the GABA content. Thus, 36 °C was set as the optimum fermentation temperature condition. Li et al. (2010) have compared and analyzed the GABA content after fermentation at 10–45 °C using *L. brevis* isolated from Paocai. It was reported that the GABA production was the best at 30–35 °C [47], similar to the results of this study. Kim et al. (2009) have compared the GABA production ability from raspberry juice through *L. brevis* fermentation at different temperatures, pH, and fermentation time [48]. In raspberry juice through liquid fermentation, the GABA was produced better at 30 °C than that at 25 or 37 °C, which was different from this study. Although different raw materials were used for fermentation, the difference between solid-phase fermentation and liquid-phase fermentation was considered a major factor affecting the GABA content.

(**A**)

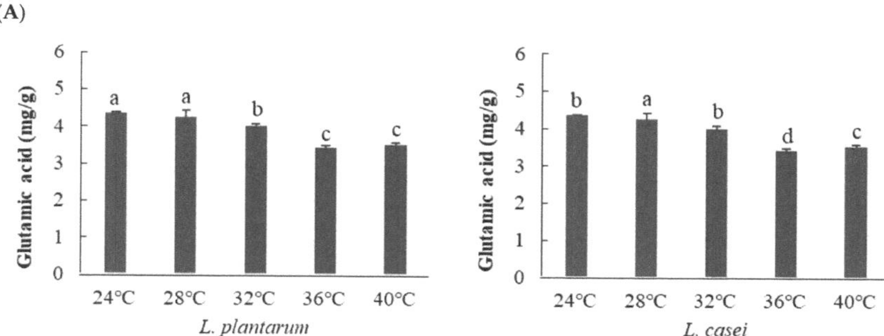

Figure 3. *Cont.*

(B)

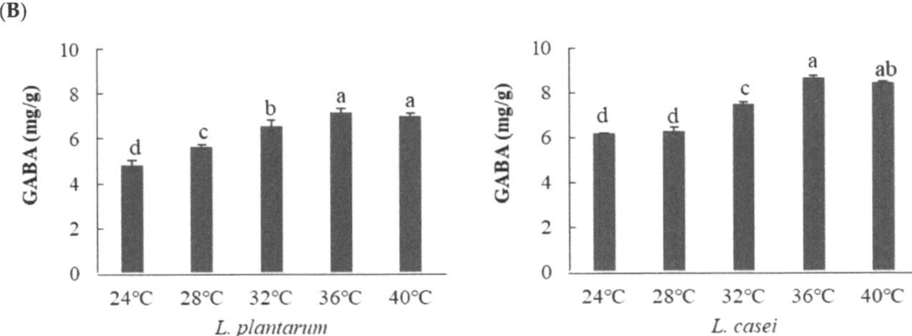

Figure 3. Glutamic acid (**A**) and GABA (**B**) contents of the *Oryza sativa* extracts depending on the fermentation temperature. Each bar is the mean ± standard deviation of the results from three different analyses (*n* = 3). The bars with different letters (a–d) indicate significant differences at $p < 0.05$ using Duncan's multiple range test.

3.2.3. Effect of Inoculated Cell Number on GABA Content

To optimize the inoculated cell number of each bacterial strain for fermentation, the fermentation was carried out by changing the bacteria concentrations (1×10^3, 1×10^5, 1×10^7, 1×10^9, and 1×10^{11} CFU/mL) (Figure 4). When fermented with *L. plantarum* and *L. casei* strains, the GABA contents were very low at 4.79 ± 0.06 mg/dry weight g and 4.65 ± 0.11 mg/ g in the experimental group inoculated with bacteria at 1×10^3 CFU/mL, respectively. In the experimental group inoculated with bacteria at 1×10^7 CFU/mL, the GABA content increased by about 35.2% and 49.0% to 6.48 ± 0.10 mg/dry weight g and 6.93 ± 0.19 mg/g, respectively. As bacterial inoculation increased, the glutamic acid content decreased but the GABA content increased significantly until the bacterial concentration was 1×10^7 CFU/mL. However, even if the concentration was higher than that, there was no statistically significant difference in the GABA production ability. Thus, the optimal concentration for bacterial inoculation was set to be 1×10^7 CFU/mL.

(A)

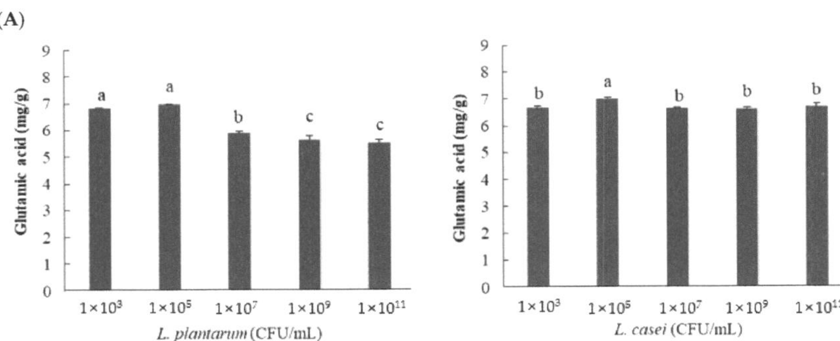

Figure 4. *Cont.*

(B)

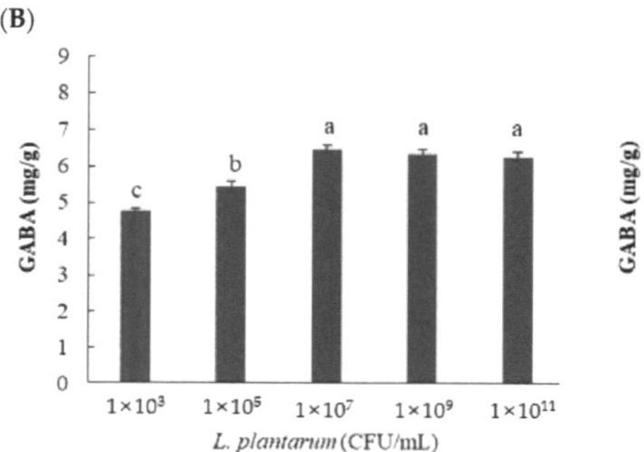
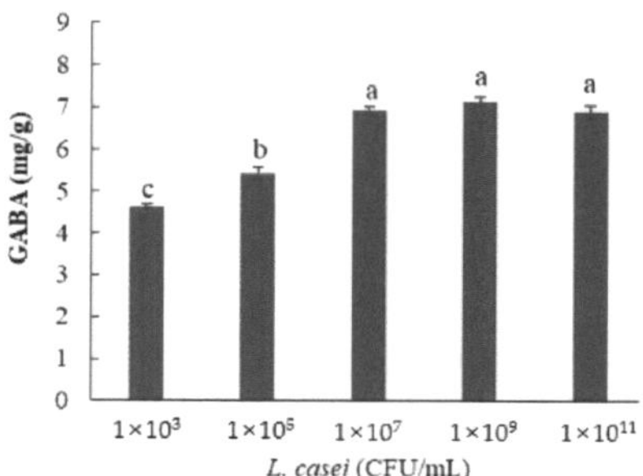

Figure 4. Glutamic acid (**A**) and GABA (**B**) contents of *Oryza sativa* extracts depending on the inoculated cell number of Lactobacillus strain. Each bar is the mean ± standard deviation of the results from three different analyses (n = 3). Bars with different letters (a–c) indicate significant differences at $p < 0.05$ using Duncan's multiple range test.

3.2.4. Effect of Fermentation Time on GABA Content

To set the optimal fermentation time, the fermentation was carried out for 2 h, 4 h, 8 h, 12 h, 24 h, 48 h, and 72 h, respectively. The glutamic acid and GABA contents were then compared (Figure 5). As the fermentation time increased, the GABA content significantly increased. With the *L. plantarum* strain, the GABA content did not increase any more at 48 h. With *L. casei*, the GABA content showed a tendency to decrease slightly after 72 h of fermentation. Li et al. (2010) have reported that the GABA content increases rapidly up to 36 h under sufficient substrate conditions, followed by almost no production thereafter [47]. In the study of Hwang and Park (2020), when lactic acid bacteria were cultured in an MRS medium including MSG, the maximum production was achieved at 40 h with no significant change after that [49]. In the case of GABA production using lactic acid bacteria, 48 h was set as the optimal fermentation condition in that the GABA content increased rapidly at the beginning of fermentation without increasing any more after a certain period.

(A)

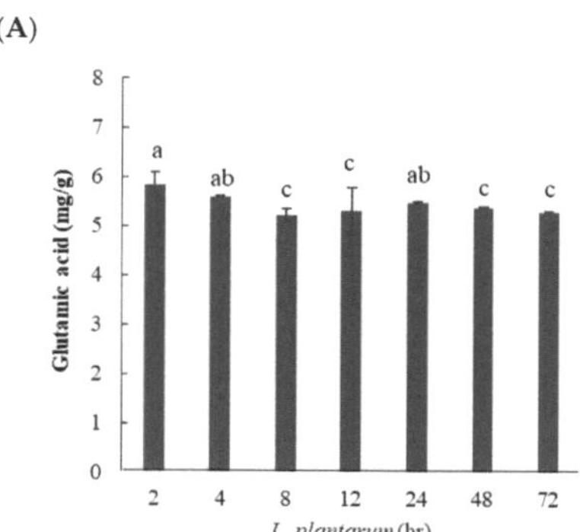
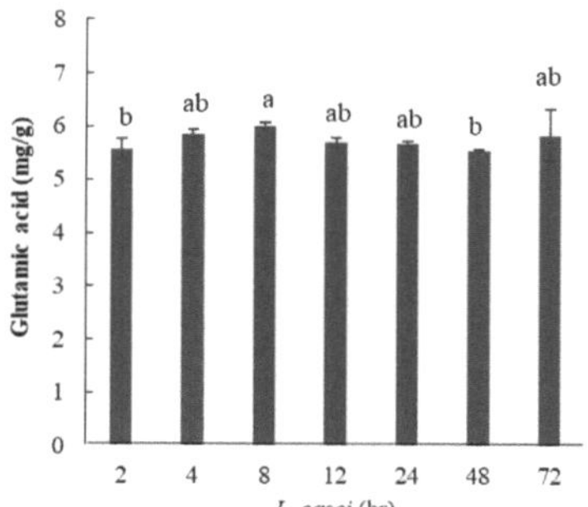

Figure 5. *Cont.*

(B)

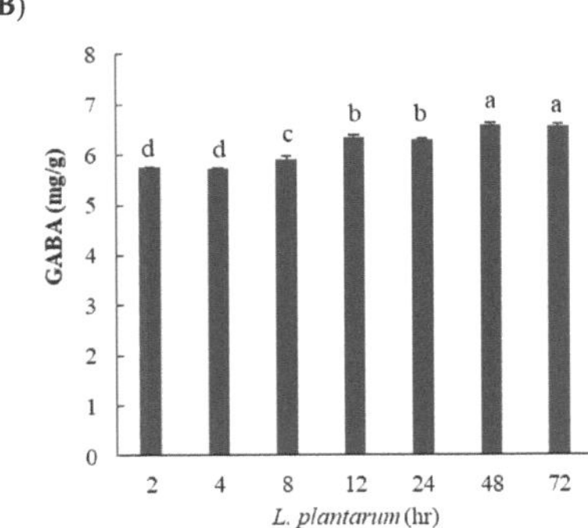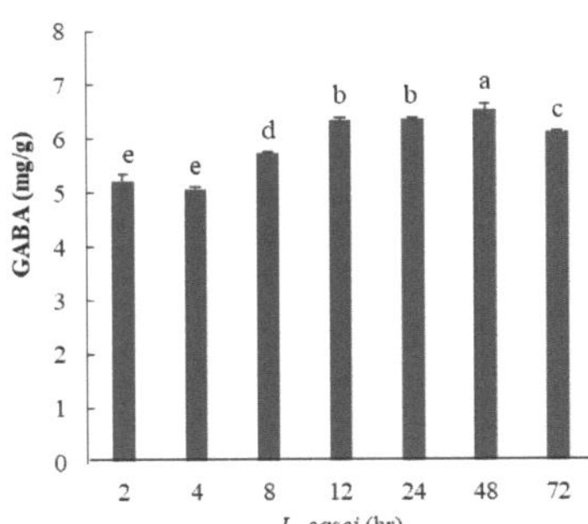

Figure 5. Glutamic acid (**A**) and GABA (**B**) contents of *Oryza sativa* extracts depending on the fermentation time. Each bar is the mean ± standard deviation of the results from three different analyses (*n* = 3). Bars with different letters (a–e) indicate significant differences at $p < 0.05$ using Duncan's multiple range test.

3.3. Analysis of GABA Content for Four Types of Fermented Grains

The GABA contents were analyzed after fermenting four kinds of grains (*Avena sativa*, *Cicer arietinum*, and red and brown *Lens culinaris*) under optimal lactic acid bacteria fermentation conditions (36 °C, 1×10^7 CFU/mL, 48 h) to enhance GABA contents (Figure 6). *Avena sativa*, commonly known as oat, is cultivated worldwide in England, France, and Germany. It is widely used as a raw material for feed, health food, and alcoholic beverages. [50]. Furthermore, *Avena sativa* are important grain crops that possess high levels of valuable nutrients such as protein, unsaturated fatty acids, soluble fiber, and minerals and are widely utilized to provide nutrient consumption for humans and livestock [51]. *Cicer arietinum*, also known as chickpea, is a grain that is evaluated as an important component of a vegetarian diet due to its high nutritional value with high contents of dietary fiber and protein [52]. Because the content of glutamic acid is the highest among amino acids contained in chickpeas, it is known as one of the grains with high potential for GABA production [53]. *Lens culinaris*, commonly known as lentils, varies in color depending on the seed and processing stage, such as green, brown, and red. It is widely used as a low-fat, high-protein, and high-fiber legume crop [54,55]. Lentils are known as one of the world's top five health foods and consumption and production are steadily increasing [56]. When *Avena sativa* was fermented with *L. plantarum* or *L. casei*, the GABA content was improved by 3% or 6%, respectively, compared to that in the control group without fermentation. The GABA contents in *Cicer arietinum* fermented with *L. plantarum* and *L. casei* were increased by 9% and 11%, respectively. In *Lens culinaris* fermented with *L. plantarum* and *L. casei*, the GABA contents were significantly increased by 8% and 11% in the red lentil and by 9% and 18% in brown lentil, respectively. It was confirmed that by using fermentation conditions established in this study, the GABA content could be significantly increased in five kinds of grains currently widely used: *Oryza sativa*, *Avena sativa*, *Cicer arietinum*, and red and brown *Lens culinaris*. The solid fermentation method has different nutritional components depending on the type of grain. Thus, the degree of increase for GABA content may vary.

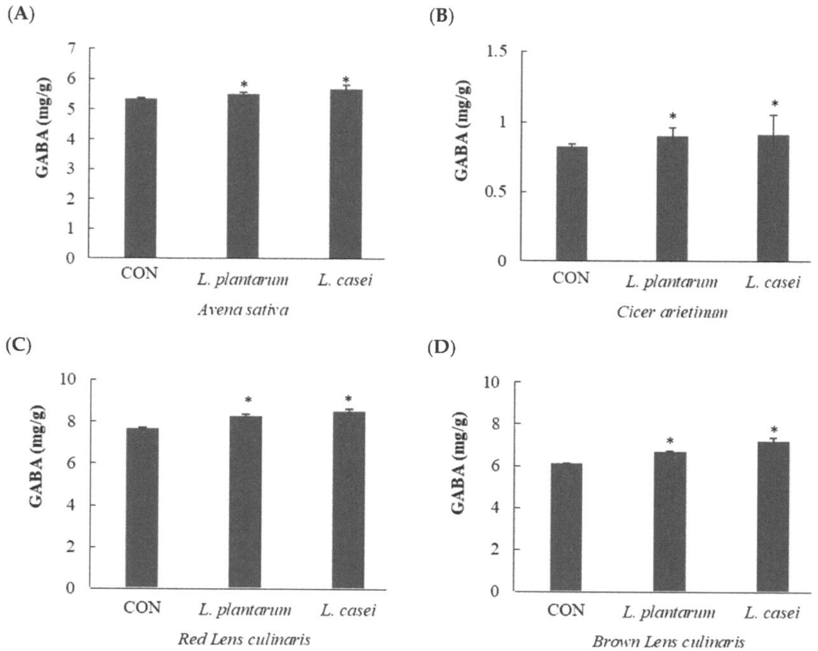

Figure 6. GABA contents of Avena sativa (**A**), Cicer arietinum (**B**), red Lens culinaris (**C**), and brown Lens culinaris (**D**) extracts. CON: Control; Each bar is the mean ± standard deviation of the results from three different analyses ($n = 3$). * $p < 0.05$, significantly different from control (Student's *t*-test).

3.4. Anti-Oxidant Activities of Five Types of Fermented Grains

The DPPH is a very stable reactive oxygen species used to measure antioxidant activity. It is used as a principle to measure the absorbance of a purple DPPH solution by changing its color to yellow after removing free radicals through hydrogen donation to substances containing hydroxyl radicals [57]. The antioxidative activity measurement using ABTS radicals is a principle of measuring absorbance by decolorizing a blue reaction solution after removing ABTS free radicals generated by the reaction with potassium persulfate by an antioxidant in the sample [58]. In recent studies, it has been reported that the antioxidant activity and phenolic contents increased when grain by-products were liquid-fermented using Lactobacillus strain [59–61]. When solid-phase fermentation was performed under optimal fermentation conditions using *L. plantarum* and *L. casei* for all five grains, it showed statistical significance and effectively removed free radicals compared to the control group without fermentation (Figure 7). Lee et al. (2010) have reported that the GABA content and antioxidant activity are increased after fermenting a kelp solution using *L. brevis* isolated from Jot-gal [23]. Jhan et al. (2015) have reported that when red bean is fermented with *B. subtilis* and *L. bulgaricus*, the contents of polyphenols and flavonoids were increased while its antioxidant activity was increased [62]. In a study by Gan et al. (2016), when mung bean (*Vigna radiata*) and soybean (*Glycine max*) were fermented using *L. plantarum* strains, the total phenolic contents and radical scavenging ability were increased [63]. Referring to the preceding studies described above, one could judge that the antioxidant activity of fermented grains was increased due to increased phenolic compounds when fermented using Lactobacillus strains.

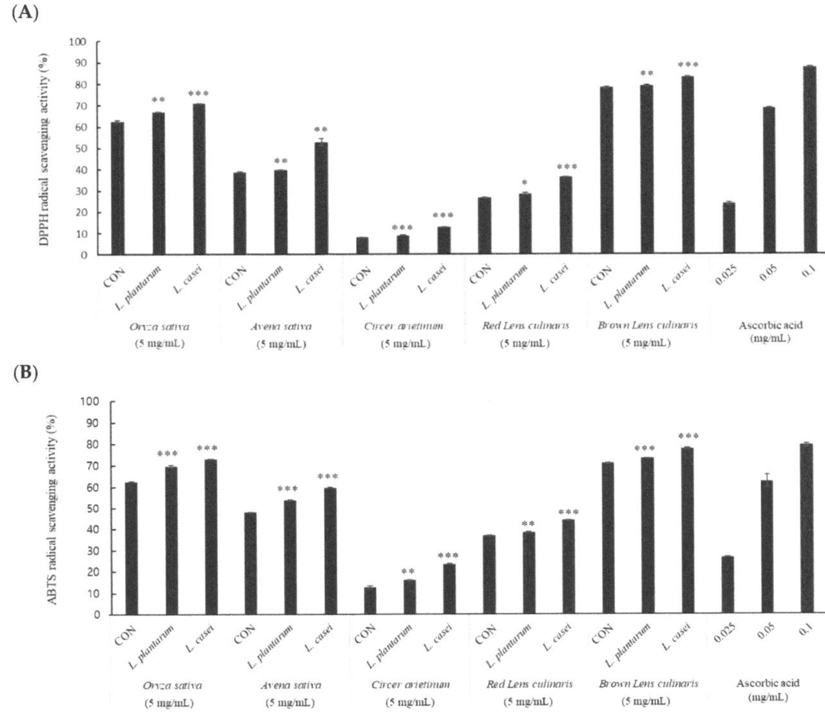

Figure 7. DPPH (**A**) and ABTS (**B**) radical scavenging activity of *Oryza sativa*, *Avena sativa*, *Circer arietinum*, red *Lens culinaris*, and brown *Lens culinaris* extracts. CON: Control; The results were shown as the mean ± SD of three independent experiments ($n = 3$). * $p < 0.05$, ** $p < 0.01$, and *** $p < 0.001$, significantly different from control (Student's *t*-test).

4. Conclusions

As a neurotransmitter in the body, GABA is attracting attention as a functional food material because it not only helps relieve nervous system diseases such as depression and stress but also has various physiologically active functions such as blood pressure improvement, liver function improvement, and alcohol metabolism promotion. It is difficult to express the functionality of GABA present in natural products only by natural ingestion in trace amounts. Therefore, various studies such as grain germination, chemical treatment, and liquid fermentation are being conducted to increase GABA contents in natural products. However, these methods cause changes in the shape or characteristics of the raw material, thus having a disadvantage in that characteristics of the existing natural product cannot be used after food manufacturing or processing. In this study, we proposed a method to improve GABA content while preserving the raw form of grains completely using a lactic acid bacteria solid-phase fermentation technique. When *Oryza sativa* was fermented, *L. plantarum* and *L. casei* showed the best GABA production ability among the four lactic acid bacteria (*L. plantarum*, *L. casei*, *L. fermentum*, and *L. rhamnosus*). As the fermentation temperature conditions were changed, the GABA production ability increased from 24 °C to 36 °C but did not increase thereafter. When the bacterial inoculum was changed during the SSF process, the GABA production increased from 1×10^3 CFU/mL to 1×10^7 CFU/mL, but, thereafter, the GABA content did not increase even as the amount of lactic acid bacteria increased. As the fermentation time increased up to 48 h, the GABA content also increased significantly and the longer fermentation times did not increase the GABA content. As a result of fermenting the grain under the optimized solid fermentation conditions in this study, it was confirmed that the GABA content increased up to 49.0% compared to the raw material. This increase was different because of the different types of grains and different nutritional components. From the DPPH and ABTS assays, it was confirmed that phenolic compounds with physiological activity could be changed in a positive direction and that the GABA content and antioxidant activity were significantly increased through fermentation. If the solid-phase fermentation technique performed in this study is combined with other previously reported techniques that can improve the GABA content, it is thought to have great potential for use in the development of health functional food materials by greatly enhancing insufficient GABA contents in natural products.

Author Contributions: Conceptualization, H.-y.K., S.-j.K., E.-m.K., Y.-d.K. and K.-t.H.; methodology, H.-y.K., J.-s.C., J.-h.U. and J.-y.L.; software, J.-s.C. and Y.-d.K.; validation, H.-y.K. and J.-s.C.; formal analysis, H.-y.K., S.-j.K., E.-m.K., B.-k.K. and Y.-d.K.; investigation, H.-y.K., J.-h.U., B.-k.K. and J.-y.L.; resources, S.-j.K. and E.-m.K.; data curation, H.-y.K. and Y.-d.K.; writing—original draft preparation, H.-y.K.; writing—review and editing, E.-m.K., J.-y.L., Y.-d.K. and K.-t.H.; visualization, H.-y.K. and J.-s.C.; supervision, J.-y.L., Y.-d.K. and K.-t.H.; project administration, J.-y.L. and Y.-d.K.; funding acquisition, J.-y.L. and Y.-d.K. All authors have read and agreed to the published version of the manuscript.

Funding: This research received no funding.

Institutional Review Board Statement: Not applicable.

Informed Consent Statement: Not applicable.

Data Availability Statement: The data supporting the findings reported here are available on reasonable request from the corresponding author.

Conflicts of Interest: The authors declare no conflict of interest.

References

1. Ueno, H. Enzymatic and structural aspects on glutamate decarboxylase. *J. Mol. Catal. B Enzym.* **2000**, *10*, 67–79. [CrossRef]
2. Checkoway, H.; Jessica, I.L.; Samir, N.K. Neurodegenerative diseases. *IARC Sci. Publ.* **2011**, *163*, 407–419.
3. Neelam, K.; Duddu, V.; Anyim, N.; Neelam, J.; Lewis, S. Pandemics and pre-existing mental illness: A systematic review and meta-analysis. *Brain Behav. Immun.* **2021**, *10*, 100177. [CrossRef] [PubMed]
4. Sengoku, R. Aging and Alzheimer's disease pathology. *Neuropathology* **2020**, *40*, 22–29. [CrossRef] [PubMed]
5. Soghomonian, J.J.; David, L.M. Two isoforms of glutamate decarboxylase: Why? *Trends Pharmacol. Sci.* **1998**, *19*, 500–505. [CrossRef] [PubMed]
6. Lin, S.D.; Jeng, L.M.; Ching, A.H. Bioactive components and antioxidant properties of γ-aminobutyric acid (GABA) tea leaves. *LWT-Food Sci. Technol.* **2012**, *46*, 64–70. [CrossRef]
7. Farahmandfar, M.; Zarrindast, M.R.; Kadivar, M.; Karimian, S.M.; Naghdi, N. The effect of morphine sensitization on extracellular concentrations of GABA in dorsal hippocampus of male rats. *Eur. J. Pharmacol.* **2011**, *669*, 66–70. [CrossRef]
8. Buzzi, A.; Chikhladze, M.; Falcicchia, C.; Paradiso, B.; Lanza, G.; Soukupova, M.; Marti, M.; Morari, M.; Franceschetti, S.; Simonato, M. Loss of cortical GABA terminals in Unverricht Lundborg disease. *Neurobiol. Dis.* **2012**, *47*, 216–224. [CrossRef]
9. Leventhal, A.G.; Wang, Y.C.; Pu, M.L.; Zhou, Y.F.; Ma, Y. GABA and its agonists improved visual cortical function in senescent monkeys. *Science* **2003**, *300*, 812–815. [CrossRef]
10. Seok, H.M. Functional Description of GABA and Possibility of New Material Development. *J. Food Technol.* **2002**, *15*, 81–85.
11. Fait, A.; Fromm, H.; Walter, D.; Galili, G.; Fernie, A.R. Highway or byway: The metabolic role of the GABA shunt in plants. *Trends Plant Sci.* **2008**, *13*, 14–19. [CrossRef] [PubMed]
12. Gironell, A.; Figueiras, F.P.; Pagonabarraga, J.; Herance, J.R.; Pascual-Sedano, B.; Trampal, C.; Gispert, J.D. GABA and serotonin molecular neuroimaging in essential tremor: A clinical correlation study. *Parkinsonism Relat. Disord.* **2012**, *18*, 876–880. [CrossRef] [PubMed]
13. Ding, J.; Yang, T.; Feng, H.; Dong, M.; Slavin, M.; Xiong, S.; Zhao, S. Enhancing contents of γ-aminobutyric acid (GABA) and other micronutrients in dehulled rice during germination under normoxic and hypoxic conditions. *J. Agric. Food Chem.* **2016**, *64*, 1094–1102. [CrossRef] [PubMed]
14. Han, S.I.; Ra, J.E.; Seo, K.H.; Park, J.Y.; Seo, W.D.; Park, D.S.; Cho, J.H.; Lee, J.H.; Sim, E.Y.; Nam, M.H. Effects of physico-chemical treatment on 'Nunkeunhukchal' (black sticky rice with giant embryo) for the enhancement of GABA (γ-aminobutyric acid) contents. *Korean J. Crop. Sci.* **2014**, *59*, 398–405. [CrossRef]
15. Zushi, K.; Naotaka, M. Salt stress-enhanced γ-aminobutyric acid (GABA) in tomato fruit. *Acta Hort.* **2007**, *761*, 431–436. [CrossRef]

16. Nejad, A.F.; Nasibi, F.; Kalantari, K.M. 24-epibrassinolide pre-treatment alleviates the salt-induced deleterious effects in medicinal pumpkin (*Cucurbita pepo*) by enhancement of GABA content and enzymatic antioxidants. *S. Afr. J. Bot.* **2019**, *124*, 111–117. [CrossRef]

17. Sun, X.; Wang, J.; Li, C.; Zheng, M.; Zhang, Q.; Xiang, W.; Tang, J. The use of γ-aminobutyric acid-producing *saccharomyces cerevisiae* SC125 for functional fermented beverage production from apple juice. *Foods* **2022**, *11*, 1202. [CrossRef]

18. Ly, D.; Mayrhofer, S.; Yogeswara, I.B.A.; Nguyen, T.-H.; Domig, K.J. Identification, classification and screening for γ-amino-butyric acid production in lactic acid bacteria from Cambodian fermented foods. *Biomolecules* **2019**, *9*, 768. [CrossRef]

19. Ueno, Y.; Hayakawa, K.; Takahashi, S.; Oda, K. Purification and characterization of glutamate decarboxylase from *Lactobacillus brevis* IFO 12005. *Biosci. Biotechnol. Biochem.* **1997**, *61*, 1168–1171. [CrossRef]

20. Higuchi, T.; Hisanobu, H.; Keietsu, A. Exchange of glutamate and gamma-aminobutyrate in a Lactobacillus strain. *J. Bacteriol. Res.* **1997**, *179*, 3362–3364. [CrossRef]

21. Park, K.B.; Oh, S.H. Production of yogurt with enhanced levels of gamma-aminobutyric acid and valuable nutrients using lactic acid bacteria and germinated soybean extract. *Bioresour. Technol.* **2007**, *98*, 1675–1679. [CrossRef] [PubMed]

22. Han, M.H.; Yang, J.M.; Joo, Y.C.; Moon, G.S. Production of functional fermented soymilk by γ-aminobutyric acid producing lactic acid bacteria. *Curr. Top. Lact. Acid Bact. Probiotics* **2019**, *5*, 59–64. [CrossRef]

23. Lee, B.J.; Kim, J.S.; Kang, Y.M.; Lim, J.H.; Kim, Y.M.; Lee, M.S.; Jeong, M.H.; Ahn, C.B.; Je, J.Y. Antioxidant activity and γ-aminobutyric acid (GABA) content in sea tangle fermented by *Lactobacillus brevis* BJ20 isolated from traditional fermented foods. *Food Chem.* **2010**, *122*, 271–276. [CrossRef]

24. Siragusa, S.; De Angelis, M.; Di Cagno, R.; Rizzello, C.G.; Coda, R.; Gobbetti, M. Synthesis of γ-aminobutyric acid by lactic acid bacteria isolated from a variety of Italian cheeses. *Appl. Environ. Microbiol.* **2007**, *73*, 7283–7290. [CrossRef] [PubMed]

25. Pandey, A. Solid-state fermentation. *Biochem. Eng. J.* **2003**, *13*, 81–84. [CrossRef]

26. Thomas, L.; Christian, L.; Ashok, P. Current developments in solid-state fermentation. *Biochem. Eng. J.* **2013**, *81*, 146–161. [CrossRef]

27. Guardado-Félix, D.; Antunes-Ricardo, M.; Rocha-Pizaña, M.R.; Martínez-Torres, A.C.; Gutiérrez-Uribe, J.A.; Saldivar, S.O.S. Chickpea (*Cicer arietinum* L.) sprouts containing supranutritional levels of selenium decrease tumor growth of colon cancer cells xenografted in immune-suppressed mice. *J. Funct. Foods* **2019**, *53*, 76–84. [CrossRef]

28. Kim, J.M.; Park, J.Y.; Kim, K.W.; Yoon, K.Y. Nutritional com- position and functionality of mixed cereals powder. *Korean J. Food Preserv.* **2014**, *54*, 388–395. [CrossRef]

29. Amarowicz, R.; Estrella, I.; Hernández, T.; Robredo, S.; Troszyńska, A.; Kosińska, A.; Pegg, R.B. Free radical-scavenging capacity, antioxidant activity, and phenolic composition of green lentil (*Lens culinaris*). *Food chem.* **2010**, *121*, 705–711. [CrossRef]

30. Lee, H.K.; Hwang, I.G.; Kim, H.Y.; Woo, K.S.; Lee, S.H.; Woo, S.H.; Lee, J.S.; Jeong, H.S. Physicochemical characteristic and antioxidant activities of cereals and legumes in Korea. *J. Korean Soc. Food Sci. Nutr.* **2010**, *39*, 1399–1404. [CrossRef]

31. Lu, Q.; Goto, K.; Nishizu, T. Study on the conditions and quality of GABA enriched brown rice. *J. JSAM* **2010**, *72*, 291–296.

32. Ma, Y.; Wang, A.; Yang, M.; Wang, S.; Wang, L.; Zhou, S.; Blecker, C. Influences of cooking and storage on γ-aminobutyric acid (GABA) content and distribution in mung bean and its noodle products. *LWT* **2022**, *154*, 112783. [CrossRef]

33. Yun, I.J. Effect of Selection Attributes of Convenience Food Instant Japgog (Mixed Cereals) Rice on Customer Satisfaction and Loyalty. Ph.D. Thesis, Catholic Kwandong University, Gangneung, Republic of Korea, 2016.

34. Nongsaro. Available online: https://www.nongsaro.go.kr/portal/ps/pst/pstb/pstbc/mngmtDtaDtl.ps?menuId=PS03213&nttSn=694&totalSearchYn=Y (accessed on 1 September 2017).

35. Kim, S.I.; Kang, J.H.; Lee, J.S.; Lee, J.S. Effect on rice image for connectedness formation and meal culture settlement. *J. Foodserv. Manag.* **2016**, *19*, 25–49.

36. Kim, H.S.; Oh, I.K.; Yang, S.K.; Lee, S.Y. A comparison of rheological measurement methods of instant cooked rice by a texture analyzer. *Food Eng. Prog.* **2018**, *22*, 381–385. [CrossRef]

37. Jung, Y.J.; Cho, Y.J.; Kim, K.W.; Yoon, K.Y. Current status and development plan of domestic cereal industry. *Food Preserv. Process. Indust.* **2013**, *12*, 31–39.

38. Panrod, K.; Anyarporn, T.; Vipaporn, P. Comparison of validated high performance liquid chromatography methods using two derivatizing agents for gamma-aminobutyric acid quantification. *Thai J. Pharm. Sci.* **2016**, *40*, 203–208.

39. Hayat, A.; Jahangir, T.M.; Khuhawar, M.Y.; Alamgir, M.; Siddiqui, A.J.; Musharraf, S.G. Simultaneous HPLC determination of gamma amino butyric acid (GABA) and lysine in selected Pakistani rice varieties by pre-column derivatization with 2-Hydroxynaphthaldehyde. *J. Cereal Sci.* **2014**, *60*, 356–360. [CrossRef]

40. Jang, G.W.; Choi, S.I.; Han, X.; Men, X.; Kwon, H.Y.; Choi, Y.E.; Kang, N.Y.; Park, B.W.; Kim, J.J.; Lee, O.H. Antioxidant and Anti-inflammatory Activities of *Phellodendron amurense* Extract Fermented with *Lactobacillus plantarum* CM. *J. Food Hyg. Saf.* **2021**, *36*, 196–203. [CrossRef]

41. Re, R.; Pellegrini, N.; Proteggente, A.; Pannala, A.; Yang, M.; Rice-Evans, C. Antioxidant activity applying an improved ABTS radical cation decolorization assay. *Free Radic. Biol. Med.* **1999**, *26*, 1231–1237. [CrossRef]

42. Radhika, D.; Bajpai, V.K.; Baek, K.H. Production of GABA (γ-aminobutyric acid) by microorganisms: A review. *Braz. J. Microbiol.* **2012**, *43*, 1230–1241.

43. Das, D.; Goyal, A. Antioxidant activity and γ-aminobutyric acid (GABA) producing ability of probiotic *Lactobacillus plantarum* DM5 isolated from Marcha of Sikkim. *LWT-Food Sci. Technol.* **2015**, *61*, 263–268. [CrossRef]

44. Wang, H.K.; Dong, C.; Chen, Y.F.; Cui, L.M.; Zhang, H.P. A new probiotic cheddar cheese with high ACE-Inhibitory activity and g-aminobutyric acid content produced with koumiss-derived *Lactobacillus casei* Zhang. *Food Technol. Biotechnol.* **2010**, *48*, 62–70.
45. Rayavarapu, B.; Tallapragada, P.; Usha, M.S. Statistical optimization of γ-aminobutyric acid production by response surface methodology and artificial neural network models using *Lactobacillus fermentum* isolated from palm wine. *Biocatal. Agric. Biotechnol.* **2019**, *22*, 101362. [CrossRef]
46. Lin, Q. Submerged fermentation of *Lactobacillus rhamnosus* YS9 for γ-aminobutyric acid (GABA) production. *Braz. J. Microbiol.* **2013**, *44*, 183–187. [CrossRef] [PubMed]
47. Li, H.; Qiu, T.; Huang, G.; Cao, Y. Production of gamma-aminobutyric acid by *Lactobacillus brevis* NCL912 using fed-batch fermentation. *Microb. Cell Fact.* **2010**, *9*, 85. [CrossRef]
48. Kim, J.Y.; Lee, M.Y.; Ji, G.E.; Lee, Y.S.; Hwang, K.T. Production of γ-aminobutyric acid in black raspberry juice during fermentation by *Lactobacillus brevis* GABA100. *Int. J. Food Microbiol.* **2009**, *130*, 12–16. [CrossRef]
49. Hwang, E.Y.; Park, J.Y. Isolation and characterization of gamma-aminobutyric acid (GABA)-producing lactic Acid Bacteria from Kimchi. *Curr. Top. Lact. Acid Bact. Probiotics* **2020**, *6*, 64–69. [CrossRef]
50. Singh, R.; De, S.; Belkheir, A. *Avena sativa* (Oat), a potential neutraceutical and therapeutic agent: An overview. *Crit. Rev. Food Sci. Nutr.* **2013**, *53*, 126–144. [CrossRef]
51. Xu, D.; Ren, G.Y.; Liu, L.L.; Zhu, W.X.; Liu, Y.H. The influences of drying process on crude protein content of naked oat cut herbage (*Avena nuda* L.). *Dry Technol.* **2014**, *32*, 321–332.
52. Jukanti, A.K.; Gaur, P.M.; Gowda, C.L.L.; Chibbar, R.N. Nutritional quality and health benefits of chickpea (*Cicer arietinum* L.): A review. *Br. J. Nutr.* **2012**, *108*, S11–S26. [CrossRef]
53. Joyce, B.; Zare, F.; Pletch, A. Pulse proteins: Processing, characterization, functional properties and applications in food and feed. *Food Res. Int.* **2010**, *43*, 414–431.
54. Jarpa, P.M. Lentil protein: A review of functional properties and food application. An overview of lentil protein functionality. *Int. J. Food Sci.* **2018**, *53*, 892–903. [CrossRef]
55. Wang, N. Optimization of a laboratory dehulling process for lentil (*Lens culinaris*). *Cereal Chem.* **2005**, *82*, 671–676. [CrossRef]
56. Thavarajah, D.; Thavarajah, P.; Sarker, A.; Vandenberg, A. Lentils (*Lens culinaris* Medikus Subspecies *culinaris*): A whole food for increased iron and zinc intake. *J. Agric. Food Chem.* **2009**, *57*, 5413–5419. [CrossRef] [PubMed]
57. Bondet, V.; Brand-Williams, W.; Berset, C. Kinetics and mechanisms of antioxidant activity using the DPPH. free radical method. *LWT-Food Sci. Technol.* **1997**, *30*, 609–615. [CrossRef]
58. van den Berg, R.; Haenen, G.R.; van den Berg, H.; Bast, A.A.L.T. Applicability of an improved Trolox equivalent antioxidant capacity (TEAC) assay for evaluation of antioxidant capacity measurements of mixtures. *Food Chem.* **1999**, *66*, 511–517. [CrossRef]
59. Ghamry, M.; Ghazal, A.F.; Al-Maqtqri, Q.A.; Li, L.; Zhao, W. Impact of a novel probiotic Lactobacillus strain isolated from the bee gut on GABA content, antioxidant activity, and potential cytotoxic activity against HT-29 cell line of rice bran. *J. Food Sci. Technol.* **2022**, *59*, 3031–3042. [CrossRef]
60. Ghamry, M.; Zhao, W.; Li, L. Impact of *Lactobacillus apis* on the antioxidant activity, phytic acid degradation, nutraceutical value and flavor properties of fermented wheat bran, compared to *Saccharomyces cerevisiae* and *Lactobacillus plantarum*. *Food Res. Int.* **2022**, *163*, 112142. [CrossRef]
61. Nisa, K.; Rosyida, V.T.; Nurhayati, S.; Indrianingsih, A.W.; Darsih, C.; Apriyana, W. Total phenolic contents and antioxidant activity of rice bran fermented with lactic acid bacteria. *IOP Conf. Ser. Earth Environ. Sci.* **2019**, *251*, 012020. [CrossRef]
62. Jhan, J.K.; Chang, W.F.; Wang, P.M.; Chou, S.T.; Chung, Y.C. Production of fermented red beans with multiple bioactivities using co-cultures of *Bacillus subtilis* and *Lactobacillus delbrueckii* subsp. *bulgaricus*. *LWT-Food Sci. Technol.* **2015**, *63*, 1281–1287. [CrossRef]
63. Gan, R.Y.; Shah, N.P.; Wang, M.F.; Lui, W.Y.; Corke, H. *Lactobacillus plantarum* WCFS1 fermentation differentially affects antioxidant capacity and polyphenol content in mung bean (*Vigna radiata*) and soya bean (*Glycine max*) milks. *J. Food Process Preserv.* **2017**, *41*, e12944. [CrossRef]

Article

Antioxidant and Functional Features of Pre-Fermented Ingredients Obtained by the Fermentation of Milling By-Products

Lorenzo Siroli [1,2,*], Barbara Giordani [3], Samantha Rossi [1], Davide Gottardi [1,2,*], Helena McMahon [4], Aleksandra Augustyniak [4], Abhay Menon [4], Lucia Vannini [1,2], Beatrice Vitali [3], Francesca Patrignani [1,2] and Rosalba Lanciotti [1,2]

1 Department of Agricultural and Food Sciences, University of Bologna, 47521 Cesena, Italy
2 Interdepartmental Centre for Agri-Food Industrial Research, University of Bologna, 47521 Cesena, Italy
3 Department of Pharmacy and Biotechnology, University of Bologna, 40127 Bologna, Italy
4 Kerry Campus, Munster Technological University, Clash, V92 HD4V Tralee, Ireland
* Correspondence: lorenzo.siroli2@unibo.it (L.S.); davide.gottardi2@unibo.it (D.G.)

Abstract: The use of milling by-products as ingredients in food formulations has increased gradually over the past years, due to their well-recognized health properties. Fermentation performed with selected microbial strains or microbial consortia is the most promising way to reduce antinutritional factors of cereals and bran, while increasing their nutritional and functional properties. This work, developed within the BBI project INGREEN, was aimed to study the functional, nutritional and technological features of a pre-fermented ingredient obtained from the fermentation of a mixture of rye bran and wheat germ by a selected microbial consortium composed of yeasts (*Kazachstania unispora* and *Kazachstania servazii*) and lactic acid bacteria (*Latilactobacillus curvatus*) using as reference the unfermented mixture and the same mixture fermented by a baker's yeast. The selected microbial consortium improved the complexity of the volatile molecules such as acids, alcohols and esters. A better retention of color parameters was maintained compared to the product fermented by a baker's yeast. In addition, the fermentation by the selected consortium showed a significant increase in short chain fatty acids (more than 5-fold), antioxidant activity (22–24%), total phenol content (53–71%), bioactive peptides (39–52%), a reduction of 20–28% in phytic acid content and an increase in prebiotic activity not only compared to the unfermented product but also compared to the preferment obtained with a baker's yeast. Overall, the fermentation by the selected microbial consortium can be considered a valuable way to valorize milling by-products and promote their exploitation as food ingredients.

Keywords: rye bran; valorisation; lactic acid bacteria; yeasts; functionality

Citation: Siroli, L.; Giordani, B.; Rossi, S.; Gottardi, D.; McMahon, H.; Augustyniak, A.; Menon, A.; Vannini, L.; Vitali, B.; Patrignani, F.; et al. Antioxidant and Functional Features of Pre-Fermented Ingredients Obtained by the Fermentation of Milling By-Products. *Fermentation* **2022**, *8*, 722. https://doi.org/10.3390/fermentation8120722

Academic Editors: Michela Verni and Carlo Giuseppe Rizzello

Received: 22 November 2022
Accepted: 8 December 2022
Published: 9 December 2022

1. Introduction

World cereal production stands at around 2.77 million tons/year [1] which makes cereals one of the main food sources for human consumption [2]. Among cereals, wheat, rice, corn, barley, rye, sorghum and oats are the most consumed ones [3]. During milling processing, several by-products are produced and classified based on particle size and endosperm content. Among them, bran and germ represent the most abundant ones [4]. Currently, cereal bran and germ are mainly used as a feed supplement, while the application in the food sector plays only a minor role [2,5,6]. Although the interest in using bran and germ in food formulation has increased gradually over the years, due to their widely recognized health properties [7,8], their current usage is too limited compared to their production rate (millions of metric tons a year), and it does not fully exploit their wide technological and functional potential [9]. For example, rye bran represents a valuable by-product due to its composition (33.4% cellulose, 5.3% hemicellulose, 3.3% lignin, 18.6% starch, 17.0% protein and 2.5% lipids) [10]. In addition, rye bran contains high amounts of bioactive compounds, including ferulic acid, characterized by antioxidant, anti-inflammatory, and anticancer properties [11]. However, its application in the food sector is still limited due to the negative impact on the sensory properties (mainly appearance, taste and texture) of food, reducing consumer acceptance to a certain extent [2,12]. Additionally, wheat germ is reported to be an excellent source of vitamins, minerals, dietary fiber, proteins, and some functional micronutrients [13,14]. However, the presence of wheat germ adversely affects the quality of flour and bran products, mainly due to the oxidation of unsaturated fatty acids [2,15].

Literature data show that fermentation and enzymatic treatments of cereal bran may increase its functionality [8,16]. In fact, some technological and health promoting properties of bran can be increased by the activity of selected microorganisms or microbial consortia [2,17]. Fermentation performed by selected microbial strains and consortia is the most promising way to reduce the phytate content of cereals and bran, increasing their nutritional values [11]. Literature data indicate that fermentation with well-characterized microbial cultures, containing yeasts and lactic acid bacteria (LAB), represents a useful tool to improve the quality, processability and functionality of fermented cereal products or high fiber ingredients, such as sourdough bread, fermented wheat bran and whole-meal flour [18]. For this reason, the interest in the selection of microbial consortia for the fermentation of milling by-products has greatly increased in recent years [2]. For example, a microbial consortium containing *Levilactobacillus brevis* and *Kazachstania exigua* increased the nutritional quality of fermented bran compared to native bran [4]. Fermentation of wheat bran with *Levilactobacillus brevis and Candida humilis* increased also the functionality of native bran due to the release of free amino acids and phenols [19]. Therefore, a sourdough-like fermentation could improve the nutritional and functional profile of rye bran and wheat germ, also compared to the use of baker's yeast, usually applied in bakery processes [20,21].

In this context, the aim of this work was to study the functional, nutritional and technological features of a pre-fermented ingredient obtained from the fermentation of a mixture of rye bran and wheat germ by a selected microbial consortium composed of yeast and LABs. The new pre-fermented ingredient was compared with an unfermented milling by-product mixture and a mixture fermented by baker's yeast (benchmark). In fact, sourdough-like fermentation could improve the nutritional and functional profile of cereal bran compared to the use of traditional baker's yeast. This study allowed us to valorize a low value by-product into a high functional food ingredient potentially suitable for bakery application.

2. Materials and Methods

2.1. Raw Material: Wheat Germ and Rye Bran

The wheat germ and rye bran were provided by Molini Pivetti S.p.A. (Ferrara, Italy) and Barilla G. e R. F.lli S.p.A. (Parma, Italy), respectively. Samples of wheat germ and rye bran were characterized for moisture, protein, ash, crude fat, soluble, insoluble, and total dietary fiber, as well as mineral content according to [22]. The fatty acid profiles and starch content were determined according to AOAC International [23] and AACC [24], respectively. Lipid fractions were analyzed by gas chromatograph combined with mass spectrometer according to AACC [24]. Phytate content was determined according to Buddrick et al. [25].

The chemical composition of the raw ingredients used in this work is summarized in Table 1.

Table 1. Chemical composition of wheat germ and rye bran used in this work.

Component %	Wheat Germ	Rye Bran
Proteins %	28.9	13.8
Humidity %	12.1	11.6
Ashes %	3.9	2.5
Soluble Fiber %	1.1	4.6
Insoluble Fiber %	11.9	18.5
Total fat %	7.5	2.9
Starch %	23.6	45.0
Phytic acid %	1.6	1.3

2.2. Microbial Consortium

A microbial consortium composed of one lactic acid bacterium (LAB) and two yeasts was used as starter for the fermentation of milling by-products. The strains were *Latilactobacillus curvatus* LS1, *Kazachstania unispora* FM2 and *Kazachstania servazii* KAZ2. All the strains belong to the collection of the Department of Agricultural and Food Sciences of Bologna University and were isolated from spontaneously fermented wheat and rye bran. This consortium was selected after a preliminary functional and technological characterization and on the basis of its rapid growth both on wheat germ and rye bran, in the framework of the EU project INGREEN. Both LAB and yeast strains were maintained as frozen stocks ($-80\ °C$), respectively, on Maltose, Man, Rogosa and Sharpe (mMRS) medium (Oxoid, Italy) and on Yeast Extract–Peptone–Dextrose (YPD) medium (Oxoid, Italy), supplemented with 25% glycerol (*w/v*). The cultures were propagated three times with about 3% (*v/v*) inoculum in MRS for *Latilactobacillus curvatus* and YPD for *Kazachstania* spp. and incubated at 25 °C for 24 h for LAB and 48 h for yeasts.

2.3. Fermentation Process

A milling by-product mixture was prepared in the amount of 1 kg using rye bran (70%) and wheat germ (30%) and placed in a commercial kneader (Major, Kenwood, Italy). The composition of the mixture was preliminarily chosen on the basis of the compositional characteristics of the by-products and the ability of the microbial consortium to have rapid fermentation kinetics and to remain stable over time. Tap water was added in proportion 2:1 to milling by-products. The three strains were previously propagated, as reported above, and then cultivated for 24 h at 25 °C in mMRS broth for *Latilactobacillus curvatus* and in YPD broth for *Kazachstania unispora* and *Kazachstania servazii*. The strains were than collected by centrifugation at 10,000 rpm, washed twice in saline solution (0.9% NaCl) and resuspended in the same solution before their use. The strains were inoculated in the hydrated milling by-product mixture in order to reach a cell load of about 6.5 log CFU/g for *Latilactobacillus curvatus* and 4.5 log CFU/g for the two yeast strains.

After mixing the inoculated ingredients with a kneader, the mixture was transferred to a sanitized tank, and the fermentation process was conducted in a static way at 25 °C for 24 h. As a benchmark, the same milling by-product mixture, prepared as described above, was inoculated with a lyophilized commercial bakery yeast (Lievital, Lesaffre Italia, Parma, Italy) at an initial level of 7.0 log CFU/g. The lyophilized yeast culture was rehydrated prior to use by adding water at 30 °C for 15 min. Eventually, an unfermented milling by-product mixture was included in the sample set. Fermentation experiments were carried out in triplicate.

Three different preferments were obtained:

1. Milling by-product mixture fermented by the microbial consortium of yeasts and LAB (FM),
2. Milling by-product mixture fermented by a commercial bakery yeast (benchmark)

3.   Unfermented milling by-product mixture (UF).

2.4. Acidification Kinetics and Plate Count

The kinetics of acidification were monitored by measuring the pH of the different preferments after 17, 20, 22 and 24 h of fermentation at 25 °C. At the beginning and at the end of fermentation, the cell loads of LAB and yeasts were determined by plate counting on selective agar media: m-MRS + 0.02% cycloheximide for the enumeration of LAB, and YPD + 0.02% chloramphenicol for the enumeration of yeasts. In addition, at the end of fermentation, total titratable acidity (TTA) was determined according to Rizzello et al. [14].

2.5. Nutritional Profile

The preferments were characterized for energy, fats, saturated fatty acids, carbohydrates, sugars, dietary fiber, proteins, and salt according to the Reg UE 1169/2011 25/10/2011 GU CE L304 22/11/2011. In particular, the energy content was evaluated according to da Rocha et al. [26]. The fat content was measured by a Soxhlet extraction method according to AOAC official methods 920.39 [27], and saturated fatty acids were analyzed according to dos Santos Oliveira et al. [28]. The carbohydrate content (%) was calculated by subtracting the contents of ash, fat, fiber and protein from 100% dry matter, according to Costantini et al. [29]. Sugar content was evaluated according to Luchese et al. [30], while dietary fiber was evaluated according to AACCI Approved Method 32–07.01, as reported by Khalid et al. [22]. Protein content was determined by the Kjeldahl method, using $N \times 5.7$ as conversion factor. Acetic acid assay kit and Lactic acid assay kit (Megazyme, Ayr, UK) were used for the quantification of acetic acid and lactic acid, respectively. The fermentation quotient (QF) was determined as the molar ratio between lactic and acetic acids.

2.6. Color Analysis

Color was assessed by a Minolta® CR-400 colorimeter (Milan, Italy), previously calibrated using a standard white ceramic tile, in standardized illuminant (C) and observation angle (0 with respect to an area of 8 mm in diameter) conditions. The CIELAB system was utilized, and the parameters of lightness (L*), redness (a*) and yellowness (b*) were used to objectively define color [31].

2.7. Volatile Molecule Profiles and Short-Chain Fatty Acids

The volatile molecule profiles and short-chain fatty acids (SCFA) were detected using a GC-MS coupled with a solid-phase microextraction technique (SPME), according to Burns et al. [32] with some modifications reported by Rossi et al. [33].

2.8. Fatty Acid Profiles

The lipid fractions of the different preferments were extracted according to the method reported by Boselli et al. [34] with some modifications described in Rossi et al. [33]. The fatty acid composition was determined as fatty acid methyl esters (FAMEs) using GC-MS. Methyl tridecanoate (Sigma, Milan, Italy) (13:0, 0.02 mg/mL) was used as internal standard, and Supelco FAME MIX 37 (Sigma) was used as external reference. The total fatty acid methyl ester profiles analyses were carried out according to Rossi et al. [30].

2.9. Peptide and Phytate Content

The peptide content was determined on the purified fractions obtained from preferment samples. Specifically, 1 g of sample was diluted in 4 mL of Tris-HCl (pH 8.8) 50 mm, stored at 4 °C for 1 h vortexing every 15 min, and centrifuged at $20,000 \times g$ for 20 min. The supernatant containing albumins and globulins was used to determine the peptide concentration by the o-phtaldialdehyde (OPA) method according to Coda et al. [35]. As a reference, a standard curve prepared with tryptone (0.1 to 1.5 mg/mL) was used. Phytate content was analyzed according to the AOAC 986.11/88 methodology [36] with a confidence interval calculated at a probability level of about 95%.

2.10. Antioxidant Activity and Phenol Content

To determine total polyphenols and antioxidant activity, 5 g of sample was previously added with 50 mL of 80% methanol. After 30 min, the mixture was centrifuged at 6000 rpm for 20 min. The evaluation of antioxidant activity was performed on the samples using different methodologies. The 2,2-diphenyl-1-picryl-hydrazyl-hydrate (DPPH) assay was used according to Rizzello et al. [37]. For the calculation of the µmol DPPH radical scavenged by the extracts, the absorbance value measured after 10 min was read at 517 nm. A blank reagent was used to verify the stability of DPPH radical dot during the test time. The kinetics of the antioxidant reaction were also determined over 30 min and compared with Trolox as an antioxidant reference. The 2,2′-azino-bis (3-ethylbenzothiazoline-6-sulfonic acid (ABTS) assay was used as reported by Miller and Rice-Evans [38]. The ABTS antioxidant reaction mixture contained 3.0 mL of ethanolic ABTS with an absorbance between 0.68 a 0.72 at 734 nm, and 30 µL of extract sample or 30 µL of ethanol/water (1:1 *v/v*) for the control. The absorbance at 734 nm was measured every 30 s for 6 min, and the Trolox equivalent was calculated using a standard curve prepared with Trolox. Triplicate determination was performed.

The analysis of total polyphenols was carried out according to the Folin–Ciocalteu methodology as reported by Slinkard and Singleton [39]. The mixture absorbance was read at 750 nm, and the concentration of total phenols was expressed as gallic acid equivalent.

2.11. Prebiotic Activity

The prebiotic activity of the samples was evaluated towards several bacterial strains (*Lactiplantibacillus plantarum* ATCC 8014, *Lacticaseibacillus rhamnosus* ATCC 7469, *Lactobacillus acidophilus* ATCC 4356, *Limosilactobacillus fermentum* ATCC 9338, *Bifidobacterium breve* DSM 20091). *Bifidobacterium angulatum* DSM 20098 and *Bifidobacterium longum* DSM 20219 were purchased from Medical-Supply Co. Ltd. (Ireland), while *Escherichia coli* ECOR 1 (ATCC 35320) was purchased in freeze-dried form from ATCC-LGC Standards (UK). The prebiotic index (PI) scores of the various strains were calculated using the following formula:

$$\text{Prebiotic Index}$$
$$= \frac{(24\text{ h increase in OD of probiote on prebiotic})-(24\text{ h increase in OD in control MRS})}{(24\text{ h increase in OD of }E.\text{ coli on feed})-(24\text{ h increase in OD in control LBA})}\frac{(24\text{ h increase in OD of probiote on glucose})-(24\text{ h increase in OD in control MRS})}{(24\text{ h increase in OD of }E.\text{ coli on glucose})-(24\text{ h increase in OD in control MRS})}$$

The formula determined the ratio of growth of Lactobacilli or Bifidobacteria using commercial prebiotics (FOS, GOS and inulin—Merck, Darmstadt, Germany) and preferments in relation to the growth of the non-probiotic *E. coli*. With the modified formula, glucose yielded a score of 1.00, and the growth of probiotic and enteric strains on known prebiotics or preferments was compared to growth on glucose. Negative or low PI scores were obtained for those strains that grew less on prebiotics or biomass than on glucose or/and exhibited less growth on the prebiotics than the enteric strains [40].

The ability of preferments to sustain the growth of *Bifidobacterium* spp. (*Bifidobacterium breve* DSM 20456, *Bifidobacterium bifidum* DSM 20213, *Bifidobacterium longum* DSM 20219 and *Bifidobacterium adolescentis* DSM 20083) was also evaluated through viable cell count. In particular, 50 mg of FM, benchmark or UF sample were dispersed in 10 mL of simulated intestinal fluid (SIF: 0.1% *w/v* pancreatin, 0.15% *w/v* Oxgall bile salt, pH 7) [41], and Bifidobacteria, previously grown in MRS, were inoculated at a final concentration of 10^6 CFU/mL. Bifidobacteria inoculated in SIF only served as control growth. Counts of viable Bifidobacteria on MRS agar plates were carried out at the inoculation time and after 3 h, 6 h and 24 h of incubation at 37 °C. Plates were then incubated at 37 °C in anaerobic jars containing GasPak EZ (Beckton, Dickinson and Co., Milan, Italy) for 24 h.

2.12. Antimicrobial Activity against Gastrointestinal Pathogens

The antibacterial activity of preferments was evaluated towards three gastrointestinal pathogens, namely *Escherichia coli* ECET, *Salmonella choleraesuis* serovar *typhimurium* and *Yersinia enterocolitica*, belonging to Department of Pharmacy and Biotechnology of

University of Bologna. The preferments (FM, benchmark and UF) were incubated in SIF (5 mg/mL) for 24 h at 37 °C. Subsequently, the samples were centrifuged ($5000\times g$ for 20 min) in order to harvest the supernatants containing the digested and soluble parts of the preferments, which were used for microbial testing.

The pathogens were cultured in nutrient broth (Difco, Detroit, MI, United States) for 24 h. One hundred microliters (100 µL) of cell suspensions in growth medium (2×10^5 CFU/mL) were then incubated inside 96-multi-well plates together with 100 µL of digested samples at 37 °C under anaerobic conditions. Wells inoculated with 100 µL of cell suspension in culture medium and 100 µL of SIF were used as growth control. Blanks, consisting of SIF and culture medium, were also included.

The growth was evaluated after 24 h of incubation by reading the absorbance at 600 nm (EnSpire Multimode Plate Reader, PerkinElmer Inc., Waltham, MA, USA), and the impact of preferments on microbial growth was expressed in percentage with respect to growth control (100%).

2.13. In Vitro Digestion, Cell Culture and Viability Assay

Tested samples were artificially digested to mimic the natural processes occurring in the human digestive system. Simulated salivary fluid (SSF), simulated gastric fluid (SGF) and simulated intestinal fluid (SIF) were prepared according to the publication of Minekus et al. [42]. Oral digestion was achieved by first mixing weighed amounts of the products with simulated salivary fluid, then adding calcium chloride and distilled water to produce a liquid sample. Gastric digestion was achieved by adding a set volume of gastric fluid to the liquid sample, followed by porcine pepsin and calcium chloride and then bringing the pH to 3 with hydrochloric acid. More distilled water was added, and the samples were placed in a shaking incubator at 37 °C for 2 h. Intestinal digestion was performed by adding simulated intestinal fluid, pancreatin, freshly made bile and calcium chloride. The pH was increased to pH 7 with sodium hydroxide. The samples were thoroughly mixed and incubated at 37 °C for 2 h in the shaking incubator. The samples were then heated to 60 °C for 20 min to inactivate the enzymes. The samples were centrifuged at 4000 rpm for 5 min. The pH was checked to ensure that it was pH 7. The supernatants were removed, filter-sterilized and stored at -20 °C in the freezer for further analysis.

The human epithelial cell line Caco-2 was purchased from European Collection of Cell Culture (ECACC, Salisbury, UK). Cells were maintained in Dulbecco's modified Eagle's medium (DMEM; Gibco™, Thermofisher, Waltham, MA, USA) supplemented with 10% fetal bovine serum, 1% non-essential amino acids solution, 100 U/mL penicillin, 100 µg/mL streptomycin and 1% GlutaMAX™ solution (Gibco™). Cells were incubated at 37 °C under 5% CO_2 atmosphere. Medium was changed every 2 days, and cells passed at 60% confluence.

The influence of tested samples on Caco-2 viability was determined using PrestoBlue™ Cell Viability Reagent (Thermofisher, Waltham, MA, USA). Caco-2 cells were seeded at a density of 10,000 cells per well in 96-well plates in 100 µL of complete culture media. Cells were allowed to adhere overnight. After 24 h, the cell culture medium was carefully removed and replaced with culture medium supplemented with tested samples. Medium only was used to treat control cells. After 24 h of incubation, PrestoBlue™ Cell Viability Reagent was added to each well of the 96-well plate to the final concentration of 10%. Plates were incubated in the dark for 2 h at 37 °C. Fluorescence was read using a 560 nm excitation/590 nm emission filter set (10 nm bandwidth) with a Thermo Scientific™ Varioskan™ LUX spectrophotometer. Fluorescence data in wells containing cells were corrected for background fluorescence using cell-free media control replicates.

2.14. Statistical Analysis

The results are expressed as the mean of three different samples from three repeated experiments on different days. The data were statistically analyzed using the one-way ANOVA procedure of Statistica 6.1 (StatSoft Italy srl, Vigonza, Italy). The differences between mean values were detected by the HSD Tukey test, and evaluations were based on a significance level of $p \leq 0.05$.

3. Results and Discussion

3.1. Fermentation Kinetics and Microbial Characteristics

In this work, a milling by-product mixture composed of rye bran (70%) and wheat germ (30%) was fermented (FM) by a microbial consortium composed of LAB (*Latilactobacillus curvatus* LS1) and yeasts (*Kazachstania unispora* FM2 and *Kazachstania servazii* KAZ2) in order to evaluate the effect of fermentation on the final nutritional, technological and functional characteristics in comparison to an unfermented mixture (UF) and a mixture fermented by commercial baker's yeast (benchmark). The composition of the milling by-products mixture and the microbial consortium were based on preliminary tests, carried out in the framework of the European project "INGREEN", in which it was observed that this specific microbial consortium was particularly suitable for fermenting rye bran in a short time and with optimal technological characteristics (fermentation kinetics, stability to backslopping, optimal LAB:yeast ratio, and sensorial features).

The fermentation of the rye bran and wheat germ mixture, with only *Saccharomyces cerevisiae* (benchmark) or with the selected microbial consortium (FM), was protracted for 24 h at 25 °C. Table 2 lists the pH and acidity at the end of fermentation, and the cell load of LAB and yeasts prior to and after fermentation.

Table 2. pH, titratable acidity, LAB and yeast cell load (pre- and post-fermentation) of the different samples after 24 h of fermentation. For the same parameter, average values lacking a common letter ([a–c]) significantly differ from each other.

| | | | Pre-Fermentation | | Post-Fermentation | |
| | | Acidity | LAB | Yeasts | LAB | Yeasts |
Sample	pH	meqNaOH	log CFU/g	log CFU/g	log CFU/g	log CFU/g
FM [1]	3.88 ± 0.16 [a]	21.10 ± 0.40 [a]	6.54 ± 0.21	4.67 ± 0.18 [a]	9.46 ± 0.19 [a]	7.38 ± 0.18 [a]
Benchmark [2]	5.69 ± 0.13 [b]	9.40 ± 0.30 [b]	<1.0	6.89 ± 0.22 [b]	2.11 ± 0.11 [b]	8.21 ± 0.21 [b]
UF [3]	6.43 ± 0.12 [c]	6.10 ± 0.10 [c]	<1.0	<1.0	-	-

[1] FM: Fermentation by microbial consortia composed of LAB and yeasts. [2] Benchmark: fermentation by a benchmark bakery yeast. [3] UF: unfermented milling by-product mixture.

The UF samples showed a pH of 6.43 and a titratable acidity of 6.10 meqNaOH. As expected, the fermentative process by the microbial consortium of yeasts and LAB led to a higher acidification of the preferment that reached a pH of 3.88 and titratable acidity of 21.2 meqNaOH after 24 h. Conversely, the fermentation by a bakery yeast resulted in a less acidic preferment. Lactic acid bacteria and yeasts were not detected in the UF sample. The benchmark sample had a starting yeast load of 6.89 log CFU/g; after 24 h of fermentation, yeasts reached 8.21 log CFU/g while LABs were present at the end of fermentation at a level of 2.11 log CFU/g. The FM samples showed an initial load of LABs and yeasts of 6.54 and 4.67 log CFU/g, respectively. At the end of fermentation, the detected loads of LABs and yeasts were 9.46 and 7.38 log CFU/g, respectively. Only the member strains in the microbial consortium used were detected in the FM preferment at the end of the fermentation process. Spontaneous fermentation was also carried out, and after 24 h of incubation, there was no significant acidification of the sample, while yeast and LAB were present at levels below 3.0 log CFU/g (data not shown).

It is evident that the ratio between yeast and LAB in FM samples at the end of fermentation was about 1:100, which is considered the required ratio for stable wheat sourdough [43–45]. The pH below 4.0 reached in FM samples strongly affected the protein solubilization, proteolysis process and protein interactions [43]. Acidification and proteolysis during sourdough fermentation typically result in an increase in amino acids, flavor precursors, and a change in dough rheology and texture [46].

3.2. Nutritional Profile and Fatty Acid Composition

The chemical composition of the preferments (FM and benchmark) and the unfermented milling by-product mixture is reported in Table 3.

Table 3. Fat, saturated fatty acids, starch, insoluble fiber, soluble fiber, proteins, salt, lactic acid and acetic acid expressed as g/100 g of dry matter (DM), of the preferment FM compared with the benchmark and the UF sample. For the same parameter, average values lacking a common letter ([a,b]) significantly differ from each other.

	FM [1]	Benchmark [2]	UF [3]
Fat g/100 g	4.0 ± 0.5 [a]	3.0 ± 0.4 [a]	3.5 ± 0.3 [a]
Saturated fatty acids g/100 g	1.2 ± 0.2 [a]	0.9 ± 0.2 [a]	0.9 ± 0.2 [a]
starch g/100 g	9.1 ± 0.4 [b]	9.3 ± 0.5 [b]	10.5 ± 0.7 [a]
Sugars g/100 g	0.3 ± 0.1 [a]	0.3 ± 0.1 [a]	2.8 ± 0.3 [b]
Insoluble fiber g/100 g	38.8 ± 1.2 [a]	41.3 ± 1.6 [a]	40.5 ± 1.3 [a]
Soluble fiber g/100 g	9.3 ± 0.4 [a]	8.5 ± 0.5 [ab]	8.3 ± 0.3 [b]
Protein g/100 g	16.6 ± 0.4 [a]	15.7 ± 0.5 [ab]	15.3 ± 0.4 [b]
Lactic acid g/100 g	1.011	0.02	-
Acetic acid g/100 g	0.104	0.06	0.02

[1] FM: Fermentation by microbial consortia composed of LAB and yeasts. [2] Benchmark: fermentation by a benchmark bakery yeast. [3] UF: unfermented milling by-product mixture.

No significant differences among samples were observed in terms of total and saturated fat content and total and insoluble fiber. However, the fermentation performed by the microbial consortium of LAB and yeasts led to a significant increase in soluble fiber compared to the UF samples but not compared to the benchmark. These data are in agreement with findings reported by Manini et al. [47] that showed a 30% increase in soluble fiber in wheat bran fermented by backslopping propagation with a stable microbiota of LAB and yeasts. A similar result was also obtained by fermenting wheat bran with a selected consortium containing *Lactobacillus bulgaricus*, *Streptococcus thermophilus* and a commercial baker's yeast [48]. An increase in fiber solubility was also observed in spontaneously fermented wheat germ and in wheat germ fermented by selected microbial consortia containing LAB and yeasts combined or not with hydrolytic enzyme treatments [4,47]. The increase in soluble fiber is very interesting from a nutritional point of view. In fact, the health benefits associated with soluble fiber are several, including longer digestive processes, increased satiety, a positive influence on postprandial glycemic response and reduction of total and LDL cholesterol [49].

Additionally, a significant reduction in starch was observed in FM and benchmark samples, which showed a starch content of 9.1 and 9.3 g/100 g DM, respectively, compared to the UF sample (10.5 g/100 g DM). As expected, the fermentation by the microbial consortium or by commercial bakery yeast reduced the sugar content to 0.3 g/100 g, compared to the native bran mixture that showed 2.8 g/100 g. These data are in agreement with findings reported by other authors that showed a reduction in starch content and sugars, such as raffinose and sucrose, after fermentation of wheat bran and germ, due to microbial metabolism [15,47].

As expected, the preferment obtained by the fermentation of the selected consortia of LAB and yeasts showed a significant increase in lactic acid and acetic acid, to 1.011 and 0.104 g/100 g, respectively, compared to the benchmark and UF. The fermentation quotient (FQ) of the FM samples was 6.5. On the other hand, the presence of *Latilactobacillus curvatus* in FM samples determined a fast acidification of the matrix due to the rapid conversion of fermentable carbohydrates into mainly lactic acid, but also other organic acids such as acetic acid, formic acid, and ethanol.

Although the amount of fat present was not significantly different as a result of the fermentation process, the fatty acid (FA) profiles of the different samples were determined and are reported in Table 4.

Table 4. Fatty acid profiles expressed as relative percentages, of each considered preferment (FM and benchmark) and unfermented milling by-product mixture (UF). For the same parameter, average values lacking a common letter ([a,b]) significantly differ from each other.

				FA Relative Percentage						
	C14:0	C15:0	C16:1	C16:0	C18:2 n-6	C18:1 Δ9	C18:3 n-3	C18:0	CL *	UL[†]
UF [1]	0.4 [a]	0.3 [a]	0.4 [a]	37.7 [b]	42.2 [a]	15.1 [a]	2.1 [a]	1.8 [a]	1722	1.06
Benchmark [2]	0.6 [a]	0.3 [a]	0.3 [a]	35.5 [b]	42.0 [a]	16.9 [a]	2.2 [a]	2.2 [a]	1725	1.08
FM [3]	0.7 [a]	0.1 [a]	0.8 [b]	30.5 [a]	40.1 [a]	22.6 [b]	2.8 [b]	2.4 [a]	1732	1.11

[1] FM: Fermentation by microbial consortia composed of LAB and yeasts. [2] Benchmark: fermentation by a benchmark bakery yeast. [3] UF: unfermented milling by-product mixture. The fatty acid relative percentages were calculated with respect to the total fatty acid methyl esters. The results are means of three repetitions of three independent experiments. The coefficients of variability, expressed in percentages as ratios between the standard deviations and the mean values, ranged between 2% and 5%. * Mean chain length calculated as (FAP * C) (where FAP is the percentage of fatty acid and C the number of carbon atoms). [†] Unsaturation level calculated as [percentage monoenes + 2(percentage dienes) + 3(percentage trienes)]/100.

Palmitic and linoleic acids were found to be the main fatty acids detected in all the samples, followed by oleic, linolenic and stearic acid. When compared to the benchmark and UF sample, the relative percentages of oleic and palmitoleic acid increased in FM, while the relative percentage of palmitic acid decreased. Therefore, fermentation by the selected microbial consortium (FM) led to an increase in the unsaturation level (UL) of the fatty acids compared to the benchmark and the UF. Other authors reported an increase in lipid content and a modification of fatty acid profiles in wheat and maize bran subjected to a sourdough-like fermentation [47,50]. The presence of unsaturated FA such as linoleic, linolenic and oleic acid in the pre-fermented ingredients, associated with a reduced content of total FA, has great industrial importance since it is associated with a nutritional, functional and health promoting effect [14]. In addition, the interest in the presence of unsaturated FA is not only related to nutritional aspects, but also to the fact that they are precursors of aromatic and antimicrobial compounds (i.e., furanones) [51].

3.3. Color Analysis

Figure 1 presents the appearance of the UF, FM and benchmark samples, while Table 5 lists the indexes of lightness (L*), redness (a*) and yellowness (b*).

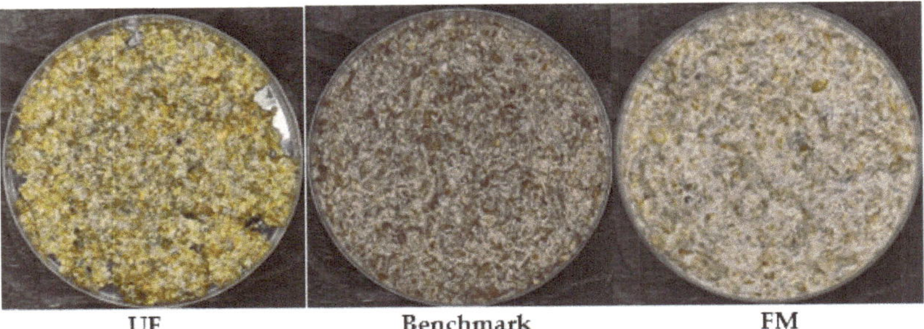

Figure 1. Appearance of unfermented rye bran and wheat germ mixture (UF) and of the preferment obtained by the fermentation of the selected microbial consortia (FM) and the benchmark after 24 h of fermentation.

As can be seen from Figure 1 and Table 5, the preferments were characterized by a very different appearance, and consequently of colorimetric indices. The fermentation by the selected microbial consortia of LAB and yeasts led to a significant increase in the L* index compared to the UF sample. In contrast, the benchmark showed rapid browning, as demonstrated by the significant reduction in the L* value compared to the other samples.

In addition, a significant increase in the a* index and decrease in the b* index were observed in the FM and benchmark samples compared to the UF sample. On the other hand, fermentation by commercial bakery yeast does not slow down the activity of browning related enzymes, while the lower pH achieved with the selected consortium is responsible for the reduction of enzymatic activity [2]. The color of a preferment, developed for applications in the bakery sector, represents an important parameter both for its acceptability and the impact on the final product. For these reasons, it is very important to prevent browning processes. Fermentation by selected microbial consortia, composed of LAB and yeasts, is from this point of view optimal, as it can increase the presence of antioxidant substances, such as ferulic acid, that can prevent color changes [2,11].

Table 5. Average lightness (L*), redness (a*) and yellowness (b*) values prior to fermentation (UF) and at the end of fermentation by the selected microbial consortia of LAB and yeasts (FM) and by a benchmark bakery yeast (benchmark). Data represent means $\pm$ SD. For the same parameter, average values lacking a common letter ([a–c]) significantly differ from each other.

	L*	**a***	**b***
UF	55.32 $\pm$ 1.07 [b]	0.85 $\pm$ 0.20 [a]	12.45 $\pm$ 0.56 [c]
Benchmark	52.63 $\pm$ 1.11 [c]	2.62 $\pm$ 0.26 [c]	8.77 $\pm$ 0.52 [a]
FM	63.15 $\pm$ 0.92 [a]	1.68 $\pm$ 0.40 [b]	10.66 $\pm$ 0.39 [b]

3.4. Volatile Molecule Profiles

Specific profiles in terms of volatile molecules were detected based on the preferments considered (Table 6). A total of 26, 50 and 45 volatile molecules were identified in the aromatic profiles of the UF, FM and benchmark samples, respectively. The fermentation process led to a qualitative and quantitative increase in the aromatic molecules present in the samples, in particular acids, alcohols, and esters. However, the type of microbial consortium used strongly influenced the aromatic profile. In fact, the use of only *Saccharomyces cerevisiae* (benchmark) led to a particularly higher amount of alcohols (ethanol 40.3 ppm) and esters (ethyl acetate 18.9 ppm). The fermentation by the selected microbial consortium of LAB and yeasts also led to an increase in alcohols and esters, but to a lesser extent than the benchmark. On the contrary, the FM samples showed an increase in acids, such as acetic acid, and short-chain fatty acids. In general, the fermentation process led to an improvement in the aromatic profile of the preferments due to the positive odor perception of the identified molecules. The greater acidic note of the FM preferment was linked to the presence of *Latilactobacillus curvatus* in the consortium used and was related to the measured lower pH and higher TTA. Other authors reported a higher increase in acetic acid and hexanoic acid in rye bran fermented only by *Latilactobacillus curvatus* [52]. Fermentation by a bakery yeast resulted in a higher amount of volatile molecules, but the volatile profiles obtained with a consortium of yeasts and LAB showed a wider spectrum of compounds. The synthesis of volatile molecules in yeasts mainly involves the conversion of pyruvate to diacetyl, the formation of esters and the Ehrlich pathway with the conversion of amino acids to volatile compounds; however, the formation of aromatic compounds by the Ehrlich pathway can be enhanced by sourdough fermentation due to the release of the precursor amino acids for the higher proteolysis level [53,54]. On the other hand, it is widely reported that the spectrum and amounts of volatile compounds are higher in sourdoughs fermented with yeasts and lactobacilli compared to baker's yeast, proving that different microbiota affect the final food volatile profile [53,55]. Additionally, in LAB, the synthesis of volatile molecule compounds is strictly dependent on the species, but in sourdough fermentation, it is well demonstrated that heterofermentative and homofermentative LAB, including *Latilactobacillus curvatus*, contribute to a higher and wider spectrum of volatile molecules with respect to the use of baker's yeast [20].

Table 6. Volatile compounds, expressed as ppm, detected through GC-MS-SPME in preferments FM, benchmark, and the mixture of unfermented milling by-products (UF). The coefficients of variability, expressed as the percentage ratios between the standard deviations and the mean values, ranged between 2% and 5%.

		UF	FM	Benchmark
	Odor Perception [a]		**ppm**	
Pentanal	Fruity	-	0.03	0.02
Hexanal	Green, Fruity	0.73	2.24	3.10
Heptanal	Green, Herbal	-	0.23	0.54
2-Hexenal	Fruity, Green	0.06	1.23	0.87
2-Nonenal, (E)-	Fatty, Green	0.05	0.18	0.83
Acetaldehyde	Ethereal, Pungent, Fruity	0.24	1.18	1.01
Benzaldehyde	Fruity, Sharp	-	0.20	0.19
Total Aldehydes		**1.08**	**5.29**	**6.56**
2,3-Butanedione	Buttery, Creamy, Pungent	-	0.13	0.15
3-Pentanone	Ethereal	-	0.44	-
Methyl Isobutyl Ketone	Green, Sharp, Herbal	0.34	0.16	0.31
2-Hexanone, 4-methyl-	Fruity, Fungal, Meaty	0.08	0.11	0.20
3-Penten-2-one, 4-methyl-	Vegetable, Pungent	0.40	0.58	0.42
4-Heptanone, 2,6-dimethyl-	Green, Fruity	0.13	0.22	0.35
2-Butanone, 3-hydroxy-	Buttery, Sweet, Creamy	-	1.23	0.28
Total Ketones		**0.96**	**2.85**	**1.71**
Ethyl Alcohol	Alcoholic, Ethereal	0.35	26.71	40.33
1-Propanol	Alcoholic, Fermented	-	0.20	0.19
1-Penten-3-ol	Green, Ethereal	-	0.21	0.18
2-Hexanol	Winey, Fruity	0.21	0.29	0.18
1-Pentanol	Fermented, Balsamic	0.61	0.76	0.21
2-Penten-1-ol, (E)-	Green	0.13	0.38	0.17
1-Hexanol	Herbal, Hethereal	1.29	8.38	4.12
1-Octen-3-ol	Earthy, Mushroom	0.19	0.49	0.09
Heptanol	Green, Musty	-	0.79	2.89
1-Hexanol, 2-ethyl-	Citrus, Floral	0.06	0.19	0.10
1-Octanol	Waxy, Green	0.06	0.53	0.47
2-Furanmethanol	Bready, Alcoholic	-	0.06	0.10
Benzyl Alcohol	Floral, Balsamic	-	0.59	0.67
1-Butanol, 3-methyl	Fruity, Sweet	-	1.62	2.88
3-Nonen-1-ol, (Z)-	Waxy, Green	-	0.15	3.18
Phenylethyl Alcohol	Floral	0.08	0.36	-
Total Alcohols		**2.96**	**41.70**	**55.67**
Acetic Acid	Acidic, Sharp	0.15	7.96	1.85
Butanoic Acid	Cheesy, Sharp	-	0.12	-
Butanoic Acid, 3-methyl-	Sweet, Mentholic	-	0.36	-
Pentanoic Acid	Cheesy, Acidic	-	0.43	-
Hexanoic Acid	Fatty, Sour	0.08	3.60	0.35
Heptanoic Acid	Cheesy, Rancid, Sour	-	0.11	-
Octanoic Acid	Fatty, Waxy, Rancid	-	0.18	-
Total Acids		**0.23**	**9.76**	**2.20**
Ethyl Acetate	Ethereal, Fruity, Sweet	-	8.45	18.90
Pentanoic Acid, Ethyl Ester	Fruity, Sweet	-	0.41	1.35
Hexanoic Acid, Ethyl Ester	Fruity	-	1.21	5.21
Heptanoic Acid, Ethyl Ester	Fruity	0.05	0.05	0.73
Octanoic Acid, Ethyl Ester	Waxy, Fruity, Winey	0.11	-	0.29
Nonanoic Acid, Ethyl Ester	Waxy, Fruity	-	-	0.51
Butanoic Acid, Ethyl Ester	Fruity, Juicy	-	-	0.27
1-Butanol, 3-methyl-, Acetate	Fruity, Sweet	-	0.15	-
1-Butanol, 3-methyl-, Formate	Green, Fruity	-	4.19	6.87
Total Esters		**0.16**	**14.46**	**34.13**

Table 6. *Cont.*

		UF	FM	Benchmark
	Odor Perception [a]		**ppm**	
Butylated Hydroxytoluene	Phenolic, Camphoreous	1.13	1.05	1.05
Cyclohexene, 1,6,6-trimethyl-	Citrus, Herbal	1.59	1.53	1.19
Cyclohexene, 3-methyl-	Citrus	-	0.07	0.11
Hexadecane	-	0.08	0.02	0.08
Furan, 2-pentyl-	Fruity, Green	-	0.02	0.07
Benzenamine	-	1.69	0.78	0.99
Total Other Compounds		**5.25**	**4.25**	**4.18**
		10.63	**81.32**	**104.45**

[a] Based on data reported in the literature and information found at: http://www.thegoodscentscompany.com/index.html (accessed on 11 November 2022).

3.5. Functionality

Several functional parameters including bioactive peptides, total SCFA, total phenols, antioxidant activity (DPPH and ABTS) and phytic acid content of the mixture of rye bran and wheat germ fermented by a yeast/LAB consortium or bakery yeast were assessed and compared with the unfermented mixture (Table 7). As shown in Table 7, The fermentation of the milling by-product mixture by the LAB/yeast consortium led to a significant increase in the SCFA content compared to both the unfermented mixture and those fermented with a benchmark baker's yeast. In fact, the SCFAs detected were 12.64 mg/kg, 2.21 mg/kg and 0.23 mg/kg, respectively, in FM, benchmark and UF sample. The greater content of SCFAs in FM can be attributed to the presence of *Latilactibacillus curvatus*. In fact, the production of SCFAs in fermented bran and sourdough is related to the metabolic activity of LAB [56]. Lactobacilli can produce SCFAs by the fermentation of carbohydrate end-products such as pyruvate, which is generated during the glycolytic pathway and also by the phosphoketolase route in heterofermenting conditions [57,58]. The detected SCFAs in the FM sample were acetic acid, propanoic acid, butanoic acid, pentanoic acid, hexanoic acid and octanoic acid. However, the dominant SCFAs were acetic acid and hexanoic acid. The beneficial properties of SCFAs are widely reported; they are easily absorbed by the host and have beneficial effects on intestinal function as well as systemic roles in insulin secretion, lipid metabolism, and inflammation, among others [59]. In addition, the contribution of SCFAs to the flavor of bread crumb is also reported [60].

Table 7. Bioactive peptides, total SCFA, phenols, antioxidant activity (ABTS and DPPH) and phytic acid content in UF, benchmark and FM. Data represent means ± SD. For the same parameter, average values lacking a common letter ([a–c]) significantly differ from each other.

	FM [1]	Benchmark [2]	UF [3]
Bioactive peptides (mg/g)	5.41 ± 0.17 [b]	3.88 ± 0.34 [a]	3.55 ± 0.28 [a]
Total SCFA (mg/kg)	12.64 ± 0.89 [c]	2.21 ± 0.50 [b]	0.23 ± 0.09 [a]
Total phenols (Gallic acid mg eq/kg DM)	451 ± 22 [b]	295 ± 16 [a]	263 ± 16 [a]
ABTS (TROLOX mg/kg DM)	432 ± 24 [b]	355 ± 16 [a]	349 ± 26 [a]
DPPH (TROLOX mg/kg DM)	229 ± 14 [b]	199 ± 9 [a]	196 ± 15 [a]
Phytic acid (g/100 g DM)	0.76 ± 0.06 [b]	0.95 ± 0.04 [a]	1.05 ± 0.05 [a]

[1] FM: Fermentation by microbial consortia composed of LAB and yeasts. [2] Benchmark: fermentation by a benchmark bakery yeast. [3] UF: unfermented milling by-product mixture.

The peptide content was found to be dependent on the sample considered. In any case, the fermentation by the yeasts and LAB consortium led to a significant increase in bioactive peptides (5.41 mg/g) compared to the concentration found in the samples fermented with a baker's yeast (3.88 mg/g) and the unfermented milling by-product mixture (3.55 mg/g). The functional importance of bioactive peptides is widely demonstrated, and activities such as mineral binding, immunomodulatory, antimicrobial, antioxidative, antithrombotic, hypocholesterolemic and antihypertensive, are attributed to them [35]. The increase in bioactive peptides observed in the FM samples can be attributed to the activities of the LABs present in the microbial consortia used. In fact, the release of various bioactive peptides (e.g., angiotensin I-converting enzyme (ACE)-inhibitory peptides) from proteins through proteolysis by LABs in various food systems is well documented [43,61,62].

The preferment FM showed a significant increase in total phenol content compared to the UF and benchmark. The increase in total phenols with respect to the UF and benchmark was found to be 70% and 50%, respectively. On the other hand, it is widely reported that sourdough fermentation increases the levels of extractable phenolic compounds [63]. These increases are mainly dependent on the breakdown of the cell wall of cereal and brans, followed by enzymatic activities that result in the release of bounded phenolic compounds [64]. This is particularly important in cereal bran, which is very rich in phenolic compounds that generally are present in esterified form linked to the cell wall matrix and consequently not easily available [64,65]. Fermentation can be considered an optimal approach to release the insoluble bounded phenolic acids. In particular, it is well documented that fermentation with LABs or consortia of yeasts and LABs, combined or not with enzymatic treatments, has a positive effect on the bioavailability of polyphenols in fermented wheat bran, wheat germ, rice bran and rye bran [2,18,64].

The increase in phenolic compounds in FM samples resulted in higher antioxidant activity, detected with ABTS and DPPH assay, compared to the UF and benchmark samples. Fermentation of the rye bran and wheat germ mixture by a benchmark baker's yeast did not lead to a significant increase in antioxidant activity compared to the unfermented mixture. On the contrary, the fermentation by the consortium showed an increase in antioxidant activity ranging between 17 and 24%, depending on the method used, compared to the unfermented milling by-product mixture. These data are in agreement with Rizzello et al. [14], who reported an increase in antioxidant activity due to phenolics and bioactive peptides in wheat germ fermented by *Lb. plantarum* and *Lb. rossiae*. An increased content of free and soluble antioxidant compounds after fermentation of wheat bran with a *Lb. rhamnosus* strain was also observed by Spaggiari et al. [66].

On the other hand, during sourdough fermentation, LABs increased the amount of extractable phenolic compounds and antioxidant peptides [44]. Several lactobacilli can reinforce their inherent cellular antioxidant defense through the secretion of antioxidant enzymes such as superoxide dismutase. In addition, LABs promote the synthesis of glutathione, the main non-enzymatic antioxidant and free-radical scavenger, and exopolysaccharides that are other biomolecules synthesized by LABs and also share antioxidant activity [67].

The FM samples also showed a significantly lower content of phytic acid compared to the UF and benchmark samples. In fact, the FM samples showed a reduction in phytate concentration of about 28% and 20% compared to the UF and benchmark, respectively. Other than bioactive compounds, cereal bran usually contains anti-nutritional molecules such as phytic acid that reduces the bioavailability of important minerals such as Ca^{2+}, Mg^{2+}, Fe^{2+}, and Zn^{2+} and amino acids. Therefore, the degradation of these compounds is fundamental in order to improve the functional and nutritional characteristics of cereal bran [66]. The results obtained in this work are in agreement with data in the literature that showed a decrease in phytic acid content in wheat bran and wheat germ fermented by LABs [14,48,66]. The hydrolysis of phytic acid is generally carried out by microbial and endogenous phytase that hydrolyzes phytic acid to a non-metal chelator compound [14]. Lactic acid bacteria are a source of phytase, and the pH reached during fermentation by LABs is more suitable to activate flour endogenous phytases [5].

3.6. Prebiotic Activity

The prebiotic activity of the samples was assessed in both aerobic and anaerobic fer-mentation models. The prebiotic index (PI) scores determined are reported in Table 8. Both the FM and UF samples showed a positive value of PI, which means all probiotic bacteria grew more than enteric strains (*E. coli*). The scores of *Lactobacillae* strains were noted to be higher than that of *Bifidobacteria*. Prebiotic indices obtained for *Limosilactobacillus fermentum* ATCC 9338 grown in UF, FM and FOS were significantly higher than values determined for the other six strains considered. The strains *Lactobacillus acidophilus* ATCC 4356, *Limosilactobacillus fermentum* ATCC 9338 and *Bifidobacterium angulatum* DSM 20098 demonstrated a significantly higher prebiotic index (PI) score for the fermented mixture (FM) in comparison with the unfermented one (UF). Data in the literature indicate that fermented cereal brans may exert prebiotic activity due to production of exopolysaccharides by LABs and in response to the presence of oligosaccharides resulting from hydrolysis of arabinoxylans present in brans [68].

Table 8. The prebiotic indices of pre-fermented ingredients in UF and FM and commercial prebiotic products, FOS and inulin. For the same bacterial strains, average values lacking a common letter ([a–d]) significantly differ from each other.

Bacterial Strains	UF [1]	FM [2]	FOS	INULIN
L. plantarum ATCC 8014	2.73 [a]	2.78 [a]	1.17 [b]	0.95 [c]
L. rhamnosus ATCC 7469	2.80 [a]	2.75 [a]	1.03 [b]	0.28 [c]
L. acidophilus ATCC 4356	2.08 [b]	2.65 [a]	0.97 [c]	0.23 [d]
L. fermentum ATCC 9338	5.60 [b]	7.86 [a]	1.95 [c]	1.13 [c]
B. angulatum DSM 20098	1.61 [b]	2.66 [a]	0.46 [c]	2.16 [ab]
B. breve DSM 20091	1.96 [b]	0.82 [c]	1.59 [bc]	5.70 [a]
B. longum DSM 20219	3.10 [a]	2.15 [b]	0.79 [d]	1.27 [c]

[1] UF: unfermented milling by-product mixture. [2] FM: fermentation by microbial consortia composed of LAB and yeasts.

The ability of pre-fermented ingredients (FM and benchmark), as well as UF one, to support the growth of *Bifidobacterium* spp. highly present in the gut microbiota (*B. breve*, *B. bifidum*, *B. longum* and *B. adolescentis*) was also confirmed through viable counts in simulated intestinal fluid (Figure 2). The viability of bifidobacteria incubated in SIF without preferments or UF strongly decreased over time and was completely abolished after 6 h (*B. bifidum* DSM 2013 and *B. longum* DMS 20219) or 24 h (*B. breve* DSM 20456 and *B. adolescentis* DSM 20083). On the contrary, bifidobacteria were able to grow in the presence of the unfermented and fermented mixtures, reaching at least 7.5 log CFU/mL after 24 h of incubation. In particular, after 3 h of incubation, the viable count of *Bifidobacterium* spp. was always lower in the presence of benchmark, while the viability in the presence of FM was equal (*B. breve* DMS 20456) or slightly lower than that observed in the presence of the UF mixture. On the contrary, after 6 h and 24 h of incubation, the highest growth was observed in the presence of the FM mixture for all bifidobacteria tested, reaching 8.31–8.96 log CFU/mL. These results underlined that bifidobacteria were able not only to survive, but also to grow in the presence of FM as substrate, confirming the prebiotic activity previously observed.

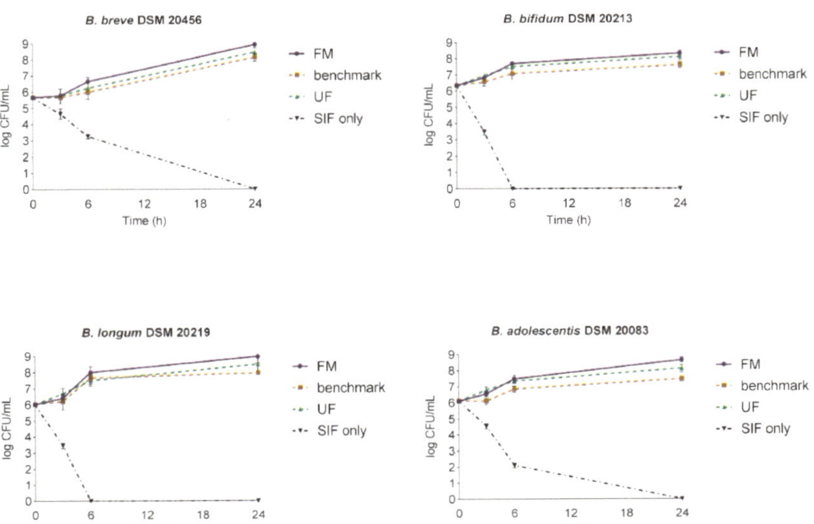

Figure 2. Viability over time of *Bifidobacterium* spp. in SIF only or in the presence of FM, benchmark and UF (mean ± SD, *n* = 3).

3.7. Antimicrobial Activity against Gastrointestinal Pathogens

The impact of preferments and UF on relevant gastrointestinal pathogens (*E. coli* ECET, *S. choleraesuis* and *Y. enterocolitica*) was also sought in order to exclude undesired stimulating effects. Results are reported in Figure 3, as growth percentages compared to the growth of pathogens in the absence of mixtures (control). The UF mixture significantly stimulated the growth of *S. choleraesuis* and *Y. enterocolitica*, by ~10–18%. *Y. enterocolitica* was also slightly stimulated by the benchmark (by ~9%, $p < 0.05$). Noticeably, the FM mixture strongly reduced the growth of gastrointestinal pathogens, with inhibition percentages against *E. coli* ECET, *S. choleraesuis* and *Y. enterocolitica* of 60.5, 48.8 and 43.9%, respectively. This behavior can be due to the lower pH and the presence of metabolites in the FM mixture that can exert antibacterial activities, such as bioactive peptides, SCFA and phenols [69–71].

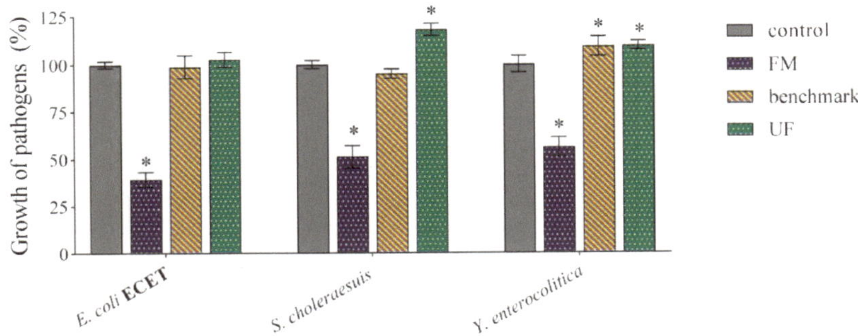

Figure 3. Effects of FM, benchmark and UF on the growth of *E. coli* ECET, *S. choleraesuis* and *Y. enterocolitica* (mean ± SD, *n* = 3). * $p < 0.05$.

3.8. Caco-2 Viability Assay

Caco-2 cells were treated with tested samples for 24 h, and the viability of the cells was measured using a fluorescent dye, PrestoBlue™. The results of the performed cytotoxicity experiment are displayed in Figure 4.

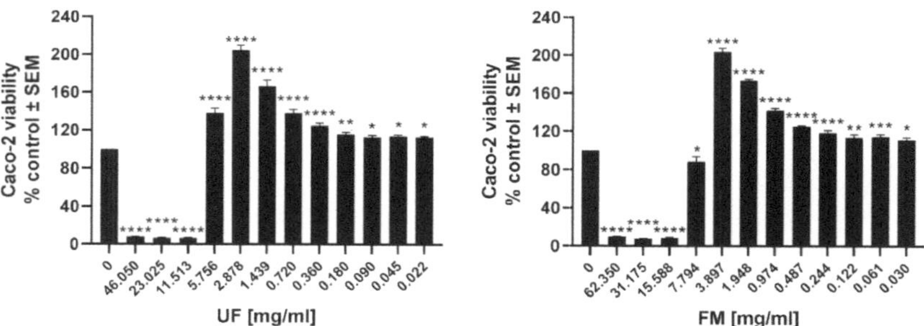

Figure 4. Effect of tested samples on Caco-2 cell proliferation. Cells were treated with different concentrations of tested samples for 24 h, and proliferation was assessed using PrestoBlue™ Cell Viability Reagent. The data are presented as the mean ± SEM. Statistical significance in comparison to the negative control (untreated cells) was assessed using one-way ANOVA followed by Dunnett's multiple comparison test; * $p < 0.05$; ** $p < 0.01$; *** $p < 0.001$; **** $p < 0.0001$.

The results show that three of the highest concentrations of the unfermented milling by-product and the four highest concentrations of sample after fermentation significantly reduced Caco-2 proliferation. All other concentrations of tested samples showed no cytotoxic effect against the analyzed cell model. Moreover, a significant increase in cell proliferation was observed for these doses of tested samples. The best results were obtained for 2.88 mg/mL UF and 3.90 mg/mL FM–Caco-2 viability was 204.77% and 203.93%, respectively. The Caco-2 cellular model is commonly used to investigate potential toxic effects of food or food metabolites in the intestinal mucosa [72]. The effect of pre-fermented cereals on growth of the Caco-2 cell line was assessed by Bruininx et al. [73]. The results of a tetrazolium-based colorimetric assay indicated that after 19 days of cell incubation with 1%, 5% and 10% of fermented cereal supernatant, the cytotoxic effect of tested samples was not observed. In the study by Galli et al. [74], 23 LAB strains were selected and used for wheat flour inoculation. The obtained sourdough extracts did not negatively affect the viability of human intestinal cells, Caco-2. Yang et al. [75] analyzed the cytotoxicity of *Lactobacillus rhamnosus* and *Bifidobacterium animalis* ssp. *Lactis*-fermented seaweed extract using an MTT assay in the Caco-2 cell line. No significant decrease in cell viability was observed. Moreover, one of the tested fermented extracts significantly increased Caco-2 viability when used at a concentration of 400 μg/mL.

4. Conclusions

The data obtained showed the excellent potential of the microbial consortium containing *Lb. curvatus*, *K. servazzii* and *K. unispora* to improve the functional and technological properties of fermented wheat germ and rye bran. First, they showed rapid fermentation kinetics on the by-products used, reaching optimal pH of 3.9 and acidity values of 21.1 meqNaOH within 24 h, with all of the consequences that this aspect has from an industrial point of view. Second, this consortium improved the complexity of the volatile molecules obtained after fermentation, mainly acetic acid and hexanoic acid. Moreover, fermentation by the selected yeasts and LAB determined an increase in SCFA (by more than 5-fold), antioxidant activity (22–24%), total phenol content (53–71%), bioactive peptides (39–52%), a reduction in phytic acid content of 20–28%, and an increase in prebiotic activity not only compared to the unfermented product but also compared to the preferment obtained with a baker's yeast. FM samples also showed a strong inhibition against gastrointestinal pathogens, ranging between 44 and 61% depending on the pathogen, and at a certain concentration, a stimulation of Caco-2 cell viability by more than 200%. This outcome confirms that a sourdough-like fermentation with a selected consortium of LAB and yeasts can provide superior technological and functional features to the final product. The characteristics of the obtained pre-fermented ingredient indicate that it is potentially suitable for application in the bakery sector, even if other sectors of application, such as the nutraceutical one, can be explored. However, further studies are required to define the optimal ratio to be used and the effects that this ingredient may provide in the final product.

Author Contributions: Conceptualization, R.L., F.P., L.V. and B.V.; methodology, L.S., B.V., D.G. and H.M.; software, L.S. and D.G.; validation, S.R., B.G. and A.A., formal analysis, S.R., A.A., B.G. and A.M.; investigation, S.R., A.A., A.M., D.G., L.S. and B.G.; resources, L.S., D.G., F.P. and R.L.; data curation, S.R., B.G., A.A. and A.M.; writing—original draft preparation, L.S., B.G., D.G., S.R. and H.M.; writing—review and editing, R.L., F.P., A.A., B.V., H.M. and L.V.; visualization, D.G., L.S., L.V., R.L. and B.V.; supervision, R.L., F.P., B.V. and L.V. All authors have read and agreed to the published version of the manuscript.

Funding: This research was performed in the context of the INGREEN project "Bio-based ingredients for sustainable industries through biotechnology". This project has received funding from the Bio Based Industries Joint Undertaking (JU) under grant agreement No 838120. The JU receives support from the European Union's Horizon 2020 research and innovation program and the Bio Based Industries Consortium.

Institutional Review Board Statement: Not applicable.

Data Availability Statement: Not applicable.

Acknowledgments: The authors would like to thank Narinder Bains (INEUVO Ltd.), the coordinator of the project "INGREEN Bio-based ingredients for sustainable industries through biotechnology", and also Nadia Morbarigazzi from Barilla G. e R. F.lli S.p.A and Simona Baraldi from Molini Pivetti S.p.A. for providing us with wheat germ and rye bran.

Conflicts of Interest: The authors declare no conflict of interest.

References

1. FAO. Cereal Supply and Demand Brief. World Food Situation. Food and Agriculture Organization of the United Nations. Available online: https://www.fao.org/worldfoodsituation/csdb/en/ (accessed on 14 November 2022).
2. Verni, M.; Rizzello, C.G.; Coda, R. Fermentation Biotechnology Applied to Cereal Industry By-Products: Nutritional and Functional Insights. *Front. Nutr.* **2019**, *6*, 42. [CrossRef] [PubMed]
3. Zamaratskaia, G.; Gerhardt, K.; Wendin, K. Biochemical Characteristics and Potential Applications of Ancient Cereals—An Underexploited Opportunity for Sustainable Production and Consumption. *Trends. Food Sci. Technol.* **2021**, *107*, 114–123. [CrossRef]
4. Coda, R.; Kärki, I.; Nordlund, E.; Heiniö, R.L.; Poutanen, K.; Katina, K. Influence of Particle Size on Bioprocess Induced Changes on Technological Functionality of Wheat Bran. *Food Microbiol.* **2014**, *37*, 69–77. [CrossRef] [PubMed]
5. Gobbetti, M.; de Angelis, M.; di Cagno, R.; Polo, A.; Rizzello, C.G. The Sourdough Fermentation Is the Powerful Process to Exploit the Potential of Legumes, Pseudo-Cereals and Milling by-Products in Baking Industry. *Crit. Rev. Food Sci. Nutr.* **2019**, *60*, 2158–2173. [CrossRef] [PubMed]
6. Onipe, O.O.; Jideani, A.I.O.; Beswa, D. Composition and Functionality of Wheat Bran and Its Application in Some Cereal Food Products. *Int. J. Food Sci. Technol.* **2015**, *50*, 2509–2518. [CrossRef]
7. Prückler, M.; Siebenhandl-Ehn, S.; Apprich, S.; Höltinger, S.; Haas, C.; Schmid, E.; Kneifel, W. Wheat Bran-Based Biorefinery 1: Composition of Wheat Bran and Strategies of Functionalization. *LWT.-Food Sci. Technol.* **2014**, *56*, 211–221. [CrossRef]
8. Coda, R.; Katina, K.; Rizzello, C.G. Bran Bioprocessing for Enhanced Functional Properties. *Curr. Opin. Food Sci.* **2015**, *1*, 50–55. [CrossRef]
9. Luithui, Y.; Baghya Nisha, R.; Meera, M.S. Cereal By-Products as an Important Functional Ingredient: Effect of Processing. *J. Food Sci. Technol.* **2019**, *56*, 1–11. [CrossRef]
10. Galanakis, C.M. Sustainable Applications for the Valorization of Cereal Processing By-Products. *Foods* **2022**, *11*, 241. [CrossRef]
11. Zhang, J.; Liu, M.; Zhao, Y.; Zhu, Y.; Bai, J.; Fan, S.; Zhu, L.; Song, C.; Xiao, X. Recent Developments in Fermented Cereals on Nutritional Constituents and Potential Health Benefits. *Foods* **2022**, *11*, 2243. [CrossRef]
12. Dziki, D. Rye Flour and Rye Bran: New Perspectives for Use. *Processes* **2022**, *10*, 293. [CrossRef]

13. Boukid, F.; Folloni, S.; Ranieri, R.; Vittadini, E. A Compendium of Wheat Germ: Separation, Stabilization and Food Applications. *Trends. Food Sci. Technol.* **2018**, *78*, 120–133. [CrossRef]

14. Rizzello, C.G.; Nionelli, L.; Coda, R.; de Angelis, M.; Gobbetti, M. Effect of Sourdough Fermentation on Stabilisation, and Chemical and Nutritional Characteristics of Wheat Germ. *Food Chem.* **2010**, *119*, 1079–1089. [CrossRef]

15. Rizzello, C.G.; Nionelli, L.; Coda, R.; di Cagno, R.; Gobbetti, M. Use of Sourdough Fermented Wheat Germ for Enhancing the Nutritional, Texture and Sensory Characteristics of the White Bread. *Eur. Food Res. Technol.* **2010**, *230*, 645–654. [CrossRef]

16. Onipe, O.O.; Ramashia, S.E.; Jideani, A.I.O. Wheat Bran Modifications for Enhanced Nutrition and Functionality in Selected Food Products. *Molecules* **2021**, *26*, 3918. [CrossRef] [PubMed]

17. Bertsch, A.; Roy, D.; LaPointe, G. Fermentation of Wheat Bran and Whey Permeate by Mono-Cultures of Lacticaseibacillus Rhamnosus Strains and Co-Culture With Yeast Enhances Bioactive Properties. *Front. Bioeng. Biotechnol.* **2020**, *8*, 956. [CrossRef]

18. Filannino, P.; di Cagno, R.; Gobbetti, M. Metabolic and Functional Paths of Lactic Acid Bacteria in Plant Foods: Get out of the Labyrinth. *Curr. Opin. Biotechnol.* **2018**, *49*, 64–72. [CrossRef]

19. Arte, E.; Rizzello, C.G.; Verni, M.; Nordlund, E.; Katina, K.; Coda, R. Impact of Enzymatic and Microbial Bioprocessing on Protein Modification and Nutritional Properties of Wheat Bran. *J. Agric. Food Chem.* **2015**, *63*, 8685–8693. [CrossRef]

20. Arora, K.; Ameur, H.; Polo, A.; di Cagno, R.; Rizzello, C.G.; Gobbetti, M. Thirty Years of Knowledge on Sourdough Fermentation: A Systematic Review. *Trends. Food Sci. Technol.* **2021**, *108*, 71–83. [CrossRef]

21. Rizzello, C.G.; Portincasa, P.; Montemurro, M.; di Palo, D.M.; Lorusso, M.P.; de Angelis, M.; Bonfrate, L.; Genot, B.; Gobbetti, M. Sourdough Fermented Breads Are More Digestible than Those Started with Baker's Yeast Alone: An In Vivo Challenge Dissecting Distinct Gastrointestinal Responses. *Nutrients* **2019**, *11*, 2954. [CrossRef]

22. Khalid, K.H.; Ohm, J.B.; Simsek, S. Whole Wheat Bread: Effect of Bran Fractions on Dough and End-Product Quality. *J. Cereal. Sci.* **2017**, *78*, 48–56. [CrossRef]

23. Official Methods of Analysis of AOAC International. WorldCat.Org. Available online: https://www.worldcat.org/it/title/official-methods-of-analysis-of-aoac-international/oclc/1085712083 (accessed on 11 November 2022).

24. AACC. *Approved Methods of the AACC*, 10th ed.; American Association of Cereal Chemists: St. Paul, MN, USA, 2000. Available online: http://www.sciepub.com/reference/241265 (accessed on 11 November 2022).

25. Buddrick, O.; Jones, O.A.H.; Cornell, H.J.; Small, D.M. The Influence of Fermentation Processes and Cereal Grains in Wholegrain Bread on Reducing Phytate Content. *J. Cereal. Sci.* **2014**, *59*, 3–8. [CrossRef]

26. da Rocha Lemos Mendes, G.; Souto Rodrigues, P.; de las Mercedes Salas-Mellado, M.; Fernandes de Medeiros Burkert, J.; Badiale-Furlong, E. Defatted Rice Bran as a Potential Raw Material to Improve the Nutritional and Functional Quality of Cakes. *Plant. Foods Hum. Nutr.* **2021**, *76*, 46–52. [CrossRef] [PubMed]

27. Aoac 920.39 2003_05 Extracto Etereo Dietileter-VSIP. INFO. Available online: https://vsip.info/aoac-92039-200305-extracto-etereo-dietileter-2-pdf-free.html (accessed on 11 November 2022).

28. Oliveira, M.D.S.; Feddern, V.; Kupski, L.; Cipolatti, E.P.; Badiale-Furlong, E.; de Souza-Soares, L.A. Changes in Lipid, Fatty Acids and Phospholipids Composition of Whole Rice Bran after Solid-State Fungal Fermentation. *Bioresour. Technol.* **2011**, *102*, 8335–8338. [CrossRef] [PubMed]

29. Costantini, L.; Lukšič, L.; Molinari, R.; Kreft, I.; Bonafaccia, G.; Manzi, L.; Merendino, N. Development of Gluten-Free Bread Using Tartary Buckwheat and Chia Flour Rich in Flavonoids and Omega-3 Fatty Acids as Ingredients. *Food Chem.* **2014**, *165*, 232–240. [CrossRef]

30. Luchese, C.L.; Gurak, P.D.; Marczak, L.D.F. Osmotic Dehydration of Physalis (*Physalis peruviana* L.): Evaluation of Water Loss and Sucrose Incorporation and the Quantification of Carotenoids. *LWT.-Food Sci. Technol.* **2015**, *63*, 1128–1136. [CrossRef]

31. Pointer, M.R. A Comparison of the CIE 1976 Colour Spaces. *Color. Res. Appl.* **1981**, *6*, 108–118. [CrossRef]

32. Burns, P.; Patrignani, F.; Serrazanetti, D.; Vinderola, G.C.; Reinheimer, J.A.; Lanciotti, R.; Guerzoni, M.E. Probiotic Crescenza Cheese Containing Lactobacillus Casei and Lactobacillus Acidophilus Manufactured with High-Pressure Homogenized Milk. *J. Dairy Sci.* **2008**, *91*, 500–512. [CrossRef]

33. Rossi, S.; Parrotta, L.; del Duca, S.; Rosa, M.D.; Patrignani, F.; Schluter, O.; Lanciotti, R. Effect of Yarrowia Lipolytica RO25 Cricket-Based Hydrolysates on Sourdough Quality Parameters. *LWT* **2021**, *148*, 111760. [CrossRef]

34. Boselli, E.; Velazco, V.; Fiorenza Caboni, M.; Lercker, G. Pressurized Liquid Extraction of Lipids for the Determination of Oxysterols in Egg-Containing Food. *J. Chromatogr. A* **2001**, *917*, 239–244. [CrossRef]

35. Coda, R.; Rizzello, C.G.; Pinto, D.; Gobbetti, M. Selected Lactic Acid Bacteria Synthesize Antioxidant Peptides during Sourdough Fermentation of Cereal Flours. *Appl. Environ. Microbiol.* **2012**, *78*, 1087–1096. [CrossRef] [PubMed]

36. AOAC 986.11-1988, Phytate in Foods Anion Exchange Method-$14.15: AOAC Official Method. Available online: http://www.aoacofficialmethod.org/index.php?main_page=product_info&cPath=1&products_id=313 (accessed on 11 November 2022).

37. Rizzello, C.G.; Coda, R.; Mazzacane, F.; Minervini, D.; Gobbetti, M. Micronized By-Products from Debranned Durum Wheat and Sourdough Fermentation Enhanced the Nutritional, Textural and Sensory Features of Bread. *Food Res. Int.* **2012**, *46*, 304–313. [CrossRef]

38. Miller, N.J.; Rice-Evans, C.A. Factors Influencing the Antioxidant Activity Determined by the ABTS Radical Cation Assay. *Free. Radic. Res.* **1997**, *26*, 195–199. [CrossRef] [PubMed]

39. Slinkard, K.; Singleton, V.L. Total Phenol Analysis: Automation and Comparison with Manual Methods. *Am. J. Enol. Vitic.* **1977**, *28*, 49–55.

40. Huebner, J.; Wehling, R.L.; Hutkins, R.W. Functional Activity of Commercial Prebiotics. *Int. Dairy. J.* **2007**, *17*, 770–775. [CrossRef]

41. Giordani, B.; Melgoza, L.M.; Parolin, C.; Foschi, C.; Marangoni, A.; Abruzzo, A.; Dalena, F.; Cerchiara, T.; Bigucci, F.; Luppi, B.; et al. Vaginal Bifidobacterium Breve for Preventing Urogenital Infections: Development of Delayed Release Mucoadhesive Oral Tablets. *Int. J. Pharm.* **2018**, *550*, 455–462. [CrossRef]

42. Minekus, M.; Alminger, M.; Alvito, P.; Ballance, S.; Bohn, T.; Bourlieu, C.; Carrière, F.; Boutrou, R.; Corredig, M.; Dupont, D.; et al. A Standardised Static in Vitro Digestion Method Suitable for Food—An International Consensus. *Food Funct.* **2014**, *5*, 1113–1124. [CrossRef]

43. Scarnato, L.; Serrazanetti, D.I.; Aloisi, I.; Montanari, ·C.; del Duca, S.; Lanciotti, R. Combination of Transglutaminase and Sourdough on Gluten-Free Flours to Improve Dough Structure. *Amino. Acids.* **2016**, *48*, 2453–2465. [CrossRef]

44. Gobbetti, M.; Rizzello, C.G.; di Cagno, R.; de Angelis, M. How the Sourdough May Affect the Functional Features of Leavened Baked Goods. *Food Microbiol.* **2014**, *37*, 30–40. [CrossRef]

45. de Luca, L.; Aiello, A.; Pizzolongo, F.; Blaiotta, G.; Aponte, M.; Romano, R. Volatile Organic Compounds in Breads Prepared with Different Sourdoughs. *Appl. Sci.* **2021**, *11*, 1330. [CrossRef]

46. Scarnato, L.; Montanari, C.; Serrazanetti, D.I.; Aloisi, I.; Balestra, F.; del Duca, S.; Lanciotti, R. New Bread Formulation with Improved Rheological Properties and Longer Shelf-Life by the Combined Use of Transglutaminase and Sourdough. *LWT* **2017**, *81*, 101–110. [CrossRef]

47. Manini, F.; Brasca, M.; Plumed-Ferrer, C.; Morandi, S.; Erba, D.; Casiraghi, M.C. Study of the Chemical Changes and Evolution of Microbiota During Sourdoughlike Fermentation of Wheat Bran; Study of the Chemical Changes and Evolution of Microbiota During Sourdoughlike Fermentation of Wheat Bran. *Cereal Chem.* **2014**, *91*, 342–349. [CrossRef]

48. Zhao, H.M.; Guo, X.N.; Zhu, K.X. Impact of Solid State Fermentation on Nutritional, Physical and Flavor Properties of Wheat Bran. *Food Chem.* **2017**, *217*, 28–36. [CrossRef] [PubMed]

49. Lattimer, J.M.; Haub, M.D. Effects of Dietary Fiber and Its Components on Metabolic Health. *Nutrients* **2010**, *2*, 1266–1289. [CrossRef]

50. Decimo, M.; Quattrini, M.; Ricci, G.; Fortina, M.G.; Brasca, M.; Silvetti, T.; Manini, F.; Erba, D.; Criscuoli, F.; Casiraghi, M.C. Evaluation of Microbial Consortia and Chemical Changes in Spontaneous Maize Bran Fermentation. *AMB Express.* **2017**, *7*, 205. [CrossRef]

51. Ndagijimana, M.; Vallicelli, M.; Cocconcelli, P.S.; Cappa, F.; Patrignani, F.; Lanciotti, R.; Guerzoni, M.E. Two 2[5H]-Furanones as Possible Signaling Molecules in Lactobacillus Helveticus. *Appl. Environ. Microbiol.* **2006**, *72*, 6053–6061. [CrossRef]

52. Kaseleht, K.; Paalme, T.; Mihhalevski, A.; Sarand, I. Analysis of Volatile Compounds Produced by Different Species of Lactobacilli in Rye Sourdough Using Multiple Headspace Extraction. *Int. J. Food Sci. Technol.* **2011**, *46*, 1940–1946. [CrossRef]

53. Xu, D.; Zhang, Y.; Tang, K.; Hu, Y.; Xu, X.; Gänzle, M.G. Effect of Mixed Cultures of Yeast and Lactobacilli on the Quality of Wheat Sourdough Bread. *Front. Microbiol.* **2019**, *10*, 2113. [CrossRef]

54. Thiele, C.; Gänzle, M.G.; Vogel, R.F. Fluorescence Labeling of Wheat Proteins for Determination of Gluten Hydrolysis and Depolymerization during Dough Processing and Sourdough Fermentation. *J. Agric. Food Chem.* **2003**, *51*, 2745–2752. [CrossRef]

55. Gänzle, M.G.; Zheng, J. Lifestyles of Sourdough Lactobacilli—Do They Matter for Microbial Ecology and Bread Quality? *Int. J. Food Microbiol.* **2019**, *302*, 15–23. [CrossRef]

56. Pérez-Alvarado, O.; Zepeda-Hernández, A.; Garcia-Amezquita, L.E.; Requena, T.; Vinderola, G.; García-Cayuela, T. Role of Lactic Acid Bacteria and Yeasts in Sourdough Fermentation during Breadmaking: Evaluation of Postbiotic-like Components and Health Benefits. *Front. Microbiol.* **2022**, *13*, 969460. [CrossRef] [PubMed]

57. LeBlanc, J.G.; Chain, F.; Martín, R.; Bermúdez-Humarán, L.G.; Courau, S.; Langella, P. Beneficial Effects on Host Energy Metabolism of Short-Chain Fatty Acids and Vitamins Produced by Commensal and Probiotic Bacteria. *Microb. Cell Fact.* **2017**, *16*, 79. [CrossRef] [PubMed]

58. Annunziata, G.; Arnone, A.; Ciampaglia, R.; Tenore, G.C.; Novellino, E. Fermentation of Foods and Beverages as a Tool for Increasing Availability of Bioactive Compounds. Focus on Short-Chain Fatty Acids. *Foods* **2020**, *9*, 999. [CrossRef] [PubMed]

59. Kasubuchi, M.; Hasegawa, S.; Hiramatsu, T.; Ichimura, A.; Kimura, I. Dietary Gut Microbial Metabolites, Short-Chain Fatty Acids, and Host Metabolic Regulation. *Nutrients* **2015**, *7*, 2839–2849. [CrossRef] [PubMed]

60. Birch, A.N.; Petersen, M.A.; Hansen, Å.S. The Aroma Profile of Wheat Bread Crumb Influenced by Yeast Concentration and Fermentation Temperature. *LWT.-Food Sci. Technol.* **2013**, *50*, 480–488. [CrossRef]

61. Tagliazucchi, D.; Martini, S.; Solieri, L. Bioprospecting for Bioactive Peptide Production by Lactic Acid Bacteria Isolated from Fermented Dairy Food. *Fermentation* **2019**, *5*, 96. [CrossRef]

62. Pessione, E.; Cirrincione, S. Bioactive Molecules Released in Food by Lactic Acid Bacteria: Encrypted Peptides and Biogenic Amines. *Front. Microbiol.* **2016**, *7*, 876. [CrossRef]

63. Liukkonen, K.-H.; Katina, K.; Wilhelmsson, A.; Myllymaki, O.; Lampi, A.-M.; Kariluoto, S.; Piironen, V.; Heinonen, S.-M.; Nurmi, T.; Adlercreutz, H.; et al. Process-Induced Changes on Bioactive Compounds in Whole Grain Rye. *Proc. Nutr. Soc.* **2003**, *62*, 117–122. [CrossRef]

64. Adebo, O.A.; Medina-Meza, I.G. Impact of Fermentation on the Phenolic Compounds and Antioxidant Activity of Whole Cereal Grains: A Mini Review. *Molecules* **2020**, *25*, 927. [CrossRef]

65. Moore, J.; Cheng, Z.; Hao, J.; Guo, G.; Liu, J.G.; Lin, C.; Yu, L. Effects of Solid-State Yeast Treatment on the Antioxidant Properties and Protein and Fiber Compositions of Common Hard Wheat Bran. *J. Agric. Food Chem.* **2007**, *55*, 10173–10182. [CrossRef]

66. Spaggiari, M.; Ricci, A.; Calani, L.; Bresciani, L.; Neviani, E.; Dall'Asta, C.; Lazzi, C.; Galaverna, G. Solid State Lactic Acid Fermentation: A Strategy to Improve Wheat Bran Functionality. *LWT* **2020**, *118*, 108668. [CrossRef]

67. Laurent-Babot, C.; Guyot, J.P. Should Research on the Nutritional Potential and Health Benefits of Fermented Cereals Focus More on the General Health Status of Populations in Developing Countries? *Microorganisms* **2017**, *5*, 40. [CrossRef] [PubMed]

68. Poutanen, K.; Flander, L.; Katina, K. Sourdough and Cereal Fermentation in a Nutritional Perspective. *Food Microbiol.* **2009**, *26*, 693–699. [CrossRef]

69. Brown, L.; Pingitore, E.V.; Mozzi, F.; Saavedra, L.M.; Villegas, J.M.; Hebert, E. Lactic Acid Bacteria as Cell Factories for the Generation of Bioactive Peptides. *Protein Pept. Lett.* **2017**, *24*, 146–155. [CrossRef] [PubMed]

70. Huang, C.B.; Alimova, Y.; Myers, T.M.; Ebersole, J.L. Short- and Medium-Chain Fatty Acids Exhibit Antimicrobial Activity for Oral Microorganisms. *Arch. Oral. Biol.* **2011**, *56*, 650–654. [CrossRef] [PubMed]

71. Guo, L.; Gong, S.; Wang, Y.; Sun, Q.; Duo, K.; Fei, P. Antibacterial Activity of Olive Oil Polyphenol Extract Against Salmonella Typhimurium and Staphylococcus Aureus: Possible Mechanisms. *Foodborne. Pathog. Dis.* **2020**, *17*, 396–403. [CrossRef]

72. Tang, A.S.; Chikhale, P.J.; Shah, P.K.; Borchardt, R.T. Utilization of a Human Intestinal Epithelial Cell Culture System (Caco-2) for Evaluating Cytoprotective Agents. *Pharm. Res.* **1993**, *10*, 1620–1626. [CrossRef]

73. Bruininx, E.M.A.M.; Koninkx, J.F.J.G.; Binnendijk, G.P.; Zandstra, T.; Heetkamp, M.J.W.; van der Peet-Schwering, C.M.C.; Gerrits, W.J.J. Effects of Preformented Cereals or the End Products of Fermentation on Growth and Metabolism of Enterocyte-like Caco-2 Cells and on Intestinal Health of Restrictedly Fed Weanling Pigs. *Animal* **2010**, *4*, 40–51. [CrossRef]

74. Galli, V.; Mazzoli, L.; Luti, S.; Venturi, M.; Guerrini, S.; Paoli, P.; Vincenzini, M.; Granchi, L.; Pazzagli, L. Effect of Selected Strains of Lactobacilli on the Antioxidant and Anti-Inflammatory Properties of Sourdough. *Int. J. Food Microbiol.* **2018**, *286*, 55–65. [CrossRef]

75. Yang, H.S.; Haj, F.G.; Lee, M.; Kang, I.; Zhang, G.; Lee, Y. Laminaria Japonica Extract Enhances Intestinal Barrier Function by Altering Inflammatory Response and Tight Junction-Related Protein in Lipopolysaccharide-Stimulated Caco-2 Cells. *Nutrients* **2019**, *11*, 1001. [CrossRef]

Article

Determination of Changes in Volatile Aroma Components, Antioxidant Activity and Bioactive Compounds in the Production Process of Jujube (*Ziziphus jujuba* Mill.) Vinegar Produced by Traditional Methods

Havva Nilgün Budak

Department of Food Processing, Egirdir Vocational School, Isparta University of Applied Sciences, Isparta TR32500, Turkey; nilgunbudak@isparta.edu.tr

Abstract: Jujube has anticancer, diabetic, antimicrobial, anti-inflammatory, cardiovascular, gastrointestinal and immune system effects. In this study, jujube juice, jujube wine and jujube vinegar were investigated in terms of chemical composition, phenolic contents, organic acid contents, volatile compound contents, and antioxidant activity. Antioxidant activity and total phenolic content of jujube vinegar produced by traditional methods were found to be higher than those of jujube juice and wine. Protocatechic acid, chlorogenic acid, phydroxybenzoic acid, caffeic acid, epicatechin, and syringic acid were detected in jujube vinegar. Moreover, oxalic acid, malic acid, tartaric acid, formic acid, ascorbic acid, lactic acid, acetic acid and some other organic acid components were determined in jujube vinegar. Volatile aroma compounds such as ester, aldehyde, alcohol, terpene, acid, and ketone were determined in jujube samples. It was seen that the antioxidant activity and bioactive compounds of jujube vinegar were very rich, and jujube vinegar, which is an alternative product with a high potential produced from jujube fruit, is an important product for the food sector due to its long shelf life. This research is the first detailed study in which the antioxidant activity and bioactive compounds determined during the production stages of jujube vinegar (jujube juice, wine, and vinegar) were evaluated in detail.

Keywords: jujube vinegar; bioactive compounds; volatile compounds; phenolic content; antioxidant activity; organic acid content; TEAC; chlorogenic acid; acetic acid; ethyl acetate

Citation: Budak, H.N. Determination of Changes in Volatile Aroma Components, Antioxidant Activity and Bioactive Compounds in the Production Process of Jujube (*Ziziphus jujuba* Mill.) Vinegar Produced by Traditional Methods. *Fermentation* **2022**, *8*, 606. https://doi.org/10.3390/fermentation8110606

Academic Editors: Michela Verni and Carlo Giuseppe Rizzello

Received: 7 October 2022
Accepted: 2 November 2022
Published: 4 November 2022

Publisher's Note: MDPI stays neutral with regard to jurisdictional claims in published maps and institutional affiliations.

1. Introduction

Jujube (*Ziziphus jujuba* Mill.) is a deciduous tree or large shrub, belonging to the *Rhamnaceae* family. Its unripe fruits are green, and its ripe fruits can be of different colors, from yellow or red to brown. It is primarily planted for edible fruit, however, it is also used as a medicinal agent. The homeland of the jujube plant is China. It is also grown in Russia, the Middle East, India, Southern Europe, Anatolia, and North Africa [1,2]. Jujube is grown intensively in the Western and Southern provinces of Turkey [3].

In the prescription prepared by Ibn Sina, the jujube plant was used as a cough suppressant, laxative, blood pressure reducer, digestive disorder remedy, and in the treatment of stomach ulcers [4]. In China, the ripe and dried fruits of the jujube fruit, called the "fruit of life", have been used as an aphrodisiac, laxative, and antidote [5]. Recent studies have shown that jujube has anticancer, antidiabetic, antimicrobial, anti-inflammatory, sedative, laxative, cardiovascular, gastrointestinal, and immune system effects, and that phytochemicals in the jujube plant are effective in the treatment of these diseases [2,6–11].

Jujube fruit has high nutritional value. Its nutritional content comprises carbohydrates, proteins, fats, vitamins, minerals, and phenolic compounds [12]. The jujube plant contains galactose, rhamnose, mannose, glucuronic acid, arabinose, glucose, fructose, sucrose, and sorbitol sugars [13]. It has been reported that the diabetic effect of jujube fruit prevents the excessive increase in blood glucose by reducing the glucose uptake of the cells during the glucose absorption stage and delays the diffusion of glucose through the dialysis membranes [9]. Also, the jujube plant is recommended for the treatment of diabetic patients due to its antioxidant capacity and significant inhibitory effect on alpha-amylase [14].

Jujube fruit contains calcium, potassium, bromine, rubidium, lanthanum, magnesium, sodium, iron, zinc, and manganese in terms of mineral substances; retinol, thiamine, riboflavin, niacin, pyridoxine, cyanocobalamin, and ascorbic acid in terms of vitamins; caffeic acid, ferulic acid, p-hydroxybenzoic acid, chlorogenic acid, gallic acid, protocatechuic acid, vanillic acid, p-coumaric acid, ellagic acid, and cinnamic acid in terms of phenolic acids; and procyanidin B2 epicatechin, quercetin-3-O-rutinoside, quercetin-3-O-galactoside, kaempferol-glycosyl-rhamnose, catechin, epicatechin, and rutin in terms of flavonoids [13,15,16]. Keleş [17] determined the organic acid content of jujube fruit to be 84.72 mg/100 g malic acid, 52.35 mg/100 g citric acid, 78.73 mg/100 g succinic acid, and 51.94 mg/100 g ascorbic acid. Vithlani and Patel [18] defined jujube vinegar as a functional vinegar due to its high-level antioxidant activity. The bioactive compounds of jujube fruit are associated with its health benefits. Therefore, jujube fruit is regarded as a potential alternative functional food product. Today, there is an increase in the demand for natural and traditional products. It has been reported that the antioxidant activity and bioactive compound levels of vinegar produced by the traditional methods (surface culture production) were very high [19]. The biochemical reactions that occur in the traditional method of vinegar production are described in two stages. In the first stage, yeast converts sugar to ethanol in an anaerobic environment, and in the second stage, acetic acid bacteria converts ethanol to acetic acid in an oxygenated (aerobic) environment [20,21]. The present study aimed to determine the antioxidant activity, volatile aroma components, organic acid and phenolic compounds of jujube vinegar produced by traditional methods, and to evaluate the potential of jujube vinegar produced from jujube fruit. The scientific studies on this subject are limited, therefore, important information will be provided for those who want to industrially produce jujube vinegar.

2. Materials and Methods

2.1. Chemicals

Chemicals were obtained from Supelco Co. (Bellefonte, PA, USA), Sigma-Aldrich Co. (St. Louis, MO, USA), and Merck Co. (Darmstadt, Germany). The wine yeast culture (*Saccharomyces cerevisiae* strain, ConFermUni V yeast) was supplied from Eaton's Begerow® Product Line Co. (Nettersheim, Germany).

2.2. Materials

Ripe jujube (20 kg) was obtained from the province of Isparta, Turkey, to be used in the production of jujube wine and vinegar. Also, two-year-old jujube vinegar produced by the traditional methods was provided for the vinegar production phase.

2.3. Preparation of Inoculums

Activation of yeast (*Saccharomyces cerevisiae*) prepared for use in wine production was carried out according to Özen et al. [22].

2.4. Production of Jujube Vinegar

Jujube vinegar was produced by the ethanol and acetic acid fermentation stages according to the traditional vinegar production technique [21]. The production flow diagram is given in Figure 1. Jujube fruits were harvested from the jujube plant. Jujube (*Ziziphus jujuba* Mill.) fruits were selected from non-bruised, intact fruits. The fruits were carefully rubbed and washed. The fruits were cut into pieces and filled into jars and the jars were filled with sterile tap water, and their lids were closed. The jars were agitated from time to time without opening their lids. The jars were kept at +10 °C for one week. Jujube juice (JJ) was obtained by filtering the mixture with cheesecloth. The °Brix was adjusted to

12 °Bx by adding sugar to the JJ. Then, alcohol fermentation was initiated by adding yeast (*Saccharomyces cerevisiae*) to the JJ. Alcohol fermentation was carried out at 25 °C for 30 days. At the end of the period, jujube wine (JW) was obtained, which was both a by-product and a raw material for the second stage of the fermentation. For acetic acid fermentation, aged vinegar was inoculated (1/3 ratio). Acetic acid fermentation lasted 60 days at 25 °C for jujube vinegar (JV). The jujube vinegar production was repeated in two parallel, and three replications. All analysis were carried out in two parallels.

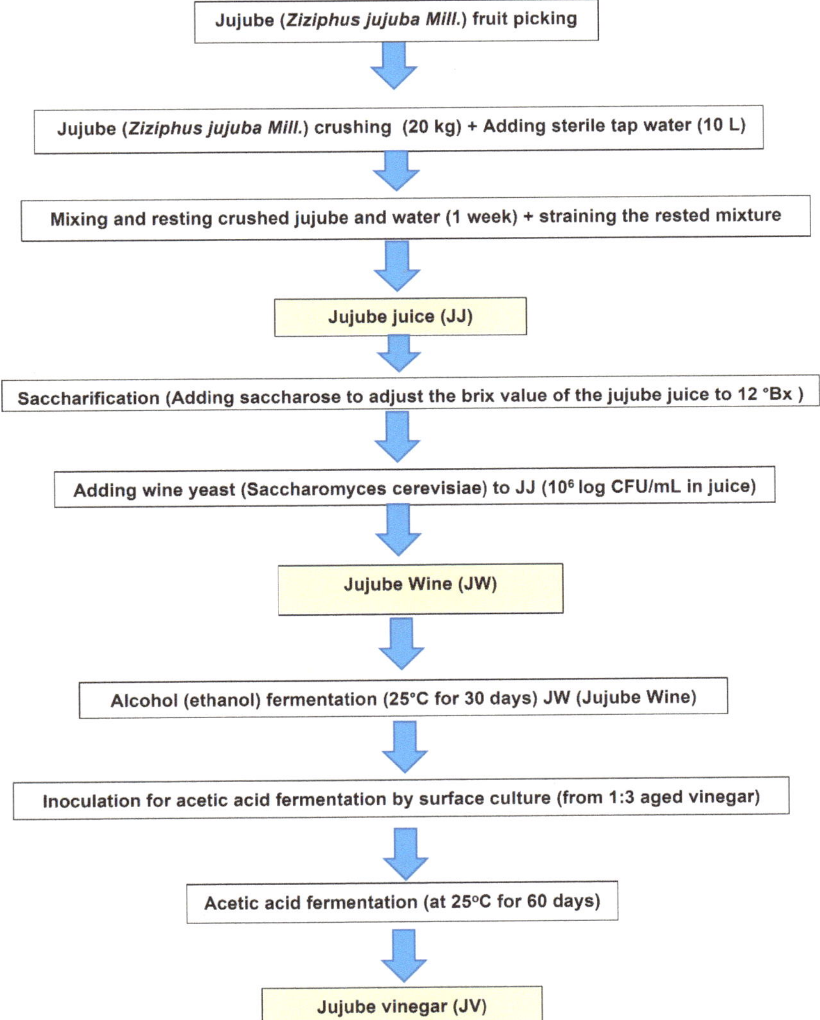

Figure 1. Jujube wine and vinegar production flow chart.

2.5. Proximate Composition Analysis

Chemical analysis applied to the samples were determined according to AOAC [23]. Total titratable acidity (TTA), total dry matter (%) and pH analysis were performed. TTA value was determined as lactic acid (%) in the fruit juice and wine samples and as acetic acid (%) in the vinegar sample. The Abbe refractometer (Bellingham Stanley Limit 60/70 Refractometer, UK) was used to determine the total soluble solids (TSS, °Brix).

2.6. Organic Acid Compound Analysis

Organic acid compound (OACs) analysis was performed using an HPLC device. The device was used effectively at the Innovative Technologies Application and Research Center (YETEM) at Süleyman Demirel University. The OACs of the samples were determined by the method decscribed by Özdemir and Budak [24]. HPLC instrument (Shimadzu SIL–20AC, Scientific Instruments, Inc., Tokyo, Japan) consists of a UV-VIS detector (SPD-10Avp), a pump (LC-20AT projection), a gas separator (DGU-20A5 projection), a column oven (CTO-10AS vp), an LC-20AT projection system controller, and a column (Technochroma TRACER EXTRASIL ODS(2) (250 × 4.6 mm) 5μ TR-016059). The supelco C18 solid phase cartridge was used in the analysis and the injection volume of the device was 20 μL [24,25].

2.7. Phenolic Compound Analysis

Phenolic compound analysis were performed using an HPLC device at the Innovative Technologies Application and Research Center (YETEM) at Süleyman Demirel University. The phenolic compounds of the samples were determined according to Özdemir and Budak [24]. The HPLC instrument (Shimadzu SCL-10A, Scientific Instruments, Inc., Tokyo, Japan) consists of a CTO-10Avp colon oven, an SPD-M 10A vp DAD detector (λmax = 278 nm), a SIL–10AD vp autosampler, an LC-10ADvp pump, a DGU-14A degasser, and a column (Agilent Eclipse XDB-C18 (250 × 4.60 mm) 5 μm; GL Sciences Inc., Southern California, CA, USA) [24,26]. The injection volume of the device was 20 μL. Process steps in HPLC are given in Table 1.

Table 1. HPLC program for phenolic compound analysis.

Time (Min)	Methanol (%; Pump B)
0.10	7
20.00	28
28.00	25
35.00	30
50.00	30
60.00	33
62.00	42
70.00	50
73.00	70
75.00	80
80.00	100
81.00	7

2.8. The Total Phenolic Content and Antioxidant Activity Analysis

The total phenolic content (TPC) of the samples was determined by the Folin–Ciocalteau method [27], while the antioxidant activity (TEAC) was determined by the 2,2′-azinobis (3-ethylbenzothiazoline)-6-sulfonic acid (ABTS) method [28]. The units of analysis of the samples were expressed as mg GAE/L for the total phenolic content and as mmol TE/L for TEAC analysis.

2.9. Volatile Aroma Component Analysis

The volatile compounds of the samples were performed using a gas chromatography-mass spectroscopy device at the Innovative Technologies Application and Research Center (YETEM) at Süleyman Demirel University. The gas chromatography-mass spectroscopy adopted the solid-phase microextraction (SPME) procedure (GC-MS-QP2010 SE; Shimadzu GC-2010 Plus Capillary, Scientific Instruments, Inc., Tokyo, Japan) [29] and modifying the method by Özen et al. [22]. Injection mode, injection temperature, column oven temperature, and mobile phase (helyum) flow rate were split (ratio; 10.0), 250.0 °C, 40 °C, and 1.50 mL/min, respectively. A Restek Rx-5Sil MS column (30 m × 0.25 mm, 0.25 μm) was used. Regarding the GC program procedure, the column oven temperature was kept at 40 °C for 2 min. Then, it reached 250 °C with an increase of 4 °C/min and was kept at this temperature for 5 min. Total time was 59.50 min. Volatile compounds were absorbed into a fused silica SPME fiber (CAR/PDMS Stable Flex, Supelco, Bellefonte, PA, USA) and analyzed based on the schedule outlined above. Compounds were identified using linear retention indices (LRIs) and by the Wiley, Nist, Tutor, and FFNSC libraries in the MS instrument [24], with component results presented in percentiles.

2.10. Statistical Analysis

The statistical analysis were performed using IBM SPSS v. 22.0 (SPSS Inc., Chicago, IL, USA) and calculating using the One-Way ANOVA test. The Duncan test was used for statistically significant results ($p < 0.05$) between fruit juice, wine, and vinegar samples. Also, the *t*-student test was carried out to reveal the statistical differences between the samples.

3. Results and Discussion

3.1. Chemical Contents of Jujube Juice, Wine, and Vinegar

Total titratable acidity (TTA), pH value, total dry matter (%), and total soluble solids (°Brix), of the JJ, JW, and JV samples are presented in Table 2.

Table 2. Chemical contents of jujube juice (JJ), jujube wine (JW) and jujube vinegar (JV).

Sample	pH-Values	Total Dry Matter (%)	Total Soluble Solids (°Brix)
JJ	3.71 ± 0.05 [a]	11.98 ± 0.04 [a]	12.00 ± 0.01 [a]
JW	3.55 ± 0.04 [b]	4.05 ± 0.66 [b]	6.10 ± 1.27 [b]
JV	3.24 ± 0.01 [c]	2.17 ± 0.49 [b]	4.75 ± 0.35 [b]

JJ; jujube juice, JW; jujube wine; JV; jujube vinegar. The results were given as mean ± standard deviation. Different lowercase letters in the same column indicate significant differences between the samples ($p < 0.05$).

The total soluble solids content (TSS) of the jujube juice was set to 12.00 °Brix at the onset of the fermentation, as described by Kim et al. [30]. In our research, it was observed that fermentations started a 12.00 °Brix in the jujube vinegar process and decreased to 6.10 °Brix at the end of the ethanol fermentation, and to 4.75 °Brix at the end of the acetic acid fermentation. The °Brix value of jujube juice was found to be considerably higher than the °Brix value of wine and vinegar samples ($p < 0.05$). Because of this, the sugar in the fruit juice was used by yeasts in alcohol fermentation and turned into ethyl alcohol [31]. Total dry matter showed a similar trend as the total soluble dry matter in this jujube sample. The total dry matter value of the JJ sample decreased from 11.98% to 4.05% in the JW sample and by 2.17% in the JV sample ($p < 0.05$). Pashazadeh et al. [32] emphasized that the total dry matter in vinegar content is based on water-soluble dry matter. In different studies, it has been stated that the total soluble solids value of jujube fruit is 8.1 (°Brix) [33] and the total soluble solids value of four different Spanish jujube fruits is in the range of 14.6–18.4 (°Brix) [13].

Total titratable acidity (TTA) values of JJ, JW, and JV samples were determined to be 0.64, 1.66, and 6.35 g/100 mL, respectively. ($p < 0.05$). Total acidity increased significantly in acetic acid fermentation ($p < 0.05$). Organic acids, and especially acetic acid content, that occur in acetic acid fermentation increases the acidity of vinegar. Vinegar samples are expected to contain at least 4% acid [19]. The pH values of jujube juice, wine and vinegar decreased significantly during the fermentation process. Changes in the pH and TTA values showed similarity in the opposite direction. However, the increase in TTA value in acetic acid fermentation was observed significantly more than the decrease in pH value. It has been stated that the changes in pH value depends on the hydrogen ion concentration [24,34].

3.2. Organic Acid Compound Profiles of Jujube Juice, Wine, and Vinegar

The organic acid content of vinegar varies according to the fruit chosen, the ripening time of the fruit, the type of fermentation, the fermentation period, and the production method of the vinegar [35]. Organic acid contents of JJ, JW, and JV samples are presented in Table 3.

Table 3. Organic acid contents (ppm) of jujube juice (JJ), jujube wine (JW) and jujube vinegar (JV).

Organic Acid Compounds	Jujube Juice (JJ)	Jujube Wine (JW)	Jujube Vinegar (JV)
Oxalic acid	56 ± 0.01 [b]	57.2 ± 0.02 [b]	75.1 ± 0.04 [a]
Tartaric acid	2232.9 ± 20.4 [a]	2075.4 ± 24.2 [a]	993.5 ± 18.0 [b]
Formic acid	865.4 ± 2.40 [a]	677.9 ± 3.24 [b]	538.5 ± 4.52 [c]
Malic acid	919.4 ± 5.65 [a]	918.8 ± 8.86 [a]	655.2 ± 6.42 [b]
Ascorbic acid	95.5 ± 2.02 [b]	323.3 ± 12.4 [a]	57.3 ± 1.04 [b]
Lactic acid	7737.8 ± 64.0 [b]	13,231.4 ± 96.2 [a]	2030.1 ± 20.4 [c]
Acetic acid	2198.5 ± 14.6 [b]	2651.2 ± 21.4 [b]	46,375.3 ± 62.4 [a]
Citric acid	391.1 ± 5.40 [b]	315.2 ± 4.86 [b]	530.7 ± 6.20 [a]
Succinic acid	794.3 ± 4.20 [b]	2288.1 ± 16.2 [a]	843.4 ± 8.20 [b]
Fumaric acid	n.d.	n.d.	3.2 ± 0.01 [a]

JJ; jujube juice, JW; jujube wine; JV; jujube vinegar. The results were given as a mean ± standard deviation. Different lowercase letters in the same column indicate significant differences between the samples ($p < 0.05$). Abbreviations: n.d., not detected.

The main organic acids in jujube juice were lactic acid, tartaric acid, and acetic acid followed by lower levels of malic acid, formic acid, succinic acid, citric acid, ascorbic acid, and oxalic acid. Organic acid values of fresh jujube fruit were determined to be 206.7 mg/100 g DW malic acid, 198.9 mg/100 g DW citric acid, and 14.8 mg/100 g DW succinic acid [36]. In another study, malic acid (219.83 mg/100 g), citric acid (108.14 mg/100 g), succinic acid (188.97 mg/100 g), and ascorbic acid (89.63 mg/100 g) were determined in organic acid results of jujube fruit [17]. Ascorbic acid is present in plant tissues bound to glucose [37]. During the jujube extraction, hygienic conditions and an oxygen-free environment were created, and yeast and mold formation were prevented. However, lactic acid formation was observed as a result of lactic acid fermentation during jujube extraction and alcohol fermentation. Ascorbic acid is released as a result of the biosynthesis of glucose in ethanol fermentation. The ascorbic acid value was determined to be 323.3 mg/L in the JW sample. Fumaric acid was not detected in the JJ or JW samples. Oxalic acid, malic acid, tartaric acid, formic acid, ascorbic acid, lactic acid, acetic acid, citric acid, succinic acid, and fumaric acid were detected in the vinegar samples. The tartaric acid, malic acid, and formic acid levels decreased during the fermentation period whereas the oxalic acid and acetic acid levels increased. It was determined that the citric acid value decreased during the alcohol fermentation, whereas it increased during the acetic acid fermentation. Xiang et al. [38] have reported that citric acid concentrations decreased during alcohol fermentation. The malic acid values of the jujube juice and wine samples were 919.4 and 918.8 mg/L, respectively. The malic acid content of the jujube vinegar sample was determined to be 655.2 mg/L. It was determined that malic acid value decreased during acetic acid fermentation. Sanarico et al. [39] have reported that citric acid, malic acid, oxalic acid, and lactic acid values decreased, and this result was associated with the oxidation by acetic acid bacteria (*Acetobacter* spp. and *Gluconobacter* spp.). The highest organic acid value in jujube vinegar was acetic acid. The acetic acid value of JV was 46,375.3 mg/L. This was an expected result since vinegar contains high levels of acetic acid.

It has been stated that organic acids in foods can be of fermentation origin as well as raw materials, processing, and aging conditions. Organic acids are formed as a result of hydrolysis, biochemical metabolism and microbial activity during fermentation. Organic acids affect the sensory and microbiological properties of fermented products [40].

3.3. Phenolic Compound Profiles of the Jujube Juice, Wine, and Vinegar

The jujube juice, wine, and vinegar were evaluated for phenolic compounds gallic acid, protocatechuic acid, catechin, p-hydroxybenzoic acid, chlorogenic acid, caffeic acid, epicatechin, syringic acid, vanillin, p-coumaric acid, benzoic acid, eriodictiol, quercetin, and campherol (Figure 2). Phenolic components that were not detected in the samples were ferulic acid, sinapinic acid, o-coumaric acid, rutin, hesperidin, rosmarinic acid, cinnamic acid, and luteolin. Gallic acid, catechin, syringic acid, p-coumaric acid, benzoic acid, eriodictiol, and quercetin were not detected whereas protocatechuic acid, p-hydroxybenzoic acid, chlorogenic acid, caffeic acid, epicatechin, vanillin, and campherol were detected in the JJ samples. Gallic acid, catechin, p-coumaric acid, eriodictiol, and quercetin were determined to be 2.1, 1.4, 0.1, 0.2, and 1.6 mg/L in the jujube wine samples, respectively. p-hydroxybenzoic acid and campherol values of the jujube wine were determined to be higher than those of the jujube juice ($p < 0.05$). Albeit statistically not significant, the chlorogenic acid, epicatechin, and vanillin values of jujube wine were determined to be higher than those of the jujube juice ($p > 0.05$). Some phenolic acids found in fruit and flower groups are dependent on fruit sugar. The use of sugar in the environment by wine yeast (*S. cerevisiae*) in alcohol fermentation releases phenolic acids bound to sugar [41].

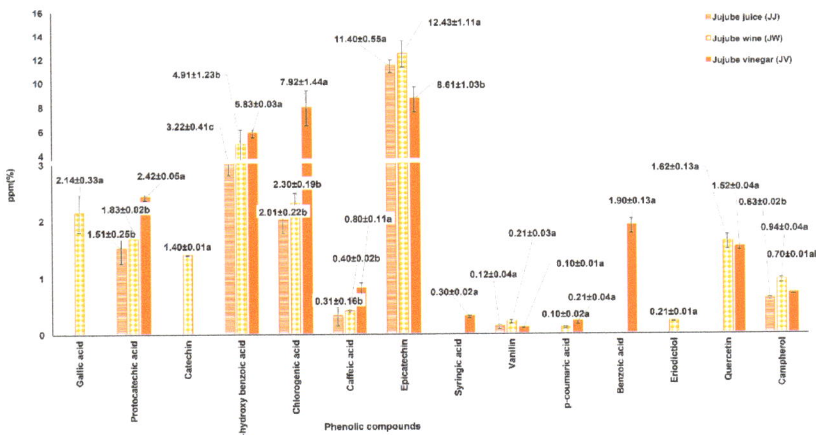

Figure 2. Phenolic compounds of jujube juice (JJ), jujube wine (JW), jujube vinegar (JV) (ppm). Different letters in the figure indicate significant differences between samples.

Gallic acid, catechin, and eriodictiol values of the jujube wine sample were also determined at very low concentrations whereas they were not detected in the JV sample. Yıkmış et al. [42] examined phenolic compounds in jujube vinegar samples and determined caffeic and ferulic acids as the most dominant phenolic compounds. It was found that some phenolic compounds in the environment such as gallic acid decreased since fermentation with oxygen takes place during acetic acid fermentation in vinegar formation [43]. Protocatechuic acid, chlorogenic acid, p-hydroxybenzoic acid, and caffeic acid values showed a gradual increase during the alcohol and acetic acid fermentations. The increase in these phenolic acids were significant in the vinegar example ($p < 0.05$). Laranjinha et al. [44] have stated that chlorogenic acid had a positive effect on health, especially on cardiovascular diseases. It has been explained that this effect occurs by inhibiting LDL oxidation. Also, it has been reported that p-hydroxybenzoic acid has antimicrobial, antialgal, antimutagenic, antiestrogenic, hypoglycemic, anti-inflammatory, anti-platelet aggregation, nematicidal, antiviral, antioxidant, etc. effects. Some 4-hydroxybenzoic acid derivatives have a direct effect on cancer molecules, inhibit acetic acid-induced edema, and are used in the treatment of sickle cell disease [45]. Protocatechuic acid has pharmacological effects in humans, both naturally and as a synthesized compound, by showing anti-inflammatory and antioxidant activity. In particular, the intake of polyphenols and flavonoid compounds in natural foods is important for health [46]. Caffeic acid and its derivatives were found to have antioxidant, anti-inflammatory, and anticarcinogenic activities in in vitro and in vivo studies [47].

3.4. Total Phenolic Content and Antioxidant Capacity of Jujube Juice, Wine, and Vinegar

The antioxidant activity (TEAC) and total phenolic substance (TPC) results of samples were shown in Figure 3A,B.

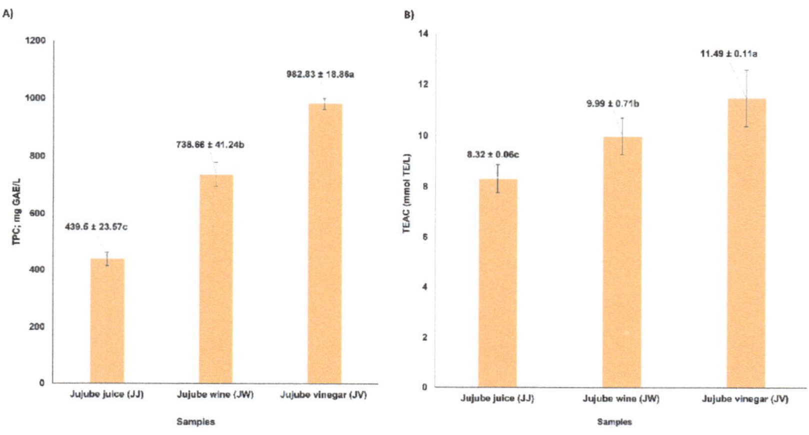

Figure 3. Total phenolic substance (**A**); and antioxidant activity (**B**) of jujube juice (JJ), jujube wine (JW) and jujube vinegar (JV). Different letters in the figure indicate significant differences between samples.

Examining the total phenolic content and antioxidant activity, there was a close relationship in the samples of jujube juice and jujube vinegar. Sorting the TPC and TEAC values from high to low, it was determined to be jujube vinegar, jujube wine, and jujube juice ($p < 0.05$). The TPC value was 982.83 mg GAE/L while the TEAC value was 11.49 mmol TE/L in the JV samples.

The total phenolic content increased with the fermentation process. Because, as a result of the use of sugar by wine yeast in the environment during alcohol fermentation, phenolic compounds that are dependent on sugar in the raw material are released and increase the phenolic compound values [48]. Vithlani and Patel [18] have stated that there was an increase in the flavonoid and flavonol contents during the fermentation stage of jujube wine and vinegar, and this increase increased the antioxidant activities of DPPH and ABTS. In the literature, the total phenolic content of jujube fruits was determined to be in the range of 275.6−541.8 mg GAE/100 g [36], while the total phenolic content of fermented jujube juice was determined to be 2663 µg mL^{-1} [49]. The ABTS scavenging activity of blackened jujube vinegar and red jujube vinegar was determined to be 0.52–0.18 mg Trolox/mL, respectively [50].

The TEAC values of different kinds of vinegar, apple cider vinegar, sour cherry vinegar, peach vinegar, hawthorn vinegar, and pomegranate vinegar were determined to be 11.90 mmol TE/mL, 8.14 µmol TE/mL, 8.05 mmol TE/L, 13.01 mmol TE/L, and 23.56 µmol TE/mL, respectively [22,34,35,51]. The high antioxidant and total phenolic content of the product obtained from the jujube fruit, which has a very short shelf-life, suggested that jujube vinegar is a valuable product.

3.5. Volatile Compound Profiles of Jujube Juice, Wine, and Vinegar

As for the aroma-related volatile compounds of the samples, the main compounds were regarded as those higher than the mean value of the percentage distribution of compounds within the sample.

In the study, a total of 42 volatile compounds including 19 esters, 9 aldehydes, 7 alcohols, 5 terpenes, 1 acid, and 1 ketone were determined. Examining the JJ (32 compounds), JW (22 compounds), and JV (19 compounds) samples together (Table 4 and Figure 4), in the JJ sample, the total ester ratio had the highest value with 66.4%, followed by the alcohols, aldehydes, and terpenes with 3.58% and 22.2%, and 1.65%, respectively. In another study, a total of 46 volatile compounds, including aldehydes, alcohols, alkenes, ketones, esters, and acids were detected in the aroma component analysis of fresh jujube fruit [52].

Table 4. The changes in volatile compounds associated with aroma during fermentation in the production of jujube vinegar.

	Compounds	JJ (Jujube Juice)		JW (Jujube Wine)		JV (Jujube Vinegar)	
		Area	Percent (%)	Area	Percent (%)	Area	Percent (%)
Acid	Acetic acid	n.d.	n.d.	n.d.	n.d.	45,730,907	60.28 ± 4.42
	Total						**60.28± 4.42**
Ketone	Diacetyl	n.d.	n.d.	n.d.	n.d.	359,304	0.47 ± 0.06
	Total						**0.47± 0.06**
Aldehyde	2-Hexenal-(E)	104,820	0.22 ± 0.05	n.d.	n.d.	n.d.	n.d.
	Acetaldehyde	152,524	0.31 ± 0.07 [a*]	n.d.	n.d.	52,781	0.07 ± 0.01 [b]
	Benzaldehyde	n.d.	n.d.	n.d.	n.d.	187,505	0.25 ± 0.09
	Phenylacetaldehyde	806,141	1.66 ± 0.42 [b]	6,665,457	10.92 ± 1.68 [a]	118,527	0.16 ± 0.08 [c]
	Decanal	n.d.	n.d.	20,651	0.03 ± 0.01	n.d.	n.d.
	Heptanal	54,258	0.11 ± 0.04	n.d.	n.d.	n.d.	n.d.
	Hexanal	418,309	0.86 ± 0.09	n.d.	n.d.	n.d.	n.d.
	Nonanal	117,790	0.24 ± 0.05 [a]	57,545	0.09 ± 0.02 [b]	n.d.	n.d.
	Octanal	86,303	0.18 ± 0.06 [a]	76,958	0.13 ± 0.03 [a]	n.d.	n.d.
	Total		**3.58± 0.48 [b]**		**11.17± 0.64 [a]**		**0.48± 0.24 [c]**
Alcohol	Ethanol	10,242,552	21.14 ± 6.84 [b]	25,782,301	42.24 ± 4.62 [a]	411,775	0.54 ± 0.14 [c]
	2-Methyl-1-butanol	104,230	0.22 ± 0.02 [b]	1,812,897	2.97 ± 0.86 [a]	310,241	0.41 ± 0.18 [b]
	Isoamyl alcohol	216,157	0.45 ± 0.12 [b]	3,630,717	5.95 ± 1.28 [a]	488,869	0.64 ± 0.34 [b]
	1-Hexanol	110,839	0.23 ± 0.01	n.d.	n.d.	n.d.	n.d.
	1-Amino-2-propanol	n.d.	n.d.	791,642	1.30 ± 0.02	n.d.	n.d.
	Phenethyl alcohol	n.d.	n.d.	80,222	0.13 ± 0.01	n.d.	n.d.
	Myristyl alcohol	75,148	0.16 ± 0.03	n.d.	n.d.	n.d.	n.d.
	Total		**22.2± 2.92 [b]**		**52.59± 3.64 [a]**		**1.59± 0.02 [c]**
Ester	2-Methylbutyl acetate	n.d.	n.d.	218,994	0.36 ± 0.06 [b]	2,330,967	3.07 ± 1.23 [a]
	Isoamyl acetate	8,850,960	18.27 ± 3.24 [a]	2,014,436	3.30 ± 1.24 [c]	7,205,463	9.50 ± 1.86 [b]
	3-Hexenyl acetate	128,237	0.26 ± 0.03	n.d.	n.d.	n.d.	n.d.
	Isobutyl acetate	n.d.	n.d.	n.d.	n.d.	765,346	1.01 ± 0.07
	Phenethyl acetate	1,676,095	3.46 ± 0.16 [b]	846,233	1.39 ± 0.48 [c]	6,352,412	8.37 ± 2.02 [a]
	Ethyl acetate	18,808,097	38.82 ± 2.86 [a]	6,350,651	10.40 ± 2.11 [b]	8,589,076	11.32 ± 4.01 [b]
	Hexyl acetate	631,438	1.30 ± 0.15 [a]	178,995	0.29 ± 0.04 [b]	n.d.	n.d.
	Methyl acetate	n.d.	n.d.	n.d.	n.d.	93,706	0.12 ± 0.01
	Propyl acetate	71,829	0.15 ± 0.01	n.d.	n.d.	n.d.	n.d.
	Ethyl phenylacetate	n.d.	n.d.	n.d.	n.d.	77,701	0.11 ± 0.02
	Ethyl benzoate	63,951	0.13 ± 0.03 [a]	114,907	0.19 ± 0.02 [a]	n.d.	n.d.
	Benzyl acetate	57,490	0.12 ± 0.02 [a]	n.d.	n.d.	78,551	0.10 ± 0.03 [a]
	Ethyl butyrate	n.d.	n.d.	58,451	0.1 ± 0.01	n.d.	n.d.
	Ethyl capronate	333,686	0.69 ± 0.24 [b]	3,134,463	5.14 ± 2.01 [a]	n.d.	n.d.
	Ethyl decanoate	597,681	1.23 ± 0.14 [b]	3,013,764	4.94 ± 1.94 [a]	n.d.	n.d.
	Ethyl laurate	166,749	0.34 ± 0.04 [a]	270,479	0.44 ± 0.21 [a]	n.d.	n.d.
	Ethyl caprylate	711,830	1.47 ± 0.04 [b]	3,972,062	6.51 ± 0.42 [a]	77,700	0.10 ± 0.01 [c]
	Ethyl palmitate	n.d.	n.d.	n.d.	n.d.	n.d.	n.d.
	Ethyl propanoate	78,836	0.16 ± 0.28	n.d.	n.d.	n.d.	n.d.
	Total		**66.4± 1.82 [a]**		**33.42± 1.02 [b]**		**33.70± 1.17 [b]**

Table 4. *Cont.*

	Compounds	JJ (Jujube Juice)		JW (Jujube Wine)		JV (Jujube Vinegar)	
		Area	Percent (%)	Area	Percent (%)	Area	Percent (%)
Terpene	α-pinene	476,664	0.98 ± 0.26	n.d.	n.d.	n.d.	n.d.
	α-Thujene	59,049	0.12 ± 0.01	n.d.	n.d.	n.d.	n.d.
	β-Phellandrene	177,922	0.28 ± 0.09	n.d.	n.d.	n.d.	n.d.
	γ-Terpinene	53,957	0.11 ± 0.08 [a]	n.d.	n.d.	47,871	0.06 ± 0.01 [b]
	p-Cymol	78,263	0.16 ± 0.07 [a]	n.d.	n.d.	53,470	0.07 ± 0.02 [b]
	Total		**1.65 ± 0.36** [a]		**n.d.**		**0.13 ± 0.01** [b]

* Statistical analysis was made according to percentages. The lowercase letters a–c denote the statistical difference between the JJ, JW, and JV samples. Bold background compounds are each sample, and therefore more effective. Abbreviations: n.d., not detected.

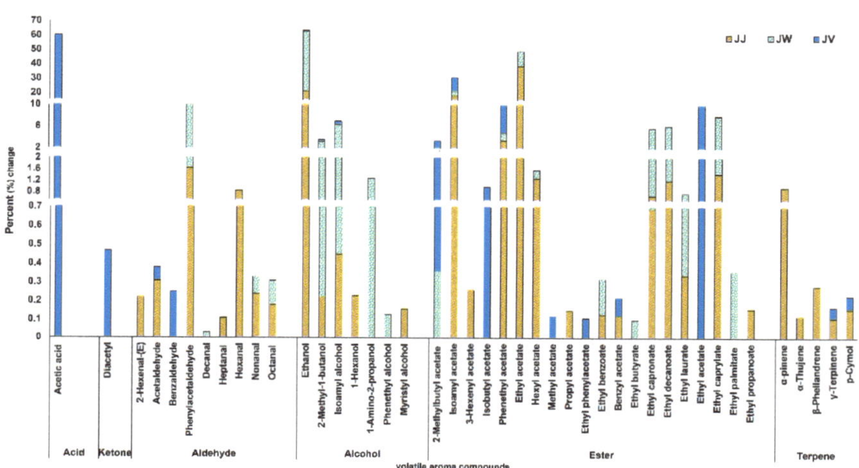

Figure 4. Volatile compounds associated with aroma of jujube juice (JJ), jujube wine (JW) and jujube vinegar (JV) (percent %).

In the present study, the total ester compound content of jujube fruit was found to be the highest. On the other hand, in the JJ sample, ethyl acetate with 38.82%, ethanol with 21.14%, isoamyl acetate with 18.27%, and phenethyl acetate with 3.46%, had high percentages. The difference between the values of these compounds was statistically significant both among each other and according to all the determined compounds ($p < 0.05$). The ethyl acetate is associated with aromatic, brandy, and grape flavors, while the ethanol is associated with alcohol flavor, the isoamyl acetate is associated with apple, banana, glue, pear flavors, and phenethyl acetate is associated with flower, honey, and rose flavors [53]. These ingredients together were considered to give the JJ sample a slightly sour taste. A good vinegar, and even a good wine, can be produced from raw material with a high content of esters, and ethyl acetate. Therefore, this fruit was considered to be a suitable raw material for vinegar fermentation in terms of the aroma related volatile components in the final product.

Other highly-effective compounds were phenylacetaldehyde with 1.66%, ethyl caprylate with 1.47%, and hexyl acetate with 1.30%. The statistical differences between these compounds were not significant whereas significant differences were found when compared to other groups. Also, among the terpene group compounds, α-pinene had a higher ratio than other terpene compounds. Regarding the JJ sample, phenylacetaldehyde is associated with berry, geranium, honey, nut, and pungent flavors, ethyl caprylate is associated with apricot, fat, floral, and pineapple, and hexyl acetate is associated with apple, banana, grass, herb, and pear flavors [53].

As for the JW sample, the total alcohol ratio was the highest with 52.59%, followed by the ester group with 33.42%, and the aldehyde group with 11.17%. No terpene compounds were found in the samples. The total alcohol content of the wine obtained from jujube fruit was found to be the highest, followed by the ester group. On the other hand, in the JW sample, the ethanol, with 42.24%, had the highest percentage ($p < 0.05$). This was followed by phenylacetaldehyde with 10.92% and ethyl acetate with 10.40% ($p < 0.05$). In particular, the ethyl acetate compounds with aromatic, brandy, and grape flavor is the desired compound in a wine [54]. However, it was thought that the close presence of the phenylacetaldehyde with the ethyl acetate might suppress the effect of the ethyl acetate. Because the phenylacetaldehyde gives wine a honey-like effect [55]. This gave the JW sample a softer flavor profile.

High levels of alcohol group content are expected in wine. However, in the present study, the ester ratio, in particular, was an interesting result. This ester content was associated with the rich ester content in the juice of the fruit used. The other highly-effective compounds following these were ethyl caprylate with 6.51%, isoamyl alcohol with 5.95%, ethyl caproate with 5.14%, and ethyl decanoate with 4.94%. The statistical differences between these compounds were not significant, whereas significant differences were found when compared to other groups ($p < 0.05$). Ethyl caprylate is associated with apricot, fatty, floral, and pineapple flavors, ethyl capronate is associated with apple peel, brandy, fruit gum, overripe fruit, pineapple flavors, and isoamyl alcohol is associated with burnt, cocoa, malt flavors, and are not desirable compounds for a wine [53,56]. In one study, a total of 182 volatile compounds, comprising 30 alcohols, 16 acids, 62 esters, 19 ketones, 18 aldehydes, 10 hydrocarbons, 7 phenols, and 20 other compounds were detected for jujube wine [57].

In the JV sample, the total acid ratio was the highest at 60.28%, and the acetic acid compound is associated with acidic, fruity, pungent, sour vinegar flavors [53] ($p < 0.05$). This showed that the vinegar fermentation was properly realized. The acid group was followed by the ester group with 33.70%, and the alcohol group with 1.59% ($p < 0.05$). As for the other compound groups, the aldehyde group and ketone groups were determined to be 0.48% and 0.47%. Also, the terpene group was determined to be 0.13%.

For the JV vinegar, the main compound was acetic acid. The acetic acid content in the vinegar samples suggested that jujube can be efficiently used in vinegar production. The other prominent compounds were ethyl acetate, associated with aromatic, brandy, and grape flavors with 11.32%, isoamyl acetate, associated with apple, banana, glue, and pear flavors with 9.50%, and phenethyl acetate, associated with flower, honey, and rose flavors, with 8.37%. The statistical differences between these compounds were not significant, whereas significant differences were found when compared to other groups ($p < 0.05$). Also, these compounds positively affected the JV sample [33,54]. The ratio of 2-methyl butyl acetate compound, associated with apple, banana, and pear flavors, was determined to be 3.07%. As was the case in the present study, high levels of acid group content are expected in vinegar and ester compounds have positive effects on the flavor formation in vinegar [22]. It has been reported that the volatile aroma compounds of jujube vinegar were determined to be acids, esters, alcohols, aldehydes, and ketones. The main components of the vinegar content were determined to be acids and esters [50].

According to the volatile compound results, the compounds with a high ratio in the aroma profile of the JJ sample were ethyl acetate, ethanol, isoamyl acetate, and phenethyl acetate. The compounds with a high ratio in the aroma profile of the JW sample were ethanol, phenylacetaldehyde, and ethyl acetate. The other highly-effective compounds in the JW sample were undesirable the compounds; ethyl caprylate, isoamyl alcohol, ethyl capronate, and ethyl decanoate. The compounds with a high ratio in the aroma profile of the JV sample were acetic acid, ethyl acetate, isoamyl acetate, and phenethyl acetate. Therefore, these results showed that jujube fruit, in terms of the aroma-related volatile component, is suitable for vinegar production, but not for wine.

4. Conclusions

In this study, aroma profile-related volatile compound analysis, antioxidant capacity, organic acid, and phenolic compounds were determined in the juice, wine, and vinegar samples, which were produced from jujube fruits. It was determined that the total phenolic compound and the TEAC antioxidant activity content of jujube vinegar were higher than those of the jujube juice and jujube wine. Also, it has been determined that the samples had important parameters that had an effect on health in terms of organic acid and phenolic compound contents. The volatile aroma compound profiles of the samples comprised the ethyl acetate, ethanol, isoamyl acetate, and phenethyl acetate for the jujube juice; the ethanol, phenylacetaldehyde, and ethyl acetate, and also the undesirable ethyl caprylate, isoamyl alcohol, and ethyl capronate for the jujube wine. As for the volatile aroma compound profiles of jujube vinegar, the most abundant compounds were acetic acid, ethyl acetate, isoamyl acetate, and phenethyl acetate. I were concluded that the use of jujube fruit, which is a healthy, pleasant-flavored, and promising alternative in the production of fruit vinegar, is important in terms of contributing to the country's economy. This study will encourage investors who want to invest in vinegar production to increase the use of jujube vinegar with increased functional properties due to its bioactive compound properties by evaluating the jujube fruit.

Funding: This research received no external funding.

Institutional Review Board Statement: Not applicable.

Informed Consent Statement: Not applicable.

Data Availability Statement: Not applicable.

Acknowledgments: I would like to thank the Innovative Technologies Application and Research Center (YETEM) in Süleyman Demirel University (Isparta/Turkey) for their support in the analysis of volatile aroma compounds, phenolic compounds and organic acid compounds. I would like to thank Nilgün Özdemir for all her support.

Conflicts of Interest: The author declares no conflict of interest.

References

1. Mengjun, L. Genetic Diversity of Chinese Jujube (*Ziziphus jujuba* Mill.). *Acta Hortic.* **2003**, *623*, 351–355. [CrossRef]
2. Kemeç-Hürkan, Y. Jujuba (*Ziziphus jujuba* Mill.) Fruit: Medical Importance from Past to Present. *J. Inst. Sci. Technol.* **2019**, *9*, 1271–1281. [CrossRef]
3. Kavas, İ.; Dalkılıç, Z. Determination of Morphological, Phenological Pomological Characteristics in Some Jujube Genotypes and Investigation of Hybridization Possibilities. *J. Adnan Menderes Univ. Agric. Fac.* **2015**, *12*, 5–72.
4. Hamedi, S.; Arian, A.A.; Farzaei, M.H. Gastroprotective Effect of Aqueous Stem Bark Extract of *Ziziphus jujuba* L. against HCl/Ethanol-Induced Gastric Mucosal Injury in Rats. *J. Tradit. Chin. Med.* **2015**, *35*, 666–670. [CrossRef]
5. Yu, L.; Jiang, B.P.; Luo, D.; Shen, X.C.; Guo, S.; Duan, J.A.; Tang, Y.P. Bioactive Compounds in The Fruits of *Ziziphus jujuba* Mill. against The İnflammatory İrritant Action of Euphorbia Plants. *Phytomedicine* **2012**, *19*, 239–244. [CrossRef]
6. Ganachari, M.S.; Kumar, S.; Bhat, K.G. Effect of *Ziziphus jujuba* Leaves Extract on Phagocytosis by Human Neutrophils. *J. Nat. Remedies* **2004**, *4*, 47–51.
7. Choi, S.H.; Ahn, J.B.; Kim, H.J.; Im, N.K.; Kozukue, N.; Levin, C.E.; Friedman, M. Changes in Free Amino Acid, Protein, and Flavonoid Content in Jujube (*Ziziphus jujube*) Fruit During Eight Stages of Growth and Antioxidative and Cancer Cell Inhibitory Effects by Extracts. *J. Agric. Food Chem.* **2012**, *60*, 10245–10255. [CrossRef]
8. Gülay, S. Investgation of The Relation Between Insulin Release and Several Zizyphus Jujuba Extracts on Pancreatic Beta Cell Lines. Master's Thesis, Erciyes University, Kayseri, Turkey, 2013.
9. Özkan, H.İ. The Examination of Some Biochemical Components of Jujube Fruit (*Zizyphus jujuba* Mill.) and Its Antibacterial, Hypoglycaemic and Total Antioxidant Activities. Master's Thesis, Balıkesir Üniversitesi, Balıkesir, Turkey, 2017.
10. Rashwan, A.K.; Karima, N.; Shishira, M.R.I.; Baoa, T.; Lua, Y.; Chena, W. Jujube fruit: A potential nutritious fruit for the development of functional food products. *J. Funct. Foods* **2020**, *75*, 104205. [CrossRef]
11. Li, G.; Yan, N.; Li, G. The Effect of In Vitro Gastrointestinal Digestion on the Antioxidants, Antioxidant Activity, and Hypolipidemic Activity of Green Jujube Vinegar. *Foods* **2022**, *11*, 1647. [CrossRef]
12. Kundi, A.H.K.; Wazir, F.K.; Abdul, G.; Wazir, Z.D.K. Physicochemical Characteristics and Organoleptic Evaluation of Different Ber (*Zizyphus jujuba* Mill.) Cultivars. *Sarhad J. Agric.* **1989**, *5*, 149–155.

13. Hernández, F.; Noguera, A.L.; Burló, F.; Wojdyło, A.; Carbonell, B.A.A.; Legua, P. PhysicoChemical, Nutritional, and Volatile Composition and Sensory Profile of Spanish Jujube (*Ziziphus jujuba* Mill.) Fruits. *J. Sci. Food Agric.* **2016**, *96*, 2682–2691. [CrossRef] [PubMed]

14. Afrisham, R.; Aberomand, M.; Ghaffari, M.; Siahpoos, A.; Jamalan, M. Inhibitory Effect of *Heracleum persicum* and *Ziziphus jujuba* on Activity of Alpha-Amylase. *J. Bot.* **2015**, 824683. [CrossRef]

15. San, B.; Yildirim, A.N. Phenolic, AlphaTocopherol, Beta-Carotene and Fatty Acid Composition of Four Promising Jujube (*Ziziphus jujuba* Miller) Selections. *J. Food Compos. Anal.* **2010**, *23*, 706–710. [CrossRef]

16. Gong, C.; Yanjing, B.; Yuying, Z. Flavonoids from *Ziziphus jujuba* Mill var. spinasa. *Tetrahedron* **2000**, *56*, 8915–8920.

17. Keleş, H. Changes of some horticultural characteristics in jujube (*Ziziphus jujube* Mill.) fruit at different ripening stages. *Turk. J. Agric. For.* **2020**, *44*, 391–398. [CrossRef]

18. Vithlani, V.A.; Patel, H.V. Production of functional vinegar from Indian jujube (*Zizyphus mauritiana*) and its antioxidant properties. *J. Food Technol.* **2010**, *8*, 143–149. [CrossRef]

19. Budak, N.H.; Aykin, E.; Seydim, A.C.; Greene, A.K.; Guzel-Seydim, Z.B. Functional Properties of Vinegar. *J. Food Sci.* **2014**, *79*, 757–764. [CrossRef]

20. Gullo, M.; Verzelloni, E.; Canonico, M. Aerobic Submerged Fermentation by Acetic Acid Bacteria for Vinegar Production: Process and Biotechnological Aspects. *Process Biochem.* **2014**, *49*, 1571–1579. [CrossRef]

21. Budak, H.N.; Guzel-Seydim, Z.B. Antioxidant Activity and Phenolic Content of Wine Vinegars Produced by Two Different Technique. *J. Sci. Food Agric.* **2010**, *90*, 2021–2026. [CrossRef]

22. Özen, M.; Özdemir, N.; Ertekin-Filiz, B.; Budak, H.N.; Kök-Taş, T. Sour Cherry (*Prunus cerasus* L.) Vinegars Produced from Fresh Fruit or Juice Concentrate: Bioactive Compounds, Volatile Aroma Compounds and Antioxidant Capacities. *Food Chem.* **2020**, *309*, 125664. [CrossRef]

23. Association of Official Analytical Chemists (AOAC). *Official Methods of Analysis*, 15th ed.; AOAC: Washington, DC, USA, 1992.

24. Özdemir, N.; Budak, H.N. Bioactive compounds and volatile aroma compounds in rose (*Rosa damascena* Mill.) vinegar during the aging period. *Food Biosci.* **2022**, *50 Pt A*, 102062. [CrossRef]

25. Krapez, K.M.; Abram, V.; Kac, M.; Ferjancic, S. Determination of Organic Acids in White Wines by RP-HPLC. *Food Technol. Biotechnol.* **2001**, *39*, 93–99.

26. Caponio, F.; Alloggio, V.; Gomes, T. Phenolic Compounds of Virgin Olive Oil: Influence of Paste Preparation Techniques. *Food Chem.* **1999**, *64*, 203–209. [CrossRef]

27. Singleton, V.L.; Orthofer, R.; Lamuela-Raventos, R.M. Analysis of Total Phenols and Other Oxidation Substrates and Antioxidants by Means of Folin-Ciocalteu Reagent. *Methods Enzymol.* **1999**, *299*, 152–178.

28. Singh, S.; Singh, R.P. In Vitro Methods of Assay of Antioxidants: An Overview. *Food Rev. Int.* **2008**, *24*, 392–415. [CrossRef]

29. Vichi, S.; Castellote, A.I.; Pizzale, L.; Conte, L.S.; Buxaderas, S.; Lopez-Tamames, E. Analysis of virgin olive oil volatile compounds by headspace solid-phase microextraction coupled to gas chromatography with mass spectrometric and flame ionization detection. *J. Chromatogr. A* **2003**, *983*, 19–33. [CrossRef]

30. Kim, S.-K.; Lee, G.-D.; Chung, S.-K. Monitoring on Fermentation of Persimmon Vinegar from Persimmon Peel. *J. Korean Soc. Food Sci. Nutr.* **2003**, *35*, 642–647.

31. Lee, S.H.; Kim, J.C. A comparative analysis for main compounds change during natural fermentation of persimmon vinegar. *J. Korean Soc. Food Sci. Nutr.* **2009**, *38*, 372–376. [CrossRef]

32. Pashazadeh, H.; Özdemir, N.; Zannou, O.; Koca, I. Antioxidant capacity, phytochemical compounds, and volatile compounds related to aromatic property of vinegar produced from black rosehip (*Rosa pimpinellifolia* L.) juice. *Food Biosci.* **2021**, *44*, 101318. [CrossRef]

33. Uddin, M.B.; Hussain, I. Development of diversified technology for jujube (*Ziziphus jujuba* L.) processing and preservation. *World J. Dairy Food Sci.* **2012**, *7*, 74–78.

34. Özdemir, G.B.; Özdemir, N.; Ertekin-Filiz, B.; Gökırmaklı, Ç.; Kök-Taş, T.; Budak, N.H. Volatile aroma compounds and bioactive compounds of hawthorn vinegar produced from hawthorn fruit (*Crataegus tanacetifolia* (lam.) pers.). *J. Food Biochem.* **2021**, *46*, e13676. [CrossRef] [PubMed]

35. Budak, H.N.; Özdemir, N.; Gökırmaklı, Ç. The changes of physicochemical properties, antioxidants, organic, and key volatile compounds associated with the flavor of peach (*Prunus cerasus* L. Batsch) vinegar during the fermentation process. *J. Food Biochem.* **2021**, *46*, e13978. [CrossRef] [PubMed]

36. Gao, Q.H.; Wu, C.S.; Yu, J.G.; Wang, M.; Ma, Y.J.; Li, C.L. Textural characteristic, antioxidant activity, sugar, organic acid, and phenolic profiles of 10 promising jujube (*Ziziphus jujuba* Mill.) selections. *J. Food Sci.* **2012**, *77*, 1218–1225. [CrossRef] [PubMed]

37. Arrigoni, O.; De Tullio, M.C. Ascorbic acid: Much more than just an antioxidant. *Biochim. Biophys. Acta* **2002**, *1569*, 1–9. [CrossRef]

38. Xiang, J.L.; Du, L.; Liu, Z.J.; Li, X.L.; Luo, L.; Zhu, W.X. Changes in biochemical parameters and organic acid content during liquid fermentation of peach vinegar. *Mod. Food Sci. Technol.* **2015**, *31*, 193–198. [CrossRef]

39. Sanarico, D.; Motta, S.; Bertolini, L.; Antonelli, A. HPLC determination of organic acids in traditional balsamic vinegar of reggio emilia. *J. Liq. Chromatogr. Relat.* **2003**, *26*, 2177–2187. [CrossRef]

40. Leejeerajumnean, A.; Duckham, S.C.; Owens, J.D.; Ames, J.M. Volatile compounds in *Bacillus*-fermented soybeans. *J. Sci. Food Agric.* **2001**, *81*, 525–529. [CrossRef]

41. Crozier, A.; Jaganath, I.B.; Clifford, M.N. Dietary Phenolics: Chemistry, Bioavailability and Effects on Health. *Nat. Prod. Rep.* **2009**, *26*, 1001–1043. [CrossRef]
42. Yıkmış, S.; Altıner, D.D.; Ozer, H.; Levent, O.; Celik, G.; Çöl, B.G. Modeling and optimization of bioactive compounds from jujube (*Ziziphus jujuba* Mill.) vinegar using response surface methodology and artificial neural network: Comparison of ultrasound processing and thermal pasteurization. *J. Food Process. Preserv.* **2022**, e17102. [CrossRef]
43. Verhoef, S.; Ruijssenaars, H.J.; de Bont, J.A.; Wery, J. Bioproduction of p-hydroxybenzoate from renewable feedstock by solvent-tolerant *Pseudomonas putida* S12. *J. Biotechnol.* **2007**, *132*, 49–56. [CrossRef]
44. Laranjinha, J.A.; Almeida, L.M.; Madeira, V.M. Reactivity of dietary phenolic acids with peroxyl radicals: Antioxidant activity upon low density lipoprotein peroxidation. *Biochem. Pharmacol.* **1994**, *48*, 487–494. [CrossRef]
45. Manuja, R.; Sachdeva, S.; Jain, A.; Chaudhary, J. A Comprehensive Review on Biological Activities of P-Hydroxy Benzoic Acid and Its Derivatives. *Int. J. Pharm. Sci.* **2013**, *22*, 109–115.
46. Kakkar, S.; Bais, S. A Review on Protocatechuic Acid and Its Pharmacological Potential. *ISRN Pharmacol.* **2014**, *12*, 952943. [CrossRef] [PubMed]
47. Espíndola, K.M.M.; Ferreira, R.G.; Narvaez, L.E.M.; Silva Rosario, A.C.R.; da Silva, A.H.M.; Silva, A.G.B.; Vieira, A.P.O.; Monteiro, M.C. Chemical and Pharmacological Aspects of Caffeic Acid and Its Activity in Hepatocarcinoma. *Front. Oncol.* **2019**, *9*, 541. [CrossRef] [PubMed]
48. Madhusoodanan, J. Wildfires Pose a Burning Problem for Wines and Winemakers. *Proc. Natl. Acad. Sci. USA* **2021**, *118*, e2113327118. [CrossRef]
49. Zhao, M.N.; Zhang, F.; Zhang, L.; Liu, B.J.; Meng, X.H. Mixed fermentation of jujube juice (*Ziziphus jujuba* Mill.) with *L. rhamnosus* GG and *L. plantarum*-1: Effects on the quality and stability. *Int. J. Food Sci. Technol.* **2019**, *54*, 2624–2631. [CrossRef]
50. Wang, F.; Song, Y.; Vidyarthi, S.K.; Zhang, R. Physicochemical properties, and volatile compounds of blackened jujube vinegar as prepared by optimized fermentation process. *Int. J. Food Prop.* **2022**, *25*, 288–304. [CrossRef]
51. Aykın, E.; Budak, H.N.; Güzel-Seydim, Z.B. Bioactive Compounds of Mother Vinegar. *J. Am. Coll. Nutr.* **2015**, *34*, 80–89. [CrossRef]
52. Wang, R.; Ding, S.; Zhao, D.; Wang, Z.; Wu, J.; Hu, X. Effect of Dehydration Methods on Antioxidant Activities, Phenolic Contents, Cyclic Nucleotides, and Volatiles of Jujube Fruits. *Food Sci. Biotechnol.* **2016**, *25*, 137–143. [CrossRef]
53. JECFA. The Joint FAO/WHO Expert Committee on Food Additives (JECFA). 2022. Available online: https://www.who.int/groups/joint-fao-who-expert-committee-on-food-additives-(jecfa) (accessed on 10 June 2022).
54. Özdemir, N.; Pashazadeh, H.; Zannou, O.; Koca, I. Phytochemical content, and antioxidant activity, and volatile compounds associated with the aromatic property, of the vinegar produced from rosehip fruit (*Rosa canina* L.). *LWT* **2022**, *154*, 112716. [CrossRef]
55. Silva Ferreira, A.C.; Guedes de Pinho, P.; Rodrigues, P.; Hogg, T. Kinetics of oxidative degradation of white wines and how they are affected by selected technological parameters. *J. Agric. Food Chem.* **2002**, *50*, 5919–5924. [CrossRef] [PubMed]
56. Lan, Y.; Wang, J.; Aubie, E.; Crombleholme, M.; Reynolds, A. Effects of Frozen Materials Other Than Grapes on Red Wine Aroma Compounds. Impacts of Harvest Technologies. *Am. J. Enol. Vitic.* **2022**, *73*, 142–155. [CrossRef]
57. Cai, W.; Tang, F.; Guo, Z.; Guo, X.; Zhang, Q.; Zhao, X.; Ning, M.; Shan, C. Effects of pretreatment methods and leaching methods on jujube wine quality detected by electronic senses and HS-SPME–GC–MS. *Food Chem.* **2020**, *330*, 127330. [CrossRef] [PubMed]

Article

Modulatory Effect of *Limosilactobacillus fermentum* grx08 on the Anti-Oxidative Stress Capacity of Liver, Heart, and Kidney in High-Fat Diet Rats

Hengxian Qu [1,2], Longfei Zhang [1,2], Xiaoxiao Liu [1,2], Yang Liu [1,2], Kaidong Mao [3], Guiqi Shen [3], Yunchao Wa [1,2], Dawei Chen [1,2], Yujun Huang [1,2], Xia Chen [1,2] and Ruixia Gu [1,2,*]

1 College of Food Science and Technology, Yangzhou University, Yangzhou 225000, China
2 Key Laboratory of Dairy Biotechnology and Safety Control, Yangzhou 225000, China
3 Shanghai HOWYOU Biotechnology Co., Ltd., Shanghai 200000, China
* Correspondence: guruixia1963@163.com

Abstract: To explore the modulating effect of *Limosilactobacillus fermentum (L. fermentum)* grx08 on anti-oxidative stress in the liver, heart, and kidney of high-fat diet in rats, a low-fat diet as a control and a high-fat diet was used to induce oxidative stress injury in rats. *L. fermentum* grx08 and its heat-inactivated bacteria were used to intervene. The results showed that the high-fat diet had caused oxidative stress injury in the liver, heart, and kidney of rats. *L. fermentum* grx08 significantly reduced the serum levels of liver, heart, and kidney injury markers (ALT, AST, LDH, CK-MB, UA, and Crea), while restoring the balance of lipid metabolism in the liver. It also enhanced the activity of antioxidant enzymes such as GSH-Px in the liver, heart, and kidney, scavenging NO radicals and reducing the content of MDA, a product of lipid peroxidation, which can regulate the anti-oxidative stress capacity of the liver, heart, and kidney to varying degrees. Among them, *L. fermentum* grx08 showed better modulating effect on kidney anti-oxidative stress, followed by liver, and the weakest modulating effect on heart. At the same time, *L. fermentum* grx08 heat-inactivated bacteria also had a partial modulatory effect as well as a similar effect profile to that of live bacteria.

Keywords: probiotics; high-fat diet; organs; anti-oxidative stress; rats

Citation: Qu, H.; Zhang, L.; Liu, X.; Liu, Y.; Mao, K.; Shen, G.; Wa, Y.; Chen, D.; Huang, Y.; Chen, X.; et al. Modulatory Effect of *Limosilactobacillus fermentum* grx08 on the Anti-Oxidative Stress Capacity of Liver, Heart, and Kidney in High-Fat Diet Rats. *Fermentation* **2022**, *8*, 594. https://doi.org/10.3390/fermentation8110594

Academic Editors: Michela Verni and Carlo Giuseppe Rizzello

Received: 27 September 2022
Accepted: 28 October 2022
Published: 1 November 2022

Publisher's Note: MDPI stays neutral with regard to jurisdictional claims in published maps and institutional affiliations.

1. Introduction

Modern dietary changes and excessive intake of high-fat and high-energy diets have led to disturbances in fat, protein, and carbohydrate metabolism in the body. Excess energy gradually accumulates in the body in the form of fat, which leads to impaired mitochondrial function and increased oxidative injury in cells [1]. The liver plays a crucial role in maintaining the metabolic balance of the body, and the metabolic disorders associated with obesity are first manifested in the liver. Excessive deposition of fat in hepatocytes leads to the development of oxidative stress injury in the liver [2]. At the same time, the disruption of glucolipid metabolism caused by a chronic high-fat diet can further lead to abnormal heart and vascular function, and the resulting lipotoxicity can also increase kidney oxidative stress levels and decrease antioxidant defense levels, leading to kidney injury [3]. In the metabolic syndrome caused by a high-fat diet, in addition to liver injury, heart and kidney injury have been found. A clinical correlation between non-alcoholic fatty liver disease (NAFLD) and heart and kidney injury has also been found [4].

Probiotics can enhance the anti-oxidative stress capacity of the liver either by themselves or by modulating intestinal microorganisms [5]. *Limosilactobacillus rhamnosus* LV108 can alleviate NAFLD in high-fat diet rats by regulating lipid metabolism and apoptosis [6]. *L. fermentum* DALI02 has been found to alleviate lipid peroxidation and improve lipid metabolism in vivo, thereby reducing oxidative stress and inflammation. They are both able to reduce lipid peroxidation products in the liver and enhance the activity of antioxidant enzymes [7]. In addition, probiotics have also been found to have a degree of protective effect on kidney function impairment in hyperuricemia, enhancing the anti-oxidative stress capacity of the kidney [8]. In addition, *Limosilactobacillus rhamnosus* GG has been found to be able to play a role in the prevention and treatment of cardiovascular disease through various mechanisms such as anti-inflammation, anti-oxidative stress, regulation of intestinal flora, and protection of the intestinal barrier [9,10]. Although the modulatory effects of probiotics on the anti-oxidative stress capacity of liver, heart, and kidney have been demonstrated, fewer studies have explored the effects of probiotics in the same model, and there is also a lack of comparative studies on the effects of probiotics on the anti-oxidative stress of the liver, heart, and kidney.

In previous studies, *L. fermentum* grx08 has been found to have strong antioxidant capacity in vitro, and its adjunctive hypolipidemic effect has also been confirmed in animal experiments [11]. We hypothesized that *L. fermentum* grx08 may have the potential to modulate the anti-oxidative stress capacity of the liver, heart, and kidney in high-fat diet rats. The aim of this work was to investigate the effect and extent of the effect of *L. fermentum* grx08 on the anti-oxidative stress capacity of the liver, heart, and kidney in high-fat diet rats. Therefore, we established a high-fat diet rat model to investigate the modulating effect of *L. fermentum* grx08 on the anti-oxidative stress capacity of the liver, heart, and kidney, and compared its modulating effect.

2. Materials and Methods

2.1. Microorganism and Preparation of Oral Samples

L. fermentum grx08 was cultured in sterile MRS broth (Hopebio., Yancheng, China) from the 3% inoculum with 18 h incubation at 37 °C (DNP-9022 incubator, Shanghai Jinghong, Shanghai, China). Cell pellets were harvested at $5000\times g$ for 10 min (H1750 centrifuge, Changsha Xiangyi, Changsha, China), resuspended with saline (repeated three times) and adjusted the concentration of the bacterium to 1×10^9 CFU/mL (plate count), which is the sample of *L. fermentum* grx08 suspension. *L. fermentum* grx08 suspension was heat inactivated at 80 °C for 20 min (TW12 water bath, JULABO, Seelbach, Germany), which was the sample of *L. fermentum* grx08 heat inactivated suspension.

2.2. Animals and Treatment

Thirty healthy male Wistar rats were purchased from Comparative Medical Center of Yangzhou University, Jiangsu, China. The rats were 5 weeks old and weighed about 145 g at the start of the experiment. All animals were housed under a 12-h light/12-h dark cycle in a controlled room with a temperature of 23 ± 3 °C and a humidity of 50% ± 10%. The animals were acclimated to their new circumstances for one week. All rats were allowed free access to food and water. All 5 rats were placed in a cage and only one cage of rats (Normal control group (NC)) was fed a low-fat diet (LFD: Flour 20%, rice flour 10%, corn 20%, drum skin 26%, soy material 20%, fish meal 2%, bone meal 2%), and other rats were fed a high-fat diet (HFD: 10% lard, 10% egg powder, 1% cholesterol and 0.2% bile salts, and 78.8% LFD) for 4 weeks. High-fat diet rats were randomly divided into 4 groups ($n = 5$): model control group (MC), positive control group (PC), *L. fermentum* grx08 suspension treatment group (Grx08), and *L. fermentum* grx08 heat inactivated suspension treatment group (Grx08H). In the next 4 weeks, all rats received the following treatments by lavage: NC and MC: physiological saline (1 mL/100 g), PC: VC solution (1 mL/100 g, 10 mg/mL), Grx08: *L. fermentum* grx08 suspension (1 mL/100 g, 1×10^9 CFU/mL), and Grx08H: *L. fermentum* grx08 heat inactivated suspension (1 mL/100 g, 1×10^9 CFU/mL). At the end of the 8th week, rats underwent 12 h of fasting prior to being anaesthetized and dissected. All rats were euthanized at the anestrus period following anesthesia under 1% sodium pentobarbital.

2.3. Body Weight and Organ Index

The body weight of rats was measured once a week and fasting weight was measured before execution. After autopsy, the liver, heart, and kidney were weighed, and the organ index was calculated.

$$\text{Organ index} = \text{organ weight (g)}/\text{body weight (100 g)}.$$

2.4. Blood Samples Handling

Blood was obtained from each group on the last day of the 8th week. Serum was incubated at 37 °C for 30 min and then centrifuged at $3000\times g$ for 15 min (H1750 centrifuge, Changsha Xiangyi, Changsha, China), the supernatant was then taken and stored at -20 °C.

2.5. Organ Samples Handling

Part of the liver/heart/kidney was put in cold physiologic saline immediately and tissue homogenate was prepared (10%, w/v). The hypothalamus homogenates were centrifuged at $4000\times g$ for 15 min at 4 °C (H1750 centrifuge, Changsha Xiangyi, Changsha, China), and the supernatant was collected and stored at -20 °C until further analyzed.

Rat liver, heart, and kidney tissues were trimmed to 0.5 cm $\times$ 0.5 cm $\times$ 0.5 cm size, fixed in 4% paraformaldehyde, and stored at 4 °C.

2.6. Indicator Testing

Liver total cholesterol (TC), triglycerides (TG), high-density lipoprotein cholesterol (HDL-C), low-density lipoprotein cholesterol (LDL-C) and blood alanine aminotransferase (ALT), aspartate aminotransferase (AST), lactate dehydrogenase (LDH), creatine kinase isoenzyme (CK-MB), uric acid (UA), and creatinine (Crea) (Ningbo Meikang, Ningbo, China) were measured by a Model 7020 fully automated biochemical analyzer (Hitachi, Tokyo, Japan); HE staining, nitric oxide (NO), malondialdehyde (MDA), superoxide dismutase (SOD), and glutathione peroxidase (GSH-Px) were measured according to the method provided in the assay kit (Nanjing Jiancheng, Nanjing, China).

2.7. Statistical Methods

Statistical analysis was performed using GraphPad Prism 9 (San Diego, CA, USA). The results are presented as the means $\pm$ standard deviation, and comparisons between different groups were assessed by analysis of variance (ANOVA) with Tukey post hoc test (one-way ANOVA-Tukey). Values of $p < 0.05$ were considered statistically significant. The graphics were created using GraphPad Prism 9 (San Diego, CA, USA).

3. Result

3.1. Effect of L. fermentum grx08 on Body Weight and Organ Index of Rats

There was no significant difference in the initial body weight of rats ($p > 0.05$). Compared with the group NC, the final body weight and liver index of rats in the group MC were significantly increased ($p < 0.05$), and then were significantly reduced after *L. fermentum* grx08 and its heat-sterilized bacteria intervention ($p < 0.05$). In contrast, there was no significant difference ($p > 0.05$) in the heart and kidney indices of rats in all intervention groups, and the high-fat diet did not cause significant changes in heart and kidney indices of rats. *L. fermentum* grx08 could alleviate the obesity induced by a high-fat diet as well as the increase of the liver index (Table 1).

3.2. Effect of L. fermentum grx08 on Organ Oxidative Stress Injuries

Compared with the group NC, the serum AST, ALT, LDH, CK-MB, UA, and Crea levels of rats in the group MC were significantly increased ($p < 0.05$), and the high-fat diet caused injury to the liver, heart, and kidney of rats. After *L. fermentum* grx08 intervention, serum AST, ALT, LDH, CK-MB, UA, and Crea were significantly decreased in rats compared with the group MC ($p < 0.05$). *L. fermentum* grx08 heat-inactivated bacteria had the same effect but had no significant effect on Crea ($p > 0.05$). *L. fermentum* grx08 had the potential to enhance the resistance to the anti-oxidative stress capacity of the liver, heart, and kidney in rats (Figure 1).

Table 1. Body weight and organ index of rats.

	Initial Weight (g)	Final Weight (g)	Liver	Heart	Kidney
NC	145.00 ± 7.97	352.80 ± 12.07 *	3.53 ± 0.04 *	0.29 ± 0.02	0.67 ± 0.03
MC	146.50 ± 12.26	415.25 ± 17.73	4.54 ± 0.05	0.28 ± 0.02	0.59 ± 0.03
PC	147.00 ± 8.52	375.50 ± 12.07 *	4.49 ± 0.05	0.32 ± 0.02	0.62 ± 0.06
Grx08	143.20 ± 4.32	358.80 ± 22.99 *	3.89 ± 0.03 *	0.28 ± 0.01	0.57 ± 0.04
Grx08H	145.40 ± 5.41	384.60 ± 37.42	4.19 ± 0.03 *	0.32 ± 0.03	0.59 ± 0.05

Same column comparison, * indicates significant difference compared to the model group ($p < 0.05$), $n = 5$.

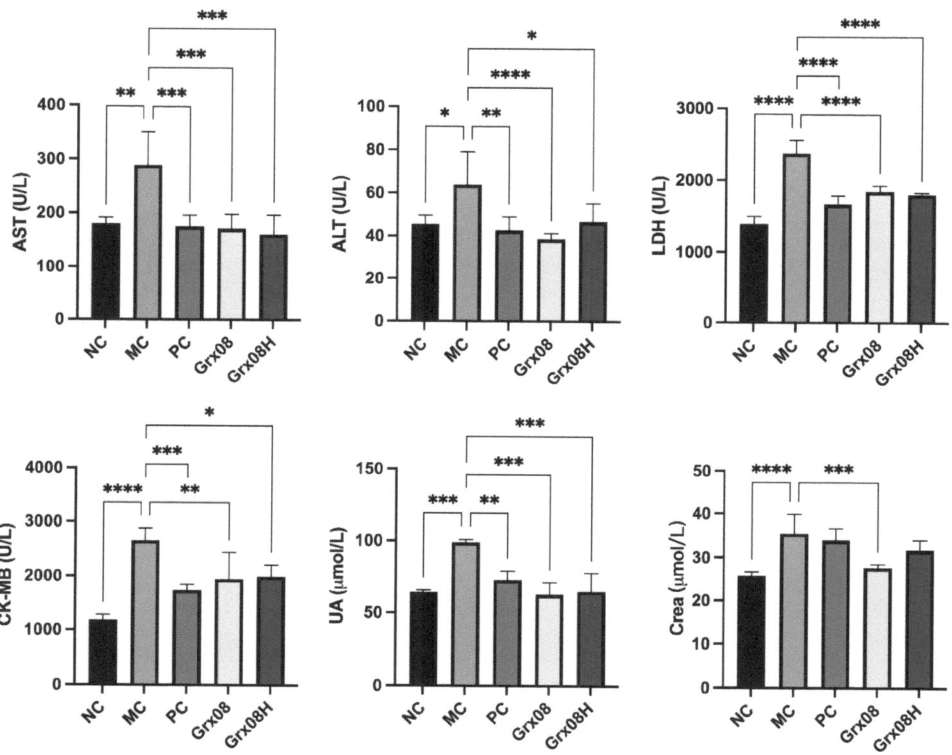

Figure 1. Effect of probiotics on serum biochemical indices of rats ($n = 5$). * indicates a significant difference between groups ($p < 0.05$), ** indicates a significant difference between groups ($p < 0.01$), *** indicates a significant difference between groups ($p < 0.001$), **** indicates a significant difference between groups ($p < 0.0001$), same as below.

3.3. Modulatory Effect of L. fermentum grx08 on Anti-Oxidative Stress of Rat Liver

Compared with the group NC, TC, TG, and LDL-C contents were significantly increased ($p < 0.05$) and HDL-C contents were significantly decreased ($p < 0.05$) in the liver of rats in the group MC. After intervention with *L. fermentum* grx08 and its heat-inactivated bacteria, TC, TG, and LDL-C contents were significantly reduced ($p < 0.05$) and HDL-C contents were significantly increased ($p < 0.05$) in the rat liver. The group PC had no significant effect on lipid accumulation in the liver ($p > 0.05$) (Figure 2a).

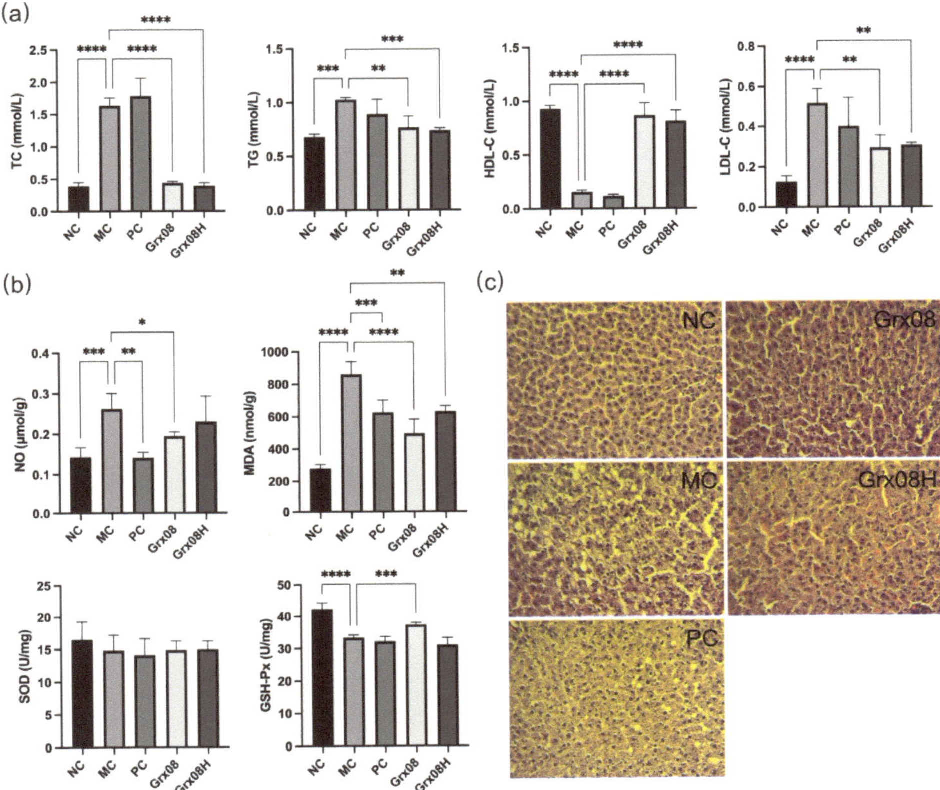

Figure 2. Effect of probiotics on anti-oxidative stress of rat liver ($n = 5$). (**a**) Lipid accumulation in rat liver, (**b**) Oxidative damage and antioxidant enzymes activity in rat liver, (**c**) Case study of rat liver ($\times 200$). * indicates a significant difference between groups ($p < 0.05$), ** indicates a significant difference between groups ($p < 0.01$), *** indicates a significant difference between groups ($p < 0.001$), **** indicates a significant difference between groups ($p < 0.0001$).

Compared with the group NC, the liver NO and MDA contents were significantly increased ($p < 0.05$) and GSH-Px contents were significantly decreased ($p < 0.05$) in the group MC. *L. fermentum* grx08 significantly reduced NO and MDA levels ($p < 0.05$), while significantly increasing GSH-Px activity ($p < 0.05$). Its heat-inactivated bacteria only significantly reduced MDA levels (Figure 2b).

Compared with the group NC, the hepatocyte structure of rat liver in the group MC was severely damaged, the cell gap was unclear, fat deposition was obvious, and a large number of hepatocyte fat vacuoles were visible. After the intervention of *L. fermentum* grx08 and its heat-inactivated bacteria, compared with the group MC, the hepatocyte structure of rats was more intact, with clearer boundaries and fewer fat vacuoles, which was closer to that of the group NC. Fat vacuoles of hepatocytes were still visible in the group PC (Figure 2c).

3.4. Modulatory Effect of L. fermentum grx08 on Anti-Oxidative Stress of Rat Heart

Compared with the group NC, there were no significant changes in NO content and SOD activity, a significant increase in MDA content ($p < 0.05$) and a significant decrease in GSH-Px activity ($p < 0.05$) in the hearts of rats in the group MC. Both *L. fermentum* grx08 and its heat-inactivated bacteria significantly reduced NO content ($p < 0.05$) and increased GSH-Px activity ($p < 0.05$) compared to the group MC, with no significant effect on MDA content ($p > 0.05$). The group PC had no significant effect on the anti-oxidative stress capacity of the heart ($p > 0.05$) (Figure 3a). The case-study observation did not reveal significant fat accumulation and lesions in the heart tissue of the group MC, as well as significant changes in the intervention of *L. fermentum* grx08 and its heat-inactivated bacteria (Figure 3b).

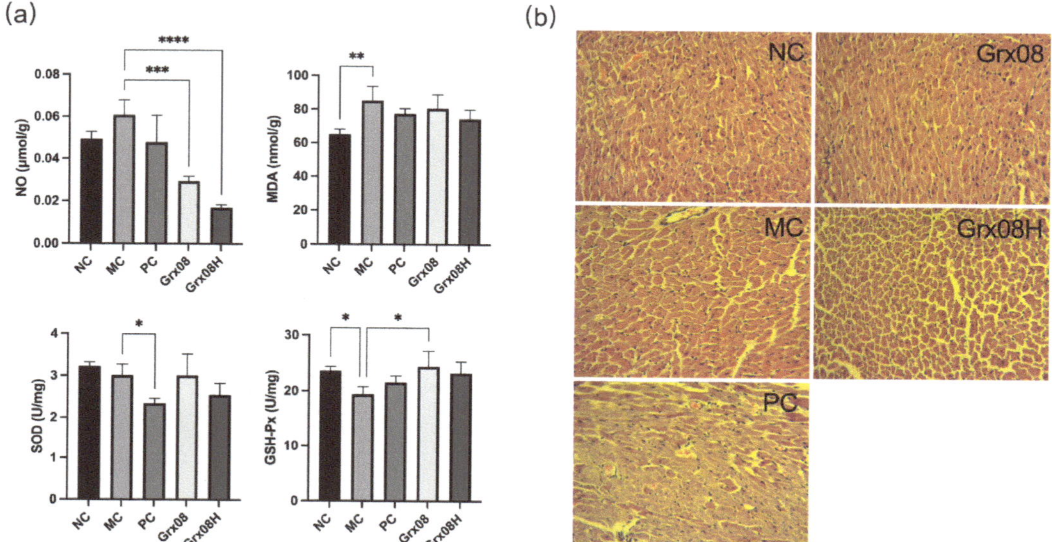

Figure 3. Effect of probiotics on anti-oxidative stress of rat heart ($n = 5$). (**a**) Oxidative damage and antioxidant enzyme activity in rat heart, (**b**) Case study of rat heart ($\times 200$). * indicates a significant difference between groups ($p < 0.05$), ** indicates a significant difference between groups ($p < 0.01$), *** indicates a significant difference between groups ($p < 0.001$), **** indicates a significant difference between groups ($p < 0.0001$).

3.5. Modulatory Effect of L. fermentum grx08 on Anti-Oxidative Stress of Rat Kidney

Compared with the group NC, the NO and MDA contents in the kidney of rats in the group MC were significantly increased ($p < 0.05$), and the SOD and GSH-Px activities were significantly decreased ($p < 0.05$). Both *L. fermentum* grx08 and its heat-inactivated bacteria significantly reduced the content of NO and MDA in rat kidney ($p < 0.05$) and significantly increased the activity of GSH-Px ($p < 0.05$), and *L. fermentum* grx08 also significantly increased the activity of SOD ($p < 0.05$). Of particular note, the levels of NO and MDA as well as the activities of SOD and GSH-Px in the kidney after *L. fermentum* grx08 intervention were restored to the level of the group NC. In addition, there was no significant effect of the group PC on both MDA and SOD in the kidney ($p > 0.05$) (Figure 4a). The case study observation did not reveal any significant fat accumulation and lesions in kidney tissue of the group MC, as well as significant changes in the intervention of *L. fermentum* grx08 and its heat-inactivated bacteria (Figure 4b).

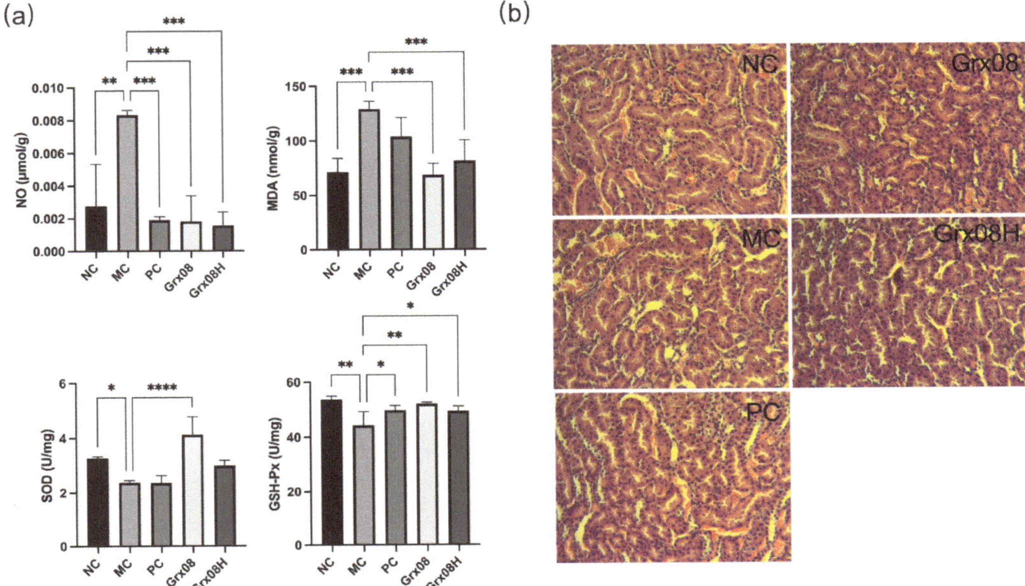

Figure 4. Effect of probiotics on anti-oxidative stress of rat kidney ($n = 5$). (**a**) Oxidative damage and antioxidant enzymes activity in rat kidney. (**b**) Case study of rat kidney ($\times 200$). * indicates a significant difference between groups ($p < 0.05$), ** indicates a significant difference between groups ($p < 0.01$), *** indicates a significant difference between groups ($p < 0.001$), **** indicates a significant difference between groups ($p < 0.0001$).

4. Discussion

A high-fat, high-energy diet results in numerous health problems. In addition to causing fatty liver, injury to the heart and kidney cannot be ignored. We used a high-fat diet to induce oxidative stress injury in the rat liver, heart, and kidney, and intervened with *L. fermentum* grx08 to investigate the modulating effect of *L. fermentum* grx08 on the anti-oxidative stress capacity of the liver, heart, and kidney in rats.

The changes in organ indices were mainly influenced by the accumulation of organ lipids. The liver is the main organ of lipid metabolism and has a regulatory role in the homeostasis of lipid metabolism in the body. Imbalance of lipid metabolism leads to the accumulation of fat in the liver first [2], which makes the liver index increase. In contrast, injury to the heart and kidney is caused by disorders of lipid metabolism [3,4] and does not result in significant accumulation of lipids, so the cardiac and renal indices did not change significantly in this study. The results of the case sections of the liver, heart, and kidney showed the same situation, with significant fat vacuoles only in the liver sections. *L. fermentum* grx08 greatly reduced fat vacuoles and the rat liver index, and alleviated liver lipid accumulation. Further measurements of lipid levels in the rat liver also revealed that the high-fat diet caused lipid accumulation in the liver, and *L. fermentum* grx08 was able to significantly reduce lipid accumulation in the liver and restore the balance of lipid metabolism in the liver.

A number of markers in the serum can more accurately reflect the extent of organ injury. Serum ALT and AST are important indicators of the extent of liver injury. When hepatocytes are damaged, cell membrane permeability increases and these two aminotransferases enter the circulation, leading to an increase in their serum concentrations [12]. LDH is the catalase for the final step of the glycolytic process, and the disruption of the cardiomyocyte membrane or abnormal permeability causes the release of LDH from the circulation [13].

CK-MB is mainly found in the myocardium and has a high sensitivity and specificity for heart injury, and is an important indicator for the diagnosis of heart injury [14].UA is derived from the breakdown of purine compounds in the body and dietary intake, and elevated serum UA is closely related to kidney injury [15]. Crea is also an important indicator for the evaluation of kidney injury, which results in a decrease in the glomerular filtration rate and, consequently, an increase in serum Crea [16]. In this study, the high-fat diet caused significant increases in ALT, AST, LDH, CK-MB, UA, and Crea in the group MC ($p < 0.05$), suggesting that although lipids only accumulated significantly in the liver, rats showed injury to the liver, heart, and kidney, which was significantly reduced after *L. fermentum* grx08 intervention ($p < 0.05$). It tentatively suggested that *L. fermentum* grx08 could enhance the anti-oxidative stress capacity of the liver, heart, and kidney in rats.

A long-term high-fat diet also causes the body to produce large amounts of free radicals through enzyme systems or non-enzyme systems. Free radicals attack unsaturated fatty acids in biological membranes and trigger lipid peroxidation, forming lipid peroxides, which deplete antioxidant substances in the body and lead to cell necrosis and apoptosis [17]. NO radicals are the main free radicals in the body [18]. MDA is the product of lipid peroxidation by free radicals, and its content reflects the degree of lipid peroxidation in the body and indirectly reflects the severity of free radical attack on cells [19]. SOD is the most important enzyme for scavenging free radicals in the body, and GSH-Px is an important peroxide-degrading enzyme that plays an important role in protecting cellular organelles from oxidative injury. Under normal conditions, antioxidants such as SOD and GSH-Px work together in the body to effectively scavenge free radicals in the body and maintain redox homeostasis [20]. We found that the high-fat diet caused different degrees of elevated NO and MDA and decreased GSH-Px in the liver, heart, and kidney, which indicated different degrees of oxidative stress injury. *L. fermentum* grx08 was able to modulate the anti-oxidative stress capacity of the liver, heart, and kidney to different degrees. *L. fermentum* grx08 had better modulating effects on the anti-oxidative stress capacity of kidney, enhancing the levels of both antioxidant enzymes SOD and GSH-Px, scavenging NO radicals and reducing lipid peroxidation products MDA, and all of them were restored to levels in the group NC. *L. fermentum* grx08 had the second highest modulatory effect on the anti-oxidative stress capacity of the liver and the weakest modulatory effect on the heart. We also found lower levels of NO and MDA in the kidney in comparison, which may also be the reason why the kidney was the first to recover after the *L. fermentum* grx08 intervention. In addition, we found that *L. fermentum* grx08 heat-inactivated bacteria also had some regulatory effects, and the regulatory effects on the liver, heart, and kidney were consistent with those of live bacteria.

5. Conclusions

In this study, we explored the regulatory effects of *L. fermentum* grx08 on oxidative stress injury in the liver, heart, and kidney of high-fat diet rats and compared its regulatory capacity. It was found that *L. fermentum* grx08 could regulate the balance of liver lipid metabolism, increase the activity of some antioxidant enzymes, scavenge free radicals, and reduce lipid oxidation products in the liver, heart, and kidney. It could also regulate the anti-oxidative stress capacity of the liver, heart, and kidney in varying degrees, especially increasing the anti-oxidative stress capacity of the kidney.

Author Contributions: Conceptualization, H.Q., L.Z. and K.M.; methodology, H.Q., X.L. and Y.L.; software, H.Q. and G.S.; validation, H.Q. and Y.W.; formal analysis, H.Q.; investigation, D.C.; resources, Y.H.; data curation, H.Q.; writing—original draft preparation, H.Q.; writing—review and editing, H.Q. and R.G.; visualization, H.Q.; supervision, R.G.; project administration, X.C.; funding acquisition, R.G. All authors have read and agreed to the published version of the manuscript.

Fermentation **2022**, 8, 594

Funding: This research was funded by the Postgraduate Research & Practice Innovation Program of Jiangsu Province (Yangzhou University) grant number KYCX20-2981, National Natural Science Foundation of China grant number 32272362 and City-school cooperation to build science and Technology Innovation Platform (YZ2020265).

Institutional Review Board Statement: The study was conducted in accordance with the U.S. National Institutes of Health guidelines for the care and use of laboratory animals (NIH Publication No. 85-23 Rev. 1985), and approved by the Animal Care Committee of the Center for Disease Control and Prevention (Jiangsu, China) (No. 202103262, 2021-03-05).

Informed Consent Statement: Not applicable.

Data Availability Statement: The data supporting the findings reported here are available on reasonable request from the corresponding author.

Conflicts of Interest: The authors declare that they have no conflict of interest.

References

1. Tian, J.L.; Gomeshtapeh, F.I. Potential roles of O-GlcNAcylation in primary cilia—Mediated energy metabolism. *Biomolecules* **2020**, *10*, 1504. [CrossRef] [PubMed]
2. Enkhmaa, B.; Anuurad, E.; Berglund, L. Lipoprotein (a): Impact by ethnicity and environmental and medical conditions. *J. Lipid Res.* **2016**, *57*, 1111–1125. [CrossRef] [PubMed]
3. Pennisi, E.M.; Garibaldi, M.; Antonini, G. Lipid Myopathies. *J. Clin. Med.* **2018**, *7*, 472. [CrossRef] [PubMed]
4. Mikolasevic, I.; Milic, S.; Wensveen, T.T.; Grgic, I.; Jakopcic, I.; Stimac, D.; Wensveen, F.; Orlic, L. Nonalcoholic fatty liver disease—A multisystem disease? *World J. Gastroenterol.* **2016**, *22*, 9488–9505. [CrossRef] [PubMed]
5. Chen, D.; Yang, Z.; Chen, X.; Huang, Y.; Yin, B.; Guo, F.; Zhao, H.; Zhao, T.; Zhenquan, Y.; Huang, J.; et al. The effect of *Lactobacillus rhamnosus* hsryfm 1301 on the intestinal microbiota of a hyperlipidemic rat model. *BMC Complement. Altern. Med.* **2014**, *14*, 386. [CrossRef] [PubMed]
6. Qu, H.; Yu, H.; Gu, R.; Chen, D.; Chen, X.; Huang, Y.; Xi, W.; Huang, Y. Proteomics for studying the effects of *L. rhamnosus* LV108 against non-alcoholic fatty liver disease in rats. *RSC Adv.* **2018**, *8*, 38517–38528. [CrossRef] [PubMed]
7. Huang, Y.; Qu, H.; Liu, D.; Wa, Y.; Sang, J.; Yin, B.; Chen, D.; Chen, X.; Gu, R. The effect of *Lactobacillus fermentum* DALI02 in reducing the oxidative stress and inflammatory response induced by high-fat diet of rats. *RSC Adv.* **2020**, *10*, 34396–34402. [CrossRef] [PubMed]
8. Garcia, F.; Gonzaga, G.; Muñoz-Jiménez, I.; Blas-Marron, M.G.; Silverio, O.; Tapia, E.; Soto, V.; Ranganathan, N.; Ranganathan, P.; Vyas, U.; et al. Probiotic supplements prevented oxonic acid-induced hyperuricemia and renal damage. *PLoS ONE* **2018**, *13*, e0202901. [CrossRef]
9. Chan, Y.K.; Brar, M.S.; Kirjavainen, P.V.; Chen, Y.; Peng, J.; Li, D.; Leung, F.C.-C.; El-Nezami, H. High fat diet induced atherosclerosis is accompanied with low colonic bacterial diversity and altered abundances that correlates with plaque size, plasma A-FABP and cholesterol: A pilot study of high fat diet and its intervention with *Lactobacillus rhamnosus* GG (LGG) or telmisartan in ApoE(-/-) mice. *BMC Microbiol.* **2016**, *16*, 264. [CrossRef]
10. Schoemaker, M.H.; Kleemann, R.; Morrison, M.C.; Verheij, J.; Salic, K.; van Tol, E.A.F.; Kooistra, T.; Wielinga, P.Y. A casein hydrolysate based formulation attenuates obesity and associated non-alcoholic fatty liver disease and atherosclerosis in LDLr-/-.Leiden mice. *PLoS ONE* **2017**, *12*, e0180648. [CrossRef] [PubMed]
11. Wa, Y.; Yin, B.; He, Y.; Xi, W.; Huang, Y.; Wang, C.; Guo, F.; Gu, R. Effects of single probiotic- and combined probiotic-fermented milk on lipid metabolism in hyperlipidemic rats. *Front. Microbiol.* **2019**, *10*, 1312. [CrossRef]
12. Al-Busafi, S.A.; Hilzenrat, N. Mild hypertransaminasemia in primary care. *ISRN Hepatol.* **2013**, *2013*, 256426. [CrossRef]
13. Van Wilpe, S.; Koornstra, R.; Den Brok, M.; De Groot, J.W.; Blank, C.; De Vries, J.; Gerritsen, W.; Mehra, N. Lactate dehydrogenase: A marker of diminished antitumor immunity. *Oncoimmunology* **2020**, *9*, 1731942. [CrossRef]
14. Dhamecha, D.; Jalalpure, S.; Jadhav, K.; Jagwani, S.; Chavan, R. Doxorubicin loaded gold nanoparticles: Implication of passive targeting on anticancer efficacy. *Pharmacol. Res.* **2016**, *113*, 547–556. [CrossRef] [PubMed]
15. Abou-Elela, A. Epidemiology, pathophysiology, and management of uric acid urolithiasis: A narrative review. *J. Adv. Res.* **2017**, *8*, 513–527. [CrossRef] [PubMed]
16. Becker, M.; Gnirck, A.-C.; Turner, J.-E. Innate lymphoid cells in renal inflammation. *Front. Immunol.* **2020**, *11*, 72. [CrossRef] [PubMed]
17. Mularczyk, M.; Michalak, I.; Marycz, K. Astaxanthin and other Nutrients from haematococcus pluvialis-multifunctional applications. *Mar. Drugs* **2020**, *18*, 459. [CrossRef] [PubMed]
18. Lorente, L.; Aller, M.-A.; Arias, J.-L.; Arias, J. Nitric oxide (NO) is a highly reactive free radical with a multitude of organ specific regulatory functions. *Ann. Surg.* **1996**, *224*, 688–689. [CrossRef] [PubMed]
19. Ayala, A.; Muñoz, M.F.; Argüelles, S. Lipid peroxidation: Production, metabolism, and signaling mechanisms of malondialdehyde and 4-hydroxy-2-nonenal. *Oxidative Med. Cell. Longev.* **2014**, *2014*, 360438. [CrossRef] [PubMed]

20. Hess, J.A.; Khasawneh, M.K. Cancer metabolism and oxidative stress: Insights into carcinogenesis and chemotherapy via the non-dihydrofolate reductase effects of methotrexate. *BBA Clin.* **2015**, *3*, 152–161. [CrossRef] [PubMed]

Article

Effects of Total Dissolved Solids, Extraction Yield, Grinding, and Method of Preparation on Antioxidant Activity in Fermented Specialty Coffee

Matúš Várady [1],*, Jan Tauchen [2], Pavel Klouček [2] and Peter Popelka [1]

1   Department of Food Hygiene, Technology and Safety, University of Veterinary Medicine and Pharmacy, Komenského 73, 041 81 Košice, Slovakia
2   Department of Food Science, Czech University of Life Sciences Prague, Kamýcká 129, 165 21 Prague, Czech Republic
*   Correspondence: varadymatus@gmail.com; Tel.: +42-19-0280-7549

Abstract: The aim of this study was to determine the effect of total dissolved solids (TDS), extraction yield (EY), and grinding on total polyphenols (TP), total flavonoids (TF), and total antioxidant capacity (TAC) in a fermented specialty coffee prepared using different methods of filtration (Hario V60, Aeropress, and the French press). The concentrations of antioxidant compounds differed between the TDS treatments and the methods of preparation. The TP and TF with Hario V60 were the highest at a TDS of 1.84%. The TP with Aeropress was at its highest at a TDS of 1.82%. TAC with the French press was at its highest at a TDS of 1.58%. EY was at its highest with fine grinding (Hario V60 > French press > Aeropress at 25.91%, 21.69%, and 20.67%, respectively). French press coffees had the highest TP ($p = 0.045$). Hario V60 coffee had the highest TF, but the TAC of the coffees remained comparable for all methods. EY and TDS influenced TP, TF, and TAC in the coffee beverages using the finest grinding size for all methods of preparation. The finer the grind, the higher the antioxidant activity of the beverages. Measuring coffee extractions should be one of the most important processes in fermented coffee preparation.

Keywords: polyphenols; antioxidant capacity; anaerobic fermentation; fermented coffee; Hario V60; Aeropress; French press

Citation: Várady, M.; Tauchen, J.; Klouček, P.; Popelka, P. Effects of Total Dissolved Solids, Extraction Yield, Grinding, and Method of Preparation on Antioxidant Activity in Fermented Specialty Coffee. *Fermentation* **2022**, *8*, 375. https://doi.org/10.3390/fermentation8080375

Academic Editors: Carlo Giuseppe Rizzello and Michela Verni

Received: 28 July 2022
Accepted: 3 August 2022
Published: 8 August 2022

Publisher's Note: MDPI stays neutral with regard to jurisdictional claims in published maps and institutional affiliations.

1. Introduction

Coffee is one of the most frequently consumed beverages in the world, and its consumption is increasing every year [1]. Coffee consumption is still affected by price, but consumers are interested in buying coffee with associated health claims [2]. Coffee production and consumption increased in the last decade with the arrival on the market of specialty coffees. Specialty coffee is made with the highest-quality green coffee beans, has a known geographical origin, uses the best postharvest treatment method (e.g., natural and washed), uses the best conditions for storing green beans, and is made using beans from the best year of the harvest [3,4]. According to the protocols of the Specialty Coffee Association of America (SCAA) and the international Q Coffee System, specialty coffee has a standardised production characterised by quality and uniqueness of origin, from the criteria used for selecting coffee plantations to brewing. The most important criterion is to achieve a cupping score of $\geq$80 points on a 100-point scale [5,6].

Many important variables affect the taste or quality of brewed coffee, such as water temperature, degree of roasting, bean origin, grinding, and extraction [7–10]. The quality of brewed coffee, however, is mainly influenced by two key factors: the amount of total dissolved solids (TDS), which is the weight fraction of soluble solids in the brew, and extraction yield (EY), which is the weight fraction of soluble solids removed from coffee grounds [11].

These factors indicate when the coffee is strong, weak, bitter, or underdeveloped. An ideal TDS in SCAA baristic practice is 1.15–1.35%, and EY is 18–22%.

Coffee beverages play an important role in human health due to their antioxidant properties, with chlorogenic acids especially responsible for the link between coffee consumption and lower incidences of various diseases [12–14]. The antioxidant activity of coffee depends on many properties, such as origin, variety, processing method, and roasting time [15–17]. Basic processing methods have recently been supplemented by new innovative processing methods, e.g., anaerobic fermentation and carbonic maceration. Anaerobic fermentation and carbonic maceration rapidly change the coffee fruit, resulting in a flavor much different from traditional fermentation methods.

Monitoring extraction for even better nutritional value and taste, however, is important for the preparation of good and healthy coffee. Specialty coffees have a higher total polyphenol (TP) than conventional coffees. The average loss of TP from green to dark-roasted conventional coffee is almost 93% [18]. Specialty coffees have a similar trend of lower TPs after roasting, but average losses are substantially lower [6,17]. The differences in TPs in specialty coffee beverages, however, also depend on the method of preparation (e.g., Hario V60, espresso, and pour-over) [19].

We hypothesised that coffees processed by anaerobic fermentation with a different TDS and EY could greatly affect the antioxidant activities in filtered specialty coffee beverages. Our aim was to determine the effect of TDS, EY, and grinding on TP, total flavonoids (TF), and TAC in a specialty fermented coffee prepared by three filtering methods (Hario V60, Aeropress, and the French press).

2. Materials and Methods

2.1. Coffee Samples

Samples of 100% *Coffee arabica* beans (Father's Coffee Roastery, Ostrava, Czech Republic) were obtained from the Cajamarca coffee area, Huabal District, in Peru. The beans were of the Catuaí variety harvested in 2020. The coffee was processed using anaerobic fermentation, where coffee is fermented in an environment without oxygen (Figure 1). The ripe cherries were harvested, washed, and then spread out on a terrace, where they were dried in the sun for 20–25 d. When this process was complete, the cherries were sealed for 24–48 h inside barrels filled with water with a one-way exit on the top. As the naturally incurring yeast interacts with the sugars of the coffee fruit, carbon dioxide is released. When oxygen is removed, yeast is forced to consume the sugars in the fruits to produce energy. This chemical reaction releases enzymes and can completely change the chemical composition and, ultimately, the flavor of the coffee.

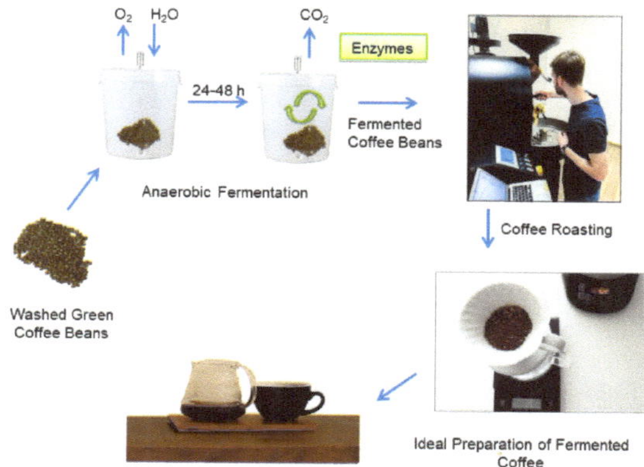

Figure 1. Life cycle of fermented coffee. Washed green coffee beans from the farm undergo anaerobic fermentation. The cherries were sealed for 24–48 h inside barrels filled with water with a one- way exit on the top. As the naturally incurring yeast interacts with the sugars of the coffee fruit, carbon

dioxide is released. When oxygen is removed, yeast is forced to consume the sugars in the fruits to produce energy. This chemical reaction releases enzymes and can completely change the chemical composition and, ultimately, the flavor of the coffee. The packaged green coffee is sent to the roastery where it is roasted lightly to preserve the original taste profile. After roasting and packaging, the coffee is finally ground and ideally prepared by various filtered methods for the most nutritious cup of coffee.

2.2. Coffee Roasting

The green beans were roasted in batches of 0.25 kg in a Probatone 5 gas roaster (Probat, Emmerich am Rhein, Germany) at a final temperature of 215 °C for 10 min, and the development time was 1.5 min. The temperature increased over time and depended on the amount of gas supplied. The temperature was measured by a probe inside the baking drum. The roasting curves are shown in Figure 2.

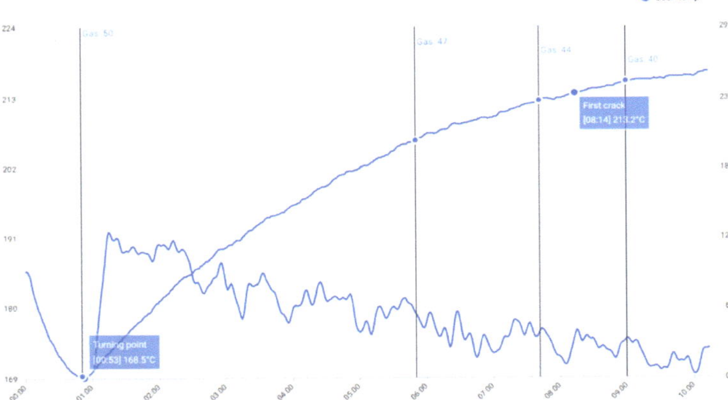

Figure 2. The roasting curves of Peruvian coffee. The green beans were roasted at a final temperature of 215 °C for 10 min, and the development time was 1.5 min. The first crack is the moment when coffee beans start to progress towards an edible state. The time after the first crack is the development time.

2.3. Coffee Grinding

The beans were ground in a Comandante C40 MK3 Nitro Blade hand coffee grinder (Comandante, Munich, Germany). We used the same grind sizes for each method of preparation: 12 clicks (fine grind), 19 clicks (medium grind), 26 clicks (medium to coarse grind), and 33 clicks (coarse grind).

2.4. Coffee Preparation

2.4.1. Hario V60 Pour-Over Method

For the Hario V60 method, a Hario filter paper (Hario, Koga-Ibaraki, Japan) was placed in a ceramic dripper and rinsed with hot water. Freshly ground beans (12 g) were added to the dripper, and water, at a temperature of 94 °C, was poured over them. An initial infusion of 30 mL of distilled water (blooming) was used to initiate the release of CO_2 from the ground beans. An additional 180 mL of water was gradually added after 30 s. The total times were: 2 min and 45 s for the fine grind, 2 min and 25 s for the medium grind, 2 min and 5 s for the medium to coarse grind, and 2 min and 10 s for the coarse grind.

2.4.2. Aeropress Immersion Method

For the Aeropress preparation, we used the classic method with one paper filter (AeroPress, Palo Alto, CA, USA). Freshly ground beans (12 g) were added to the Aeropress, and water, at a temperature of 94 °C, was poured over them. An initial infusion of 30 mL of distilled water (blooming) was used to initiate the release of CO_2 from the ground beans. An additional 180 mL of water was gradually added after 30 s. We started to push the Aeropress after 2 min, and the total time for each grind size was 2 min and 30 s.

2.4.3. French Press-Immersion Method

For the French press preparation, we used a classic French press (Bodum, Triengen, Switzerland). Freshly ground beans (12 g) were added to the French press, and water, at a temperature of 94 °C, was poured over them. An initial infusion of 30 mL of distilled water (blooming) was used to initiate the release of CO_2 from the ground beans. An additional 180 mL of water was gradually added after 30 s. The total time for each grind size was 3 min.

2.5. Total Dissolved Solids

TDSs were measured for each grind, and a digital refractometer (Atago, Tokyo, Japan) was used as the method of preparation. Each sample was measured until the temperature and TDS stabilised.

2.6. Coffee Extraction

The coffee extractions were calculated using two equations [20].
Pour-over method:

$$E\% = (TDS \times \text{weight of brewed coffee}/\text{weight of dry grounds}) + TDS$$

Immersion method:

$$E\% = (TDS \times \text{weight of brewed coffee}/\text{weight of dry grounds}) - TDS$$

2.7. Antioxidant Activity

2.7.1. Total Polyphenols

TPs were measured using a modified method for the analysis of phenols [21,22]. A sample of 100 µL (1:1000 coffee:water) was added to the wells of a 96-well microtiter plate. The range of concentrations of gallic acid (used as a reference compound) was 128–0.015625 mg/mL. Twenty-five microlitres of pure Folin–Ciocalteu reagent was then added. The reaction was initiated by adding 75 mL of 20% Na_2CO_3. The mixtures were kept in the dark at 37 °C for 2 h. The absorbance was measured at 700 nm on a Synergy H1 microplate reader (BioTek, Winooski, VT, USA). The results were expressed as gallic acid equivalents (µg GAE/mg coffee).

2.7.2. Total Flavonoids

TFs were measured by modifying the previously described method [23]. A sample of 100 µL of coffee was added to the wells of a 96-well microtiter plate and mixed with 100 µL of 10% aluminium chloride. The concentration range of quercetin (used as a reference compound) was 256–0.25 mg/mL. Afterward, the solution was incubated in the dark at room temperature for 60 min. The absorbance was read at 420 nm (Synergy H1, Biotek). The results were expressed as quercetin equivalents (µg QE/mg coffee).

2.7.3. Total Antioxidant Capacity–Radical-Scavenging Assay

The antioxidant effect of coffee was established by a slightly modified DPPH (2,2-diphenyl-1-picrylhydrazyl) radical-scavenging assay [24]. A two-fold serial dilution of each sample was prepared in absolute methanol (100 mL) in a 96-well microtiter plate. Subsequently, 100 µL of a freshly prepared 1 mM DPPH solution in methanol was added to each well (final volume 200 mL), creating a range of concentrations of 33.333–32.552 mg/mL. Trolox was used as a positive control (range of concentration: 512–0.5 mg/mL). The mixture was kept in the dark at room temperature for 30 min. The absorbance was read at 517 nm (Synergy H1, BioTek). The results were expressed as Trolox equivalents (μg TE/mg coffee).

2.8. Statistical Analysis

All of the analyses were performed in three independent tests, each in triplicate. Results were expressed as mean ±SD (standard deviation). The acquired data were analysed using GraphPad Prism 8.3.0 (538) 2019 (GraphPad Software, Inc., San Diego, CA, USA) using an ordinary one-way ANOVA and Tukey's multiple-comparisons test. The results were considered significant at $p < 0.05$.

3. Results and Discussion

Brewed coffee has antioxidant activity, which affects the sensory profile and can be modified by various TDS and methods of preparation. The concentrations of antioxidant compounds differed between the TDS treatments and the methods of preparation (Table 1).

Table 1. Effect of TDS, grind size, and EY (%) on TP (μg GAE/mg), TF (μg QE/mg), and TAC (μg TE/mg) in specialty coffees prepared by different methods (means ± SDs, n = 7).

	Hario V60				*p*
TDS	**1.84%**	**1.62%**	**1.45%**	**1.26%**	
TP	33.3 ± 3.76 [c]	28.9 ± 5.49 [b,c]	25.3 ± 3.75 [a,b]	21.3 ± 3.69 [a]	<0.001
TF	0.73 ± 0.213 [b]	0.60 ± 0.198 [a,b]	0.50 ± 0.164 [a,b]	0.42 ± 0.172 [a]	0.015
TAC	7.72 ± 0.506 [a]	11.52 ± 0.558 [b]	10.15 ± 0.563 [b]	7.93 ± 0.894 [a]	0.002
Grind size	Fine	Medium	Medium/Coarse	Coarse	
EY	25.91	22.82	20.06	17.64	
	Aeropress				
TDS	**1.82%**	**1.44%**	**1.32%**	**1.22%**	
TP	34.9 ± 3.74 [b]	29.3 ± 1.93 [a]	28.0 ± 1.53 [a]	27.1 ± 1.54 [a]	<0.001
TF	0.45 ± 0.141	0.42 ± 0.160	0.37 ± 0.134	0.32 ± 0.143	0.200
TAC	8.85 ± 1.674	9.21 ± 1.358	9.54 ± 2.182	10.7 ± 1.311	0.565
Grind size	Fine	Medium	Medium/Coarse	Coarse	
EY	20.67	16.35	15.05	14.67	
	French press				
TDS	**1.58%**	**1.36%**	**1.17%**	**1.10%**	
TP	34.0 ± 3.98	33.6 ± 4.13	29.8 ± 4.21	29.6 ± 5.18	0.044
TF	0.56 ± 0.155 [b,c]	0.47 ± 0.134 [a,b]	0.40 ± 0.140 [a,b]	0.35 ± 0.155 [a]	0.036
TAC	39.1 ± 2.02 [c]	15.8 ± 1.54 [b]	9.35 ± 1.07 [a]	12.3 ± 0.52 [a,b]	<0.001
Grind size	Fine	Medium	Medium/Coarse	Coarse	
EY	21.69	17.16	15.84	14.44	

TDS: total dissolved solids; TP: total polyphenols; TF: total flavonoids; TAC: total antioxidant capacity; EY: extraction yield; GAE: gallic acid; QE: quercetin; TE: Trolox.; [a,b,c] different letters within a row indicate significant differences at $p < 0.05$.

For Hario V60, TP ($p < 0.001$) and TF ($p = 0.015$) were the highest at a TDS of 1.84%, and TAC was at its highest ($p < 0.002$) at a TDS of 1.62% and 1.45%. For Aeropress, TP was at its highest at a TDS of 1.82% ($p < 0.001$), reaching 34.9 μg GAE/mg. For the French press, TF ($p = 0.036$) and TAC ($p < 0.001$) were the highest at a TDS of 1.58%. The antioxidant qualities of the fermented brewed coffee were strongly correlated with TDS and EY [25]. A few studies, however, have addressed the effect of EY on the sensory quality of coffee. Batali et al. [10] examined the role of TDS using time fractionation of drip infusions and found that attributes such as hot, sour, and smoky flavours were positively correlated with TDS, and attributes such as sweet, fruity, and floral flavours were negatively correlated with TDS. Frost et al. [26] characterised the effect of TDS and EY on the sensory quality of drip coffee, concluding that coffee with a higher TDS was more acidic and that coffee with a lower TDS was sweeter. Brewing temperatures, however, had little effects on the sensory profile of drip coffee when sensory attributes were investigated with brewing and TDS parameters [10]. TDS may, thus, decrease with increasing particle size [27,28]. Similar results were reported for the high-pressure extraction of whole beans, which increased TDS from 1.57 to 2.05%, but with minimal changes for ground samples [29]. Studies comparing the effect of TDS and EY on antioxidant activities in filtered fermented specialty coffee beverages, however, are limited.

Physicochemical characteristics, such as EY, TDS, TP, pH, and titratable acidity, are strongly influenced by the level of grinding [30]. Reducing the size of roasted beans by grinding is essential for controlling extraction and dispersion. Fine grinding in our study produced the highest extraction values, with EY in the order Hario V60 > French press > Aeropress (25.91%, 21.69%, and 20.67%, respectively). The formation of small particles by grinding to produce large surface areas is essential for the rapid release of CO_2, the reduction in the diffusion distance for soluble substances during extraction, and the improved transfer of colloidal substances to the liquid phase [31]. Fine grinding, therefore, produced the highest percentage of EY because coarser grinding reduced extractions due to the reduced contact area of the beans during extraction. Hario V60 extraction had a higher EY, even with the medium (22.82%) and medium-coarse grinds (20.06%), which implies that this method protected the coffee better during preparation before reducing the diffusion distance for soluble substances during extraction [32]. The level of grinding, however, can slightly affect the quality of coffee prepared by filtration (American), brewing (Turkish), and extraction under pressure (espresso), but these methods strongly affect antioxidant properties [33]. Fine grinding produces a large volume of ground coffee where the surfaces of the beans consist mainly of broken cells. In many brewing techniques, the extraction of soluble beans from individual beans is very useful because identifying the effect of specific ingredients on taste is possible [34]. Coffee extraction is a basic process that determines the brewing properties because water-soluble components, such as chlorogenic acids, caffeine, polyphenols, flavonoids, volatile hydrophilic compounds, and water-insoluble lipids, are extracted during this process.

We combined all measured TDSs for the given TP, TF, and TAC to determine the potential impact of the method of preparation on bioactive components. Aeropress and French press coffees had the highest TPs ($p = 0.045$). Hario V60 coffee had the highest TF ($p = 0.004$), but TAC for the coffees remained comparable for all methods (Table 2).

Table 2. Effect of method of coffee preparation on TP (μg GAE/mg), TF (μg QE/mg), and TAC (μg TE/mg) in specialty coffees (means ± SDs, n = 7).

	Method of Preparation			p
	Hario V60	**Aeropress**	**French Press**	
TP	27.2 ± 5.10 [a]	29.8 ± 3.49 [a,b]	31.7 ± 2.40 [b]	0.045
TF	0.562 ± 0.135 [b]	0.391 ± 0.059 [a]	0.444 ± 0.093 [a]	0.004
TAC	9.33 ± 1.83	9.58 ± 0.819	19.1 ± 13.59	0.298

TP, total polyphenols; TF, total flavonoids; TAC, total antioxidant capacity; GAE, gallic acid equivalent; QE, quercetin equivalent; TE, Trolox equivalent. [a,b] Different letters within a row indicate significant differences at $p < 0.05$.

The fermented coffee prepared by Aeropress and the French press had the highest TPs (29.8 and 31.7 μg/GAE mg), whereas Hario V60 coffee had the highest TF, reaching 0.562 μg QE/mg. More than 70% of coffee antioxidants are extracted during the first 8 s with the espresso method, but extraction starts later after about 75 s and has higher efficiency with filtered coffees, as in our methods, especially for less polar antioxidants [35]. Organic coffees contain substantially more TPs than conventional coffees [16,36], and the content depends mainly on the origin of the beans, the roasting process, and the brewing technique [37,38].

In contrast, brewing times and water temperatures (between 86°C and 90 °C) have smaller influences on TP [7]. Of the three methods of preparation (Hario V60, espresso, and poured), Hario V60 had the highest TP in samples of specialty coffee from Ethiopia, ranging from 32.0 to 46.8 mg GAE/g [19]. In a comparison of the contents of a cup of coffee prepared by filtration (American), brewing (Turkish), and extraction under pressure (espresso), American coffee had higher TAC and TP than did espresso and Turkish coffee [33]. TP in Rwandan coffee was high when using the cold-brew method for 9 h (i.e., >7.5 mg GAE/g) and in roasted coffee using hot-water preparation (95 °C, 10 min, >7.0 mg GAE/g) [38]. Our results indicated that the brewing method had a large effect on TP and TF, consistent with the previous results [15,19,35,39], and that the hot-brewing method generally produced coffee with higher antioxidant activities [38].

Some studies have reported that TF increases with higher roasting temperatures, coffee quality, grind size, and method of storage [40,41]. TF (8.6 mg/100 mL) with light roasting (186.5 °C) was at its highest after an extraction time of 8 min compared to 4 min for organic Peruvian *C. arabica* [36]. TF depended on the method used (Hario V60, Aeropress, and the French press, at 0.562 µg QE/mg, 0.391 µg QE/mg, and 0.444 µg QE/mg, respectively). The mean TF was lower with the Aeropress method (Table 2) without a significant TDS effect (Table 1). EY was also lowest in the Aeropress method. Producing coffee that can retain its bioactive content is, therefore, extremely important for guaranteeing the quality of coffee and protecting the interests of consumers.

TAC did not differ significantly between the methods ($p = 0.298$), but the relatively high average SD for the French press indicated a potentially strong differential effect between TDS for the French press. TAC, however, depends on the balance of compounds formed and degraded during roasting and on coffee type and method of preparation and is strongly correlated with TP and TF [42]. Hot-brewed coffees have higher TACs, and cold infusions have lower TACs [43]. Our results indicated that Hario V60 and the French press had a large effect on the amount of antioxidant compounds and produced the highest EY, but antioxidant activities remained comparable for all methods. Much of the new coffee equipment (e.g., Hario V60, Kalita Wave, and Aeropress) has responded to changing coffee trends in the last decade. Classic espresso has been replaced in several countries by filtered coffee due to the difference in aroma and the greater sensory experience. With the advent of new coffee machines, however, questions arise about the ideal extraction of coffee using new methods. Our goal is to drink the most delicious, but also nutritious, coffee. Choosing a quality, traceable specialty coffee, and paying attention to its preparation and the parameters that may affect it, is therefore very important. This study should lead to a better understanding of coffee preparation in the coffee industry and its impact on beneficial bioactive substances. We should not forget, however, the impact of anaerobic fermentation on coffee, which is a brand-new processing method. The real question will be how the specialty coffee market, as well as the customers, will respond to the taste of fermented coffee.

4. Conclusions

The results of our study lead to the conclusion that EY and TDS affected TP, TF, and TAC in filtered fermented beverages, which also depended on the method of coffee preparation. The most nutritious beverages were prepared using the finest grind size for all methods of preparation. TAC, however, remained comparable amongst the methods. We thus conclude that the finer the grind, the higher the content of bioactive substances in the coffee beverage and that measuring coffee extraction should be one of the most important processes in coffee preparation.

Author Contributions: Conceptualization, M.V. and J.T.; methodology, M.V. and J.T.; validation, M.V., J.T. and P.K.; formal analysis, M.V.; investigation, M.V.; resources, M.V.; data curation, P.K.; writing—original draft preparation, M.V.; writing—review and editing, P.K.; supervision, P.P.; project administration, P.P.; funding acquisition, P.P. All authors have read and agreed to the published version of the manuscript.

Funding: This research received no external funding.

Institutional Review Board Statement: Not applicable.

Informed Consent Statement: Not applicable.

Data Availability Statement: All data are presented in this manuscript.

Acknowledgments: This study was supported by funds from The Scientific Grant Agency of the Ministry of Education of the Slovak Republic and the Slovak Academy of Sciences (VEGA 1/0073/22) and from the METROFOOD-CZ research infrastructure project (MEYS Grant No.: LM2018100). The authors are grateful to Project FOX Roastery, Prague, Czech Republic, for providing a roasting machine and to Fathers Coffee Roastery, Ostrava, Czech Republic, for their donation of the coffee samples.

Conflicts of Interest: The authors declare no conflict of interest.

References

1. Vegro, C.L.R.; de Almeida, L.F. Global coffee market: Socio-economic and cultural dynamics. In *Consumer Sci & Strat Market, Coffee Consumption and Industry Strategies in Brazil*; de Almeida, L.F., Spers, E.E., Eds.; Woodhead Publishing: Cambridge, UK, 2020; pp. 3–19.
2. Samoggia, A.; Riedel, B. Consumers' perceptions of coffee health benefits and motives for coffee consumption and purchasing. *Nutrients* **2019**, *11*, 653. [CrossRef]
3. Poole, R.; Kennedy, O.J.; Roderick, P.; Fallowfield, J.A.; Hayes, P.C.; Parkes, J. Coffee consumption and health: Umbrella review of meta-analyses of multiple health outcomes. *BMJ* **2018**, *360*, k194. [CrossRef]
4. Zarebska, M.; Stanek, N.; Barabosz, K.; Jaszkiewicz, A.; Kulesza, R.; Matejuk, R.; Andrzejewski, D.; Biłos, Ł.; Porada, A. Comparison of chemical compounds and their influence on the taste of coffee depending on green beans storage conditions. *Sci. Rep.* **2022**, *12*, 2674. [CrossRef]
5. Lingle, T.R.; Menon, S.N. Chapter 8—Cupping and Grading—Discovering Character and Quality. In *The Craft and Science of Coffee*; Folmer, B., Ed.; Academic Press: Cambridge, MA, USA, 2017; pp. 181–203.
6. Bolka, M.; Emire, S. Effects of coffee roasting technologies on cup quality and bioactive compounds of specialty coffee beans. *Food Sci. Nutr.* **2020**, *8*, 6120–6130. [CrossRef]
7. Klotz, J.A.; Winkler, G.; Lachenmeier, D.W. Influence of the brewing temperature on the taste of espresso. *Foods* **2020**, *9*, 36. [CrossRef]
8. Zakidou, P.; Plati, F.; Matsakidou, A.; Varka, E.-M.; Blekas, G.; Paraskevopoulou, A. Single origin coffee aroma: From optimized flavor protocols and coffee customization to instrumental volatile characterization and chemometrics. *Molecules* **2021**, *26*, 4609. [CrossRef]
9. Uman, E.; Colonna-Dashwood, M.; Colonna-Dashwood, L.; Perger, M.; Klatt, C.; Leighton, S.; Miller, B.; Butler, K.T.; Melot, B.C.; Speirs, R.W.; et al. The effect of bean origin and temperature on grinding roasted coffee. *Sci. Rep.* **2016**, *6*, 24483. [CrossRef]
10. Batali, M.E.; Ristenpart, W.D.; Guinard, J.-X. Brew temperature, at fixed brew strength and extraction, has little impact on the sensory profile of drip brew coffee. *Sci. Rep.* **2020**, *10*, 16450. [CrossRef]
11. Lockhart, E.E. The soluble solids in beverage coffee as an index to cup quality. In *Pan American Coffee Bureau*; Coffee Brewing Institute, National Coffee Association of USA: New York, NY, USA, 1957.
12. Yamagata, K. Do coffee polyphenols have a preventive action on metabolic syndrome associated endothelial dysfunctions? An assessment of the current evidence. *Antioxidants* **2018**, *7*, 26. [CrossRef]
13. Tajik, N.; Tajik, M.; Mack, I.; Enck, P. The potential effects of chlorogenic acid, the main phenolic components in coffee, on health: A comprehensive review of the literature. *Eur. J. Nutr.* **2017**, *56*, 2215–2244. [CrossRef]
14. Farah, A.; de Paula Lima, J. Consumption of chlorogenic acids through coffee and health implications. *Beverages* **2019**, *5*, 11. [CrossRef]
15. Bobková, A.; Hudáček, M.; Jakabová, S.; Belej, L.; Capcarová, M.; Čurlej, J.; Bobko, M.; Árvay, J.; Jakab, I.; Čapla, J.; et al. The effect of roasting on the total polyphenols and antioxidant activity of coffee. *J. Environ. Sci. Health* **2020**, *55*, 495–500. [CrossRef]
16. Król, K.; Gantner, M.; Tatarak, A.; Hallmann, E. The content of polyphenols in coffee beans as roasting, origin and storage effect. *Eur. Food Res. Technol.* **2020**, *246*, 33–39. [CrossRef]
17. Várady, M.; Slusarczyk, S.; Boržíková, J.; Hanková, K.; Vieriková, M.; Marcinčák, S.; Popelka, P. Heavy-metal contents and the impact of roasting on polyphenols, caffeine, and acrylamide in specialty coffee beans. *Foods* **2021**, *10*, 1310. [CrossRef]
18. Farah, A.; Donangelo, C.M. Phenolic compounds in coffee. *Braz. J. Plant Physiol.* **2006**, *8*, 23–36. [CrossRef]
19. Várady, M.; Hrušková, T.; Popelka, P. Effect of preparation method and roasting temperature on total polyphenol content in coffee beverages. *Czech J. Food Sci.* **2020**, *38*, 417–421. [CrossRef]
20. Gagné, J. The physics of filter coffee. In *Scott Rao Coffee Books*; Gagné, J., Ed.; Scott Rao Coffee Consulting: Huntington Beach, CA, USA, 2021; p. 249.
21. Singleton, V.L.; Orthofer, R.; Lamuela-Raventos, R.M. Analysis of total phenols and other oxidation substrates and antioxidants by means of folin-ciocalteu reagent. *Meth. Enzymol.* **1999**, *299*, 152–178. [CrossRef]
22. Tauchen, J.; Maršík, P.; Kvasnicová, M.; Maghradze, D.; Kokoška, L.; Vaněk, T.; Landa, P. In vitro antioxidant activity and phenolic composition of Georgian, Central and West European wines. *J. Food Composit. Anal.* **2015**, *41*, 113–121. [CrossRef]

23. Chandra, S.; Khan, S.; Avula, B.; Lata, H.; Yang, M.H.; Elsohly, M.A.; Khan, I.A. Assessment of total phenolic and flavonoid content, antioxidant properties, and yield of aeroponically and conventionally grown leafy vegetables and fruit crops: A comparative study. *Evid. Based Complement Alternat. Med.* **2014**, *2014*, 253875. [CrossRef]
24. Sharma, O.P.; Bhat, T.K. DPPH antioxidant assay revisited. *Food Chem.* **2009**, *113*, 1202–1205. [CrossRef]
25. Liang, J.; Chan, K.C.; Ristenpart, W.D. An equilibrium desorption model for the strength and extraction yield of full immersion brewed coffee. *Sci. Rep.* **2021**, *11*, 6904. [CrossRef]
26. Frost, S.C.; Ristenpart, W.D.; Guinard, J. Effect of basket geometry on the sensory quality and consumer acceptance of drip brewed coffee. *J. Food Sci.* **2019**, *84*, 2297–2312. [CrossRef]
27. Fuller, M.; Rao, N.Z. The effect of time, roasting temperature, and grind size on caffeine and chlorogenic acid concentrations in cold brew coffee. *Sci. Rep.* **2017**, *7*, 17979. [CrossRef]
28. Cordoba, N.; Pataquiva, L.; Osorio, C.; Moreno, F.L.M.; Ruiz, R.Y. Effect of grinding, extraction time and type of coffee on the physicochemical and flavour characteristics of cold brew coffee. *Sci. Rep.* **2019**, *9*, 8440. [CrossRef]
29. Zhang, L.; Wang, X.; Manickavasagan, A.; Lim, L.-T. Extraction and physicochemical characteristics of high pressure-assisted cold brew coffee. *Future Foods* **2022**, *5*, 100113. [CrossRef]
30. Cordoba, N.; Fernandez-Alduenda, M.; Moreno, F.L.; Ruiz, Y. Coffee extraction: A review of parameters and their influence on the physicochemical characteristics and flavour of coffee brews. *Trends Food Sci. Technol.* **2020**, *96*, 45–60. [CrossRef]
31. Baggenstoss, J.; Perren, R.; Escher, F. Water content of roasted coffee: Impact on grinding behaviour, extraction, and aroma retention. *Eur. Food Res. Technol.* **2008**, *227*, 1357–1365. [CrossRef]
32. Olechno, E.; Puścion-Jakubik, A.; Zujko, M.E.; Socha, K. Influence of various factors on caffeine content in coffee brews. *Foods* **2021**, *10*, 1208. [CrossRef]
33. Derossi, A.; Ricci, I.; Caporizzi, R.; Fiore, A.; Severini, C. How grinding level and brewing method (Espresso, American, Turkish) could affect the antioxidant activity and bioactive compounds in a coffee cup. *J. Sci. Food Agric.* **2018**, *98*, 3198–3207. [CrossRef]
34. Moroney, K.M.; Lee, W.T.; O'Brien, S.B.; Suijver, F.; Marra, J. Coffee extraction kinetics in a well mixed system. *J. Math. Ind.* **2016**, *7*, 3. [CrossRef]
35. Ludwig, I.A.; Sanchez, L.; Caemmerer, B.; Kroh, L.W.; De Peña, M.P.; Cid, C. Extraction of coffee antioxidants: Impact of brewing time and method. *Food Res. Int.* **2012**, *48*, 57–64. [CrossRef]
36. Górecki, M.; Hallmann, E. The Antioxidant content of coffee and its in vitro activity as an effect of its production method and roasting and brewing time. *Antioxidants* **2020**, *9*, 308. [CrossRef]
37. Vignoli, J.A.; Viegas, M.C.; Bassoli, D.G.; Benassi, M.T. Roasting process affects differently the bioactive compounds and the antioxidant activity of arabica and robusta coffees. *Food Res. Int.* **2014**, *61*, 279–285. [CrossRef]
38. Muzykiewicz-Szymańska, A.; Nowak, A.; Wira, D.; Klimowicz, A. The effect of brewing process parameters on antioxidant activity and caffeine content in infusions of roasted and unroasted arabica coffee beans originated from different countries. *Molecules* **2021**, *26*, 3681. [CrossRef]
39. Dybkowska, E.; Sadowska, A.; Rakowska, R.; Dębowska, M.; Świderski, F.; Świąder, K. Assessing polyphenols content and antioxidant activity in coffee beans according to origin and the degree of roasting. *Rocz. Państw. Zakł. Hig.* **2017**, *68*, 347–353.
40. Odžaković, B.; Džinić, N.; Kukrić, Z.; Grujić, S. Effect of roasting degree on the antioxidant activity of different Arabica coffee quality classes. *Acta Sci. Pol. Technol. Aliment.* **2016**, *15*, 409–417. [CrossRef]
41. Merecz, A.; Marusinska, A.; Karwowski, B.T. The content of biologically active substances and antioxidant activity in coffee depending on brewing method. *Pol. J. Nat. Sci.* **2018**, *33*, 267–284.
42. Samsonowicz, M.; Regulska, E.; Karpowicz, D.; Leśniewska, B. Antioxidant properties of coffee substitutes rich in polyphenols and minerals. *Food Chem.* **2019**, *278*, 101–109. [CrossRef]
43. Rao, N.Z.; Fuller, M. Acidity and antioxidant activity of cold brew coffee. *Sci. Rep.* **2018**, *8*, 16030. [CrossRef]

Article

Enhancement of Antioxidant Activities in Black Soy Milk through Isoflavone Aglycone Production during Indigenous Lactic Acid Bacteria Fermentation

Benediktus Yudo Leksono [1,2], Muhammad Nur Cahyanto [1], Endang Sutriswati Rahayu [1,3], Rini Yanti [1] and Tyas Utami [1,*]

[1] Department of Food and Agricultural Product Technology, Faculty of Agricultural Technology, Universitas Gadjah Mada, Flora Street, Bulaksumur, Yogyakarta 55281, Indonesia; benediktus.leksono@uajy.ac.id (B.Y.L.); mn_cahyanto@ugm.ac.id (M.N.C.); endangsrahayu@ugm.ac.id (E.S.R.); riniyanti@ugm.ac.id (R.Y.)
[2] Faculty of Biotechnology, Universitas Atma Jaya Yogyakarta, Yogyakarta 55281, Indonesia
[3] Center for Food and Nutrition Studies, Universitas Gadjah Mada, Yogyakarta 55281, Indonesia
* Correspondence: tyas_utami@ugm.ac.id

Abstract: Black soybeans contain high antioxidant compounds such as isoflavone but mainly in glucoside form, with low antioxidant activities. Fermentation by lactic acid bacteria (LAB) can enhance the antioxidant properties, but its ability is strain-dependent. This study aims to study the ability of Indonesian indigenous LAB, *Lactiplantibacillus plantarum* WGK 4, *Streptococcus thermophilus* Dad 11, and *Lactiplantibacillus plantarum* Dad 13, to enhance the antioxidant properties during black soy milk fermentation. Fermentation was carried out at 37 °C for 24 h. Viable cell, acid production, Folin–Ciocalteu assay, antioxidant activity (DPPH), isoflavone aglycone daidzein and genistein, and β-glucosidase activity were measured every six hours. All LAB strains could grow well during the fermentation of black soy milk. *Lactiplantibacillus plantarum* WGK 4 produced the highest acid (1.50%). All three LAB strains could enhance antioxidant activity (DPPH) from 24.90% to 31.22–38.20%, followed by increased isoflavone aglycone. All strains could increase daidzein and genistein content, ranging from 61% to 107% and 81% to 132%, respectively. All three Indonesian indigenous LAB enhanced antioxidant properties of black soy milk relatively at the same level and potentially could be used as a starter culture of black soy milk fermentation.

Keywords: antioxidant activity; black soy milk; daidzein; fermentation; genistein; lactic acid bacteria

Citation: Leksono, B.Y.; Cahyanto, M.N.; Rahayu, E.S.; Yanti, R.; Utami, T. Enhancement of Antioxidant Activities in Black Soy Milk through Isoflavone Aglycone Production during Indigenous Lactic Acid Bacteria Fermentation. *Fermentation* **2022**, *8*, 326. https://doi.org/10.3390/fermentation8070326

Academic Editors: Michela Verni and Carlo Giuseppe Rizzello

Received: 15 June 2022
Accepted: 8 July 2022
Published: 13 July 2022

Publisher's Note: MDPI stays neutral with regard to jurisdictional claims in published maps and institutional affiliations.

1. Introduction

Black soybeans (*Glycine max* (L.) Merr.) provide health benefits due to their high protein, fiber, vitamin B, and mineral content [1–4]. Black soybeans also contain many bioactive compounds, such as anthocyanin and isoflavones, which can act as antioxidants [5,6]. This points out the potential of black soybean to be developed into various functional food products. Unfortunately, the utilization of black soybeans is still not diverse.

Like yellow soybean, black soybean could be processed into soy milk. Nevertheless, isoflavones in soy milk are mostly in glucoside form [7]. In glucoside form, isoflavones are attached to glucose. Isoflavone glucosides are more water-soluble and polar compared to their aglycone form. Thus, it is difficult to pass intestinal epithelium and be absorbed [8]. Moreover, isoflavone aglycone, without glucose moiety, exhibits higher antioxidant activity. Its hydroxyl group can react with free radicals, act as an oxygen donor, and terminate the chain reaction of free radicals [9].

Several studies have proven that soy milk fermentation by lactic acid bacteria (LAB) can enhance antioxidant activities by releasing isoflavone aglycone. Lactic acid bacteria need carbon sources for cell growth and their metabolic activity. Glucose moiety in isoflavone glucoside might be used as a carbon source. Isoflavone glucoside can be hydrolyzed by β-glucosidase into glucose and isoflavone aglycone, thus increasing the antioxidant activities [10]. Zhao and Shah [11] reported that fermentation by several LAB increased the antioxidant level of soy milk due to deglycosylation of isoflavones glucoside. In addition, Lee et al. [12] stated that *Streptococcus thermophilus* S10 produced β-glucosidase during fermentation and hydrolyzed isoflavone glucoside into its aglycone form, increasing the antioxidant activities of soy milk. Furthermore, Hati et al. [7] studied the ability of six different strains of *Lactobacillus rhamnosus* and *Lactobacillus casei* to produce β-glucosidase and bioconvert isoflavone glucoside into isoflavone aglycone during fermentation and found out that each strain produced significantly different bioconversion of the glucoside isoflavones. This points out the potential of black soybean to be developed into fermented non-dairy milk that possesses functional properties. Nevertheless, these findings indicated that the ability of LAB to enhance antioxidant activity by producing β-glucosidase to hydrolyze isoflavone glucoside is strain-dependent.

Even though previous research already reported that some LAB strains could enhance antioxidant activity, their ability differs among strains [7] and fermentation conditions [10,13]. *Lactiplantibacillus plantarum* WGK 4, *Streptococcus thermophilus* Dad 11, and *Lactiplantibacillus plantarum* Dad 13 are some indigenous LAB strains that have been isolated in previous studies from various sources. *Lactiplantibacillus plantarum* WGK 4 was isolated from red lima bean soaking water in tempeh production [14], while *Streptococcus thermophilus* Dad 11 and *Lactiplantibacillus plantarum* Dad 13 was isolated from dadih, Indonesian traditional fermented buffalo milk [15]. These LAB could grow well in milk and jack bean milk [14–16]. The ability of these three indigenous LAB to grow and enhance their antioxidant properties by increasing the isoflavone aglycone content during the fermentation of black soy milk is still unknown. Additionally, most of the previous research studied the fermentation of yellow soy milk. Studies on the fermentation of black soy milk are still limited. Black soybeans have a different nutritional composition and might affect the growth of LAB. Moreover, previous studies have not investigated sugar utilization during fermentation, whereas simple sugar availability might affect isoflavone degradation. Therefore, this study aims to investigate the ability of three Indonesian indigenous LAB strains, *L. plantarum* WGK 4, *S. thermophilus* Dad 11, and *L. plantarum* Dad 13, to grow and enhance antioxidant properties during the fermentation of black soy milk.

2. Materials and Methods

2.1. Materials

Lactiplantibacillus plantarum WGK 4 (previously identified as *Lactobacillus plantarum* WGK 4) culture was obtained from the Laboratory of Biotechnology, Department of Food and Agricultural Product Technology, Faculty of Agricultural Technology, Universitas Gadjah Mada, Yogyakarta, Indonesia. Meanwhile, *Streptococcus thermophilus* Dad 11 and *Lactiplantibacillus plantarum* Dad 13 (previously identified as *Lactobacillus plantarum* Dad 13) cultures were obtained from Food and Nutrition Culture Collection (FNCC), Center for Food and Nutrition Studies, Universitas Gadjah Mada, Yogyakarta, Indonesia. Black soybean seed Detam-1 variety was purchased from UPBS Balitkabi, Malang Regency, Indonesia. De Mann, Rogosa, and Sharpe (MRS) medium was purchased from Merck (Darmstadt, Germany), and all the chemical reagent was purchased from Sigma-Aldrich (Burlington, MA, USA).

2.2. Inoculum Preparation

The inoculum preparation was performed according to Yudianti et al. [15]. The stock culture was maintained in a sterile sucrose (20% $w \cdot v^{-1}$) and skim milk (10% $w \cdot v^{-1}$) mixture (1:1) at $-20\ ^\circ$C. For working culture, the culture was incubated in MRS broth at 37 $^\circ$C for 24 h and maintained in MRS deep tube agar, stored at 4 $^\circ$C. Culture from working culture was activated in MRS broth and incubated at 37 $^\circ$C for 24 h twice as a starter culture. The viable cell of LAB in starter culture was measured before inoculation into black soy milk and expressed as a colony-forming unit CFU·mL^{-1}.

2.3. Preparation of Black Soy Milk and Fermented Black Soy Milk

The black soy milk and fermented black soy milk preparation were conducted according to Yudianti et al. [15]. Black soybeans were washed and soaked overnight at room temperature before being blended in a blender (Matsushita, Japan) with water (80 °C, 1:2) for four minutes. The slurry was then filtered twice by a filter cloth, placed in a sterile 100 mL glass bottle, pasteurized at 65–70 °C for 30 min, and let cool before inoculating with LAB. Black soy milk was immediately analyzed for chemical properties such as proximate compounds, Folin–Ciocalteu assay, and antioxidant activities. Meanwhile, for isoflavone aglycone (daidzein and genistein) concentration, sugar and amino acid profile, and minerals (iron (Fe), zinc (Zn), magnesium (Mg), and manganese (Mn)) content, the sample was freeze-dried (Modulyo, Edwards, UK) and stored in a freezer (−20 °C) until further analysis. The fermentation of black soy milk was conducted by inoculating the black soy milk with a single culture of either *L. plantarum* WGK 4, *S. thermophilus* Dad 11, or *L. plantarum* Dad 13 (1% v·v^{-1}). Fermentation was conducted in an incubator (Sanyo MIR-262, Osaka, Japan) at 37 °C for 24 h. Fermented black soy milk was analyzed every six hours. Viable cell, titratable acidity, pH, Folin–Ciocalteu assay, antioxidant activity (DPPH), and β-glucosidase activity were immediately analyzed. Meanwhile, for the lactic and acetic acid concentration, sugar profile, and isoflavone aglycone concentration, fermented black soy milk was freeze-dried and stored in a freezer (−20 °C) until further analysis.

2.4. Proximate Composition Analysis

The proximate composition of black soy milk was analyzed according to AOAC [17]. The total fat of black soy milk was determined by the Mojonnier method and the crude protein by the Kjeldahl method according to AOAC 989.05 (Fat in Milk—Modified Mojonnier Ether Extract) and 991.20 (Nitrogen (Total) in Milk—Kjeldahl Methods), respectively. Moreover, black soy milk's water and ash content was established gravimetrically according to AOAC 925.23 (Solids (total) in Milk) and 945.46 (Ash of Milk—Gravimetric Method), respectively. Meanwhile, the total carbohydrates were quantified by calculating the percentage remaining after all the other components were measured (by difference).

2.5. Mineral Element Analysis

Mineral contents (Fe, Zn, Mg, and Mn) in black soy milk samples were carried out using Inductively Coupled Plasma-Optical Emission Spectrometry (Agilent Technologies 700 Series ICP-OES, Santa Clara, CA, USA) described by AOAC 2011.14 (calcium, copper, iron, magnesium, manganese, potassium, phosphorus, sodium, and zinc in Fortified Food Products. Microwave Digestion and Inductively Coupled Plasma-Optical Emission Spectrometry) [17]. Approximately 0.5 g of freeze-dried black soy milk was weighed into a vessel and digested using a 10 mL concentrated nitric acid (HNO$_3$). Digestion was conducted using microwave digestion under the following conditions: temperature—150 °C; ramp time—10 min; hold time—15 min. The digested sample was cooled down and transferred into a 50 mL volumetric flask. As an internal standard, 100 mg·L^{-1} of yttrium (Y) was added to the digested sample and diluted using double distilled water till the mark. The solution was filtered using filter paper before being analyzed. Inductively coupled plasma optical emission spectrometry was performed using concentric glass as nebulizer type, the intensity of torch alignment >1,000,000 for horizontal and vertical, and the position of torch alignment −1 to 1 for horizontal and vertical. The absorbance of Fe, Zn, Mg, Mn, and Y was measured at 238 nm, 213 nm, 285 nm, 257 nm, and 371 nm, respectively. The concentration of each mineral was measured using a calibration curve of its respective standard solution.

2.6. Determination of Amino Acids Profile

The amino acid determination was measured using Acquity Ultra Performance Liquid Chromatography (UPLC) H-Class Amino Acid Analysis from Waters (Queenstown, Singapore) according to the kit's procedure (AccQ-Tag Ultra Chemistry Kit number 176001235). One gram of freeze-dried black soy milk was hydrolyzed using hydrochloric acid (6 N) solution. The hydrolyzed sample was transferred into a 50 mL volumetric flask and diluted using double distilled water till the mark. The diluted sample was filtered using a 0.2 μm syringe filter (Sartorius Minisart, Germany) and added with a 2.5 mM internal standard. Alpha-Aminobutyric acid (AABA) was used as an internal standard. The protein hydrolysate was derivatized using AccQ Tag Ultra reagent containing borate buffer, 6-aminoquinolyl-N-hydroxysuccinimidyl carbamate (AQC), and acetonitrile as reagent diluent based on the procedure listed on the kit. The derivatized solution was analyzed using Acquity UPLC H-Class (Waters) system equipped with Quaternary Solvent Manager (QSM), Sample Manager—Flow Waters Corporation (SM-FTN), Column heater module (CH-A), ACQUITY UPLC BEH C18 (2.1 × 100 mm, i.d., 1.7 μm) column and ACQUITY UPLC photodiode array detector (PDA), from Waters. The mobile phase is composed of 100% AccQ Tag Ultra eluent A concentrate (A), 90:10 water: AccQ Tag Ultra eluent B (B), 100% HPLC-grade water (C), and 100% AccQ Tag Ultra eluent B (D). The samples were injected into the UPLC system under the following conditions: column and sample temperatures were 49 °C and 20 °C, respectively, the injection volume was 1 μL, pressure around 13,000 psi, and the amino acids were detected at 260 nm. The mobile phase flow rate and gradient system are shown in Table 1. The quantification of individual amino acids was based on the ratio of samples or amino acid standard area and internal standard area.

Table 1. Mobile phase composition for amino acids determination.

Time (min)	Flow Rate (mL·mL^{-1})	% Mobile Phase A	% Mobile Phase B	% Mobile Phase C	% Mobile Phase D
Initial	0.700	10.0	0.0	90.0	0.0
0.29	0.700	9.9	0.0	90.1	0.0
5.49	0.700	9.0	80.0	11.0	0.0
7.10	0.700	8.0	15.6	57.9	18.5
7.30	0.700	8.0	15.6	57.9	18.5
7.69	0.700	7.8	0.0	70.9	21.3
7.99	0.700	4.0	0.0	36.3	59.7
5.89	0.700	4.0	0.0	36.3	59.7
8.68	0.700	10.0	0.0	90.0	0.0
10.20	0.700	10.0	0.0	90.0	0.0

2.7. Determination of Sugars Profile

The sugar compounds were determined using High-Performance Liquid Chromatography (Waters, Singapore) connected to Waters 2414 Refractive Index (RI) Detector according to AOAC 980.13 (Fructose, Glucose, Lactose, Maltose, and Sucrose in Milk Chocolate—Liquid Chromatographic Method) [17]. A freeze-dried black soy milk sample (0.5 g) was weighed into a 25 mL volumetric flask and diluted with double distilled water to half the mark. The solution was sonicated (Eyela Sonicator Cleaner, Singapore) for 15 min, and 1 mL of carrez I and II solution were added and shaken until homogeneous. Double distilled water was added to the mark and shaken until homogeneous. The solution was transferred into a 2 mL tube and centrifuged (Thermo Fisher Scientific, Waltham, MA, USA) at 14,000 rpm for 3 min. The samples were filtered using GHP/RC 0.45 μm syringe tube into a 2 mL vial before being injected into the HPLC system. The chromatographic separation of sugars was achieved on a Carbohydrate column (5 μm, 250 × 4.6 mm). The injection volume was 10 μL, and the mobile phase was 80% acetonitrile at a flow rate of 1 mL·min^{-1} under isocratic conditions with ambient temperature. Sugar quantification was carried out using the calibration curve of an external mixed standard solution.

2.8. Determination of Lactic and Acetic Acid Content

Lactic and acetic acid in black soy milk was measured based on the procedure described by [18] using Acquity Ultra Performance Liquid Chromatography (UPLC) equipped with ACQUITY UPLC photodiode array detector (PDA) from Waters (Singapore). First, 1 gram of freeze-dried black soy milk samples was weighed into a 25 mL volumetric flask, and 5 milliliters of 20 mM H_3PO_4 buffer solution were added. The mixture was vortexed and sonicated for 10 min at room temperature. Thereafter, 20 mM H_3PO_4 buffer was added to the mark and shaken until homogeneous. The solution was transferred into a 15 mL conical centrifuge tube and centrifuged at 4000 rpm for 10 min. The supernatant was cleaned up using SPE-C18. The SPE-C18 was conditioned by eluting 1.5 mL methanol and equilibrated by eluting 1 mL of 20 mM H_3PO_4 buffer twice before being used. A total of 1 milliliter of samples was loaded into the SPE C-18, and the eluate was discarded. The process was repeated twice using 2.5 mL samples, and the eluate was collected. Samples were filtered with a 0.2 μm syringe filter into a 2 mL vial before being injected into the UPLC-PDA system. The conditions and parameters of the instrument were as follows: stationary phase—C18 column (100 × 3.0 mm); mobile phase—20 mM H_3PO_4; flow rate—0.425 mL·min^{-1} under isocratic conditions; run time—10 min; injection volume 10 μL; needle wash—10% acetonitrile. Lactic and acetic acid quantification was determined using an external standard calibration curve.

2.9. Viable Cell, Titratable Acidity, and pH Assay

The viable cell, titratable acidity, and pH were determined by the method described by Yudianti et al. [15]. The viable cell of LAB was determined by serial dilution and pour plate method using MRS media containing 0.5% calcium carbonate ($CaCO_3$) and 1.5% bacteriological agar. Titratable acidity was measured by titrating 5 mL samples with 0.1 N sodium hydroxide (NaOH) using phenolphthalein (PP) as the indicator. Meanwhile, pH was examined using a pH meter (HANNA HI 2210, UK).

2.10. Preparation of Crude Phenolic Extract

The extractions of phenolic compounds in black soy milk were performed according to Ulyatu et al. [19] for phenolic content and antioxidant activity assay. A total of 2 milliliters of fermented black soy milk samples were extracted with 10 mL methanol (70%) using a water bath shaker (Sibata WS-240, Japan) at 120 rpm for 72 min at room temperature, followed by 24 h maceration at 4 °C in a dark room. Then, the extracts were centrifuged at 3000× *g* for 15 min, and the supernatant was filtered using Whatman paper no. 42. The natant was extracted again using the same method. Crude extract from the first and second extraction was mixed, and volume was measured and stored at −20 °C before being analyzed.

2.11. Folin-Ciocalteu Assay

Folin–Ciocalteu assay was analyzed colorimetrically using the Folin–Ciocalteu method described by Ulyatu et al. [19]. Crude extracts were diluted four times before analysis. Two milliliters of samples were added with one milliliter of Folin-Ciocalteu reagent. The sample was vortexed and allowed to stand for one minute. Then, four milliliters of 15% sodium carbonate (Na_2CO_3) were added, and the samples were vortexed again. Samples were allowed to stand in a dark room for 2 h at room temperature. The absorbance was measured at 760 nm using a UV-Vis spectrophotometer (Thermo Fisher Scientific Genesys 150, Waltham, MA, USA). Gallic acid was used as the standard, and methanol was used as blank. The results are expressed in mg Gallic Acid Equivalent (GAE) per 100 mL fermented black soy milk.

2.12. Antioxidant Activity Assay

The antioxidant activity of fermented black soy milk was determined colorimetrically by the ability of the extracts to scavenge the 2,2′-diphenyl-1-picrylhydrazyl (DPPH) radical described by Ulyatu et al. [19]. One milliliter of crude extracts was added with three milliliters of 0.1 mM DPPH solution and was incubated for 30 min in a dark room at room temperature. The absorbance was measured using the UV-Vis spectrophotometer at 515 nm. Methanol was used as a control, and the radical scavenging activity was determined using the following equation.

$$Radical\ Scavenging\ Activity\ (RSA) = \left(\frac{control\ absorbance - sampel\ absorbance}{control\ absorbance} \right) \times 100\% \qquad (1)$$

2.13. Isoflavone Aglycone Daidzein and Genistein Analysis

Determination of isoflavone aglycone daidzein and genistein was performed according to Sulistyowati et al. [20] using High-Performance Liquid Chromatography. Fermented black soy milk was freeze-dried before analysis. One gram of freeze-dried samples was extracted with 10 mL methanol (50%). Extraction was performed using an ultrasonic bath for 30 min. The samples were then centrifuged at 3000 rpm for 15 min, and the supernatant was filtered two times using Whatman paper no.1 and a syringe filter of 0.45 μm before being injected into HPLC systems. Samples were analyzed using HPLC (LC-20AD, Shimadzu, Japan) equipped with an autosampler (SIL-20A HT, Shimadzu, Japan), quaternary pump, PDA detector (CTO-20A, Shimadzu, Japan), degassing unit (DGU-20A SR, Shimadzu, Japan). The column used was Sun Fire TMC reverse phase C-18 column (150 mm × 4.6 mm, 5 μm). Meanwhile, the mobile phase consists of methanol (solvent A) and 0.1% acetic acid in water (solvent B). The flow rate was set isocratic at 1 mL·min^{-1} with the ratio of solvent A:B = 53:47 for 15 min at 30 °C. The detector was set at 254 nm, respectively. The quantity of daidzein and genistein was calculated based on a standard curve.

2.14. Determination of β-Glucosidase Activity

The β-glucosidase activity was determined colorimetrically by the rate of hydrolysis of substrate p-nitrophenyl-α-D-glucopyranoside (pNPG), according to Djafaar et al. [21]. Fermented black soy milk was centrifuged at 4000 rpm, 4 °C for 15 min, and the supernatant was used as the crude enzyme. Then, 500 μL crude enzyme was added into 1 mL of 5 mM pNPG prepared in a 100 mM sodium phosphate buffer (pH 7) and incubated at 37 °C for 30 min. One milliliter of cold sodium carbonate was added to stop the reaction. The absorbance was measured using a spectrophotometer UV-Vis at 401 nm, and p-nitrophenol was used as standard. One unit of the enzyme was defined as the amount of enzyme releasing one μmol of p-nitrophenol from the substrate p-NPG per min under assay conditions.

2.15. Data Analysis

The experiment was performed in two trials, each with two replicates of analysis. Data are expressed as mean ± standard deviation (SD). Data were analyzed using one-way ANOVA followed by Duncan's Multiple Range Test with a significance level of $p < 0.05$ and performed using IBM SPSS Statistic 20 software (IBM, Armonk, NY, USA) [22]. The p-values below 0.05 were considered statistically significant.

3. Results

3.1. Chemical Properties of Black Soy Milk

The chemical properties such as proximate, minerals, sugar profile, gallic acid equivalent, isoflavone aglycone, and amino acid composition of black soy milk were studied before fermentation was conducted. Table 2 shows the chemical properties of black soy milk. Black soy milk contains a high amount of protein and sucrose. This protein and sugar might be useful for LAB growth as nitrogen and carbon sources, respectively. Some minerals such as iron, zinc, magnesium, and manganese, which are needed for LAB growth, are also available in black soy milk. Moreover, black soy milk already contains isoflavone aglycone daidzein and genistein. Table 3 presents the amino acid profile of black soy milk. Glutamic acid, L-arginine, and L-aspartic acid were the major amino acids in black soy milk. Essential amino acid L-arginine was found higher compared to other essential amino acids. Moreover, essential amino acids L-phenylalanine and L-leucine were also found in an abundant amount.

Table 2. Chemical properties of black soy milk.

Parameters	Black Soy Milk
Moisture (%)	92.45 ± 0.03
Crude fat (g·100 mL^{-1})	0.18 ± 0.03
Crude protein (g·100 mL^{-1})	3.78 ± 0.05
Ash (g·100 mL^{-1})	0.41 ± 0.01
Iron (µg·mL^{-1})	6.11 ± 0.03
Zinc (µg·mL^{-1})	4.66 ± 0.02
Magnesium (µg·mL^{-1})	246.98 ± 0.04
Manganese (µg·mL^{-1})	2.98 ± 0.01
Carbohydrate (g·100 mL^{-1})	95.61 ± 0.02
Fructose (g·100 mL^{-1})	0.48 ± 0.01
Glucose (g·100 mL^{-1})	n.d. [1]
Sucrose (g·100 mL^{-1})	3.93 ± 0.01
Maltose (g·100 mL^{-1})	n.d. [1]
Lactose (g·100 mL^{-1})	n.d. [1]
Gallic acid equivalent (mg GAE·100 mL^{-1})	33.26 ± 3.55
Daidzein (µg·mL^{-1})	10.39 ± 0.07
Genistein (µg·mL^{-1})	6.14 ± 0.01

[1] n.d. = not detected. Limit of detection (LOD) = 0.06 g·100 mL^{-1}. Values are expressed as mean ± SD ($n = 4$).

Table 3. Amino acid composition of black soy milk.

Amino Acids	Concentration (g·100 mL^{-1})
L-Serine	0.38 ± 0.01
Glutamic acid	0.99 ± 0.01
L-phenylalanine	0.46 ± 0.01
L-Isoleucine	0.25 ± 0.01
L-Valine	0.25 ± 0.01
L-Alanine	0.21 ± 0.01
L-Arginine	0.60 ± 0.01
Glycine	0.28 ± 0.01
L-Lysine	0.27 ± 0.01
L-Aspartic acid	0.56 ± 0.01
L-Leucine	0.44 ± 0.01
L-Tyrosine	0.28 ± 0.01
L-Proline	0.28 ± 0.01
L-Threonine	0.28 ± 0.01
L-Histidine	0.22 ± 0.01
L-Cysteine	0.09 ± 0.01
L-Methionine	0.06 ± 0.01
L-Tryptophan	0.05 ± 0.01

Values are expressed as mean ± SD ($n = 2$).

3.2. Fermentation of Black Soy Milk by Indonesian Indigenous LAB

Black soy milk contains an abundant nutrient that has potential as a carbon and nitrogen source and growth factor for LAB growth. Figure 1 shows the growth and acid production of Indonesian LAB during black soy milk fermentation. All strains could grow well in black soy milk medium, reaching 9 log CFU·mL^{-1}. Both *L. plantarum* strains showed a similar growth pattern that reached the stationary phase after 18 h of fermentation. Meanwhile, *S. thermophilus* Dad 11 reached the stationary phase after 12 h of fermentation. The metabolic activities during LAB growth led to the production of acid. An increase in titratable acidity and decreased pH value indicated the acid production during fermentation of black soy milk. All strains exhibited an increase in titratable acidity throughout fermentation even though the pH value did not significantly increase after 18 h for *L. plantarum* WGK 4 and 12 h for *S. thermophilus* Dad 11 and *L. plantarum* Dad 13. *L. plantarum* WGK 4 produces the highest acid, 1.50%, compared to the other two strains.

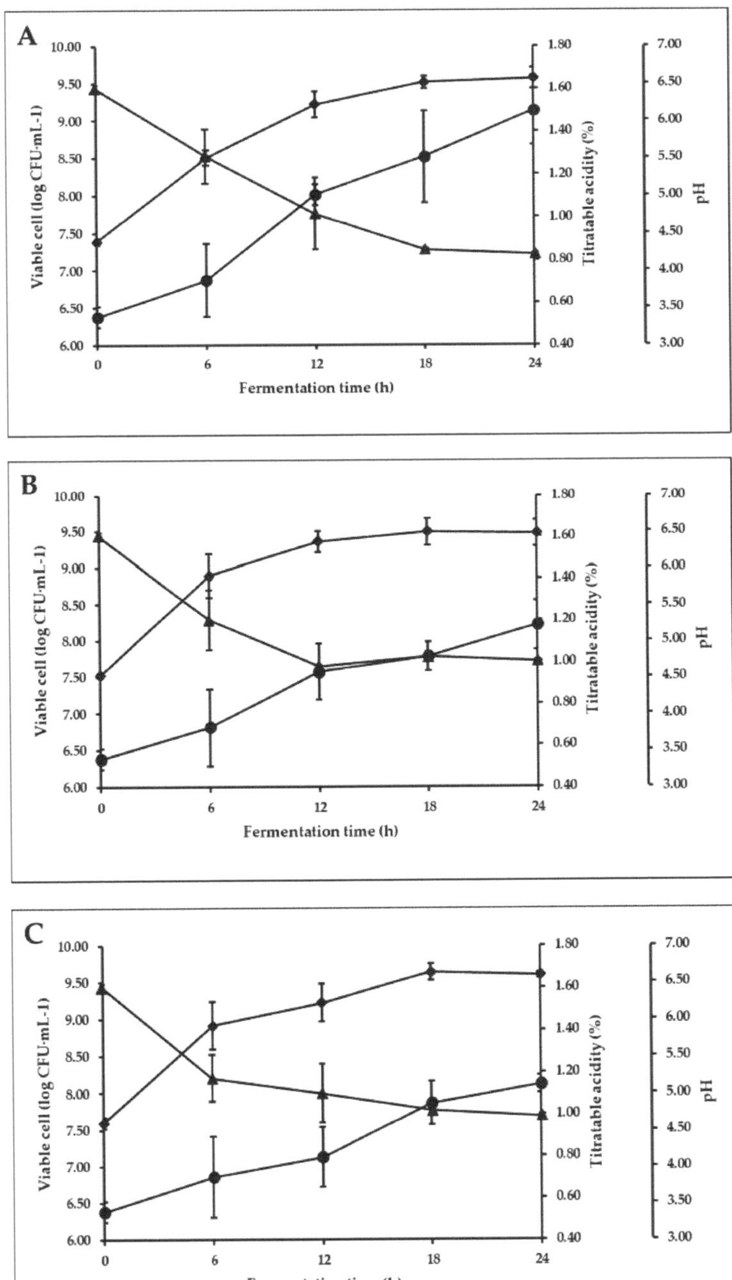

Figure 1. Growth and acid production (*n* = 4) of: (**A**) *L. plantarum* WGK 4, (**B**) *S. thermophilus* Dad 11, and (**C**) *L. plantarum* Dad 13 during fermentation of black soy milk (24 h, 37 °C).

The organic acid composition (lactic and acetic acid) was further analyzed during 18 h of fermentation of black soy milk since the pH value at 24 h of fermentation was not significantly different compared to the previous hour. Table 4 describes the lactic and acetic acid content in black soy milk during fermentation by Indonesian indigenous LAB. Lactic acid was the dominant acid produced. Acetic acid was not detected until 18 h of fermentation.

Table 4. Lactic and acetic acid of black soy milk during fermentation by Indonesian indigenous LAB (24 h, 37 °C).

Strains	Fermentation Time (h)	Lactic Acid (mg·100 mL^{-1})	Acetic Acid (mg·100 mL^{-1})
L. plantarum WGK 4	0	n.d. [1]	n.d. [1]
	6	0.20 ± 0.01	n.d. [1]
	12	0.72 ± 0.01	n.d. [1]
	18	1.14 ± 0.01	0.03 ± 0.01
S. thermophilus Dad 11	0	n.d. [1]	n.d. [1]
	6	0.40 ± 0.01	n.d. [1]
	12	0.73 ± 0.01	n.d. [1]
	18	0.93 ± 0.01	0.07 ± 0.01
L. plantarum Dad 13	0	n.d. [1]	n.d. [1]
	6	0.36 ± 0.01	n.d. [1]
	12	0.52 ± 0.01	n.d. [1]
	18	1.03 ± 0.01	0.05 ± 0.01

[1] n.d. = not detected. Limit of detection (LOD) = 0.41 µg·100 mL^{-1} (lactic acid) and 0.04 µg·100 mL^{-1} (acetic acid). Values are expressed as mean ± SD (n = 2).

One of the requirements for LAB to grow is sufficient carbon sources and the ability to utilize those carbon sources. Table 5 displays the sugar composition throughout black soy milk fermentation. A decrease in sucrose showed that Indonesian indigenous LAB could utilize this sugar. Sucrose was broken down into fructose and glucose. There was also a decrease in fructose content during fermentation. This points out that Indigenous LAB utilizes this sugar as a carbon source for their growth and metabolic activities. Glucose was mainly not detected during fermentation. It might be because the LAB rapidly consumed it.

Table 5. Sugars profile of black soy milk during fermentation by Indonesian indigenous LAB (18 h, 37 °C).

Strains	Fermentation Time (h)	Fructose (g·100 mL^{-1})	Glucose (g·100 mL^{-1})	Sucrose (g·100 mL^{-1})	Maltose (g·100 mL^{-1})	Lactose (g·100 mL^{-1})
L. plantarum WGK 4	0	0.48 ± 0.01	n.d. [1]	3.94 0.01	n.d. [1]	n.d. [1]
	6	0.57 ± 0.01	0.70 ± 0.01	2.18 ± 0.01	n.d. [1]	n.d. [1]
	12	0.32 ± 0.01	n.d. [1]	0.54 ± 0.01	n.d. [1]	n.d. [1]
	18	n.d. [1]	n.d. [1]	n.d. [1]	n.d. [1]	n.d. [1]
S. thermophilus Dad 11	0	0.48 ± 0.01	n.d. [1]	3.94 ± 0.01	n.d. [1]	n.d. [1]
	6	0.43 ± 0.01	n.d. [1]	1.71 ± 0.01	n.d. [1]	n.d. [1]
	12	n.d. [1]	n.d. [1]	0.46 ± 0.01	n.d. [1]	n.d. [1]
	18	n.d. [1]	n.d. [1]	n.d. [1]	n.d. [1]	n.d. [1]
L. plantarum Dad 13	0	0.48 ± 0.01	n.d. [1]	3.94 ± 0.01	n.d. [1]	n.d. [1]
	6	n.d. [1]	n.d. [1]	1.69 ± 0.01	n.d. [1]	n.d. [1]
	12	0.28 ± 0.01	n.d. [1]	0.73 ± 0.01	n.d. [1]	n.d. [1]
	18	n.d. [1]	n.d. [1]	n.d. [1]	n.d. [1]	n.d. [1]

[1] n.d. = not detected. Limit of detection (LOD) = 0.06 g·100 mL^{-1}. Values are expressed as mean ± SD (n = 2).

3.3. Antioxidant Properties of Black Soy Milk during Fermentation by Indonesian Indigenous LAB

Plant-based milk, such as black soy milk, is a primary source of phenolic compounds. Phenolic compounds are known to have beneficial effects, such as antioxidants. Table 6 shows the Folin–Ciocalteu assay and antioxidant activity of fermented black soy milk. Black soy milk fermented with either *L. plantarum* WGK 4 or Dad 13 significantly increased the gallic acid equivalent until 18 h of fermentation. It decreased after 24 h of fermentation, while *S. thermophilus* Dad 11 exhibited increased gallic acid equivalent content until 24 h of fermentation but not as much as two other strains. Nevertheless, for antioxidant activity, fermented black soy milk by *L. plantarum* WGK 4 and *S. thermophilus* Dad 11 showed a similar pattern in enhancing antioxidants, while *L. plantarum* Dad 13 exhibited increased DPPH-scavenging activity until 18 h but decreased afterward.

Table 6. Folin–Ciocalteu assay and DPPH-scavenging activity of black soy milk during fermentation by Indonesian indigenous LAB (24 h, 37 °C).

Strains	Fermentation Time (h)	Folin–Ciocalteu Assay (mg GAE·mL^{-1})	DPPH-Scavenging Activity (%)
L. plantarum WGK 4	0	33.26 ± 3.55 [a]	24.90 ± 2.50 [a]
	6	32.04 ± 0.84 [a]	30.37 ± 1.30 [b]
	12	40.17 ± 3.05 [b]	29.51 ± 0.26 [b]
	18	44.44 ± 2.56 [c]	36.37 ± 0.59 [c]
	24	38.27 ± 1.89 [b]	38.20 ± 0.55 [c]
S. thermophilus Dad 11	0	33.26 ± 3.55 [ab]	24.90 ± 2.50 [ab]
	6	30.17 ± 0.37 [a]	24.11 ± 0.72 [ab]
	12	31.25 ± 0.89 [a]	22.83 ± 2.35 [a]
	18	33.70 ± 2.25 [ab]	27.15 ± 1.56 [b]
	24	36.11 ± 4.38 [b]	32.04 ± 3.44 [c]
L. plantarum Dad 13	0	33.98 ± 3.97 [a]	24.90 ± 2.50 [a]
	6	32.82 ± 0.70 [a]	25.74 ± 0.78 [a]
	12	32.41 ± 0.82 [a]	29.27 ± 2.44 [b]
	18	37.66 ± 1.03 [b]	36.87 ± 0.68 [c]
	24	32.20 ± 2.61 [a]	31.22 ± 2.93 [b]

Values are expressed as mean ± SD (n = 4). Values of each strain with different superscripts ([a,b,c]) are significantly different ($p < 0.05$) by Duncan's multiple range test.

3.4. Isoflavone Aglycone Liberation throughout Black Soy Milk Fermentation by Indonesian Indigenous LAB

The increased antioxidant activity of black soy milk during fermentation might be due to the liberation of isoflavone aglycone from its glucoside form in β-glucosidase. The current results showed an increased daidzein and genistein concentration in black soy milk during fermentation by all LAB strains, along with an increase in β-glucosidase activity (Table 7). Daidzein concentration was increased by about 61%, 107%, and 103% for black soy milk fermented with *L. plantarum* WGK 4, *S. thermophilus* Dad 11, and *L. plantarum* Dad 13, respectively. Meanwhile, genistein concentration was increased by about 81%, 132%, and 108% for black soy milk fermented with *L. plantarum* 4, *S. thermophilus* Dad 11, and *L. plantarum* Dad 13, respectively. Nevertheless, isoflavone aglycone decreased after 24 h fermentation, even though the β-glucosidase activity still increased in black soy milk fermented with *S. thermophilus* Dad 11.

Table 7. Isoflavone aglycone (daidzein and genistein) and β-glucosidase activity of black soy milk during fermentation by Indonesian indigenous LAB (24 h, 37 °C).

Strains	Fermentation Time (h)	Daidzein (µg·mL^{-1})	Genistein (µg·mL^{-1})	β-Glucosidase Activity (mU·mL^{-1})
L. plantarum WGK 4	0	10.39 ± 0.07 [a]	6.14 ± 0.01 [a]	-
	6	13.97 ± 0.67 [b]	7.08 ± 0.01 [b]	25.28 ± 0.40 [a]
	12	16.04 ± 0.01 [cd]	9.72 ± 0.02 [c]	27.43 ± 3.44 [a]
	18	15.81 ± 0.05 [c]	10.56 ± 0.01 [d]	24.09 ± 1.86 [a]
	24	16.69 ± 0.12 [d]	11.03 ± 0.01 [e]	32.57 ± 1.34 [b]
S. thermophilus Dad 11	0	10.39 ± 0.07 [a]	6.14 ± 0.01 [a]	-
	6	17.03 ± 0.09 [b]	10.21 ± 0.07 [b]	23.62 ± 2.90 [a]
	12	20.34 ± 0.01 [d]	13.14 ± 0.06 [d]	36.69 ± 5.18 [b]
	18	21.55 ± 0.11 [e]	14.24 ± 0.09 [e]	34.84 ± 1.41 [b]
	24	19.48 ± 0.01 [c]	11.40 ± 0.01 [c]	48.32 ± 1.04 [c]
L. plantarum Dad 13	0	10.39 ± 0.07 [a]	6.14 ± 0.01 [a]	-
	6	16.94 ± 0.65 [b]	10.62 ± 0.02 [b]	25.18 ± 5.81 [a]
	12	18.77 ± 0.09 [c]	11.30 ± 0.02 [c]	36.55 ± 3.03 [b]
	18	20.09 ± 0.04 [d]	12.75 ± 0.01 [e]	45.05 ± 4.53 [c]
	24	18.89 ± 0.02 [c]	12.20 ± 0.10 [d]	35.37 ± 1.00 [b]

Values are expressed as mean ± SD (n = 4). Values of each strain with different superscripts ([a,b,c,d,e]) are significantly different ($p < 0.05$) by Duncan's multiple range test.

4. Discussion

The main aim of the current study was to investigate the ability of Indonesian LAB to enhance the antioxidant properties of black soy milk during fermentation. The enhancement of the antioxidant activities through fermentation can be developed into a functional food product to diversify black soybean utilization. The current LAB needs to grow well in a black soy milk medium to enhance its antioxidant properties. Lactic acid bacteria growth depends on the nutrients available in the medium and the LAB's ability to utilize those nutrients [23].

Carbon source is one of the primary nutrients needed for LAB growth. Generally, the major sugar in soybean is sucrose [24]. This finding is in agreement with the current study. Moreover, sucrose concentration in black soy milk is higher than glucose concentration in the MRS medium, which is 20 g·L^{-1}. LAB could use sucrose as a carbon source, supported by the decreased sucrose concentration during fermentation of black soy milk (Table 6). Additionally, Baú et al. [25] also found that stachyose was the second major sugar found in soy milk. Meanwhile, the major sugar in hazelnut milk and *kerandang* milk was glucose [26] and raffinose [27], respectively. It is possible that LAB could also utilize these sugars after utilizing simple sugars such as sucrose. Singh et al. [28] stated that besides sucrose, *L. plantarum* C6 could utilize other sugars, such as stachyose and raffinose, as a substrate for its growth by producing α-galactosidase. Moreover, the reduction of sucrose was faster than stachyose and raffinose. In addition, Baú et al. [25] found that sucrose was reduced at an early stage of fermentation, while stachyose and raffinose were decreased after the sucrose was not detected. In addition to carbon sources, nitrogen sources are also an essential nutrient for LAB growth.

This study also found that black soy milk contains a high protein compared to other milk. Chalupa-Krebzdak et al. [29] reported that the protein content in whole bovine milk, soy milk, coconut milk, and cashew milk is 3.15%, 2.28%, 1.28%, and 1.31%, respectively. Additionally, Kundu et al. [30] found that the protein content of almond milk was 1.308%. Lactic acid bacteria might utilize small peptides and free amino acids as a nitrogen source. Singh et al. [28] mentioned that several lactobacilli have proteolytic activity by producing extracellular proteinase to hydrolyze soy protein in soy milk. Moreover, Boulay et al. [31] stated that *S. thermophilus* LMD-9 could hydrolyze soy milk protein, and the cell wall protease PrtS plays an important role. Additionally, LAB also needs some specific amino acids for their growth. Arginine, leucine, valine, glutamic acid, and cysteine are needed for *L. plantarum* growth; meanwhile, cysteine or methionine, and histidine are needed for *S. thermophilus* growth [32]. This study showed that black soy milk contained those amino acids, especially glutamic acid and arginine. For its metabolic activity, some LAB needs minerals for the enzymatic reaction. The minerals usually required are Fe, Mg, Mn, and Zn, but this requirement is basically strain-dependent [33]. This study found that black soy milk has these minerals. Furthermore, the Mg and Mn concentrations in black soy milk were on par with the MRS medium. The findings indicate that the nutrient in black soy milk might be enough for Indonesian indigenous LAB growth.

Indonesian indigenous LAB demonstrated good growth in black soy milk medium. Results showed that *S. thermophilus* Dad 11 was on par with the viable cell of *S. thermopilus* S10, which increased 2 log cycles, reaching 9–9.5 log CFU·mL^{-1}, after 24 h fermentation of black soy milk, reported by Lee et al. [12]. Even the current study strain, *S. thermophilus* Dad 11, has a higher viable cell in fermented black soy milk (Figure 1) compared to the viable cell of *S. thermophilus* ATCC BAA-250 after 24 h of soy milk fermentation, which was 8.58 log CFU·mL^{-1}, demonstrated by Liu et al. [34]. For *L. plantarum*, the results (Figure 1) were higher than the viable cell of *L. plantarum* 70810, 8.57 log CFU·mL^{-1}, after fermenting soy milk for 12 h [35] and on par but reached faster with the cell counts of *L. plantarum* LMG6940 during fermentation of soy milk, which was increased from 7.63 log CFU·mL^{-1} to 9.52 log CFU·mL^{-1} after 48 h [36]. Additionally, *L. plantarum* WGK 4 showed higher black soy milk growth than jack bean milk, which only increased 1 log cycle after 24 h of fermentation at 37 °C [15]. Furthermore, *L. plantarum* Dad 13 and *S. thermophilus* Dad 11 showed better growth in black soy milk than in skimmed milk, which only increased 1 log cycle after 24 h of fermentation at 37 °C [14].

Not only do they grow well, but Indonesian indigenous LAB also showed good metabolic activity, proven by the acid production. These LAB strains have a higher acidification rate than previous studies, especially *L. plantarum* WGK 4 (Figure 1). Kuda et al. [37] developed several *L. plantarum* strains from plants of the coastal Satoumi regions as starter culture in soy milk fermentation. The pH and lactic acid concentrations range around 5.1–5.4 and 3.0–5.7 mg·mL^{-1}, respectively. Kim et al. [38] introduced soy yogurt fermented by *Leuconostoc mesenteroides* and *L. plantarum* from kimchi, wherein the pH and titratable acidity after 24 h were around 4.5% and 0.8%. Moreover, Liu et al. [34] demonstrated the acidification rate of *S. thermophilus* ATCC BAA-250 during soy milk fermentation. The pH did not decrease much at 12 to 24 h, around 5.22–5.18. Wardani et al. [16] found that milk's titratable acidity and pH after fermentation using *L. plantarum* Dad 13 at 37 °C were 0.46% and 5.27, respectively. This research found that *L. plantarum* WGK 4 has higher acid production than *S. thermophilus* Dad 11 and *L. plantarum* Dad 13 (Figure 1). It might be because *S. thermophilus* Dad 11 and *L. plantarum* Dad 13 were isolated from Dadih (Indonesian fermented buffalo milk). Therefore, the ability to utilize the nutrient in plant-based milk was lower than *L. plantarum* WGK 4, isolated from red lima bean soaking water in tempeh production. Indonesian indigenous LAB showed better growth and acid production due to the nutrients available in black soy milk. In addition to utilizing the available sugars in the black soybean milk, there is also a possibility that LAB used glucose moiety in isoflavone glucoside and, therefore, might enhance the antioxidant properties.

The gallic acid equivalent and antioxidant activity increased during fermentation of black soy milk by Indonesian LAB. Consistent with current results, Ulyatu et al. [19] also noted an increase in gallic acid equivalent content and DPPH-scavenging activity during fermentation of sesame milk using *L. plantarum* Dad 13. A similar increment of both gallic acid equivalent content and antioxidant activity was described by Lee et al. [12] in fermented black soy milk by *S. thermophilus* S10. Gallic acid equivalent content and DPPH scavenging activity of black soy milk increased significantly throughout 24 h fermentation using single culture of *S. thermophilus* S10. Moreover, the current study showed higher DPPH-scavenging activity than the finding of Zhao et al. [11], which reported that soy milk fermented with *L. acidophilus* CSCC 2400, *L. paracasei* CSCC 279, *L. zeae* ASCC 15820, and *L. rhamnosus* WQ2 until the pH reached 4.55 have DPPH-scavenging activity around 15–20%. Salar et al. [39] stated that lactic acid bacteria could produce β-glucosidase during fermentation. This enzyme can hydrolyze glucoside phenolic into free phenols, thus, increasing the gallic acid equivalent content of black soy milk. Moreover, Fitrotin [13] mentioned that during sesame milk fermentation, lactic acid bacteria could produce β-glucosidase to hydrolyze isoflavone glucoside into its aglycone form. Aglycone phenolic compound has more reactivity to reduce both phosphomolybdate phosphotungstate complex in Folin–Ciocalteu reagent and DPPH radical due to the more available hydroxyl groups; therefore, increasing the gallic acid equivalent content and antioxidant activity.

Nevertheless, this study also discovered a decrease in gallic acid equivalent and antioxidant activity at 24 h of fermentation in some strains. The gallic acid equivalent content of black soy milk after 24 h of fermentation by *L. plantarum* WGK 4 and *L. plantarum* Dad 13 exhibited a lower concentration than 18 h of fermentation. Phenolic compounds have been reported to have an inhibitory effect on LAB, and *L. plantarum* possessed the ability to degrade those phenolic compounds as a stress response [40]. It is possible that at 24 h of fermentation, *L. plantarum* strains shifted to degrade phenolic compounds due to stress response. Along with the decreased gallic acid equivalent, this study found a decreased DPPH-scavenging activity for black soy milk fermented by *L. plantarum* Dad 13 but not for *L. plantarum* WGK 4. It points out the possibility of other antioxidant compounds released during black soy milk fermentation, such as bioactive peptides. Yusuf et al. [41] discovered an increased antioxidant activity of 10 *Lactobacillus* spp. isolated from Indonesian kefir grains due to the liberation of the bioactive peptide by proteolytic activities during milk fermentation.

Fermentation by Indonesian indigenous LAB enhances the antioxidant properties of black soy milk. This study demonstrated increased isoflavone aglycone daidzein and genistein concentration and β-glucosidase activity. Lee et al. [42] also discovered an increased daidzein and genistein concentration of several cultivars of fermented soybean powder hydrolysate milk by *L. plantarum* P1201 for 48 h at 35 °C. Current findings were higher than the results reported by Hati et al. [7], which found the daidzein and genistein concentration in soy milk fermented with several *L. rhamnosus* and *L. casei* strain for 12 h at 37 °C was around 0.81–0.98 mg·100 mL^{-1} and 1.13–1.93 mg·100 mL^{-1}, respectively. They also stated that the β-glucosidase activity was varied among strains which corresponds to these findings. The results were also higher than the daidzein and genistein concentration in soy milk fermented with *L. plantarum* for 24 h, 0.423 mg·100 mL^{-1} and 0.753 mg·100 mL^{-1}, respectively [10]. Moreover, Lee et al. [12] also reported increased β-glucosidase activity, around 60 UA·g^{-1}, along with increased daidzein and genistein content, reaching 119.39 mg·100 g^{-1} and 191.38 mg·100 g^{-1}, respectively, during fermentation of black soy milk by *S. thermophilus* S10 at 37 °C. The β-glucosidase activity in fermented black soy milk by indigenous LAB was higher than the β-glucosidase activity in *kerandang* extract for *L. plantarum* T33, which was 20 mU·mL culture^{-1}, but lower than the β-glucosidase activity from *L. plantarum-pentosus*, T14 558 mU·mL culture^{-1} [21]. Nevertheless, the current study also found a decrease in isoflavone aglycone concentration at 24 h fermentation even though the β-glucosidase activity was still increasing in black soy milk fermented by *S. thermophilus* Dad 11. Cairns et al. [43] mentioned that β-glucosidase also catalyzes glycosylation reaction or reverse hydrolysis. Thus, reverse conversion might happen. The current study demonstrated that

Indonesian indigenous LAB could produce β-glucosidase to break down isoflavone glucoside into isoflavone aglycone and glucose. Glucose was utilized as a carbon source for LAB's growth and metabolic activity. Meanwhile, the isoflavone aglycone enhances antioxidant properties.

5. Conclusions

Indonesian indigenous LAB, *L. plantarum* WGK 4, *S. thermophilus* Dad 11, and *L. plantarum* Dad 13 grow well in black soy milk medium. These LAB could utilize the nutrient contained in black soy milk for its growth and metabolic activity. Additionally, they could produce β-glucosidase to hydrolyze isoflavone glucoside into glucose and isoflavone aglycone daidzein and genistein. Thus, it enhanced the antioxidant properties, such as gallic acid equivalent content and DPPH-scavenging activity, of black soy milk. It points out that Indonesian indigenous LAB has a promising potential to be used as a starter culture to develop fermented non-dairy milk, such as black soy milk, as functional food products. Further studies are necessary to investigate other antioxidant properties and other antioxidant compounds that might be released during fermentation. Product formulation and sensory evaluation, including flavor, taste, and physical properties, are also needed to develop fermented black soy milk into a functional food product.

Author Contributions: Conceptualization, M.N.C., E.S.R., R.Y. and T.U.; methodology, B.Y.L., M.N.C. and T.U.; software, B.Y.L.; validation, M.N.C., E.S.R., R.Y. and T.U.; formal analysis, B.Y.L.; investigation, B.Y.L., M.N.C. and T.U.; resources, M.N.C., E.S.R., R.Y. and T.U.; data curation, B.Y.L.; writing—original draft preparation, B.Y.L. and T.U.; writing—review and editing, M.N.C., E.S.R. and R.Y.; visualization, B.Y.L. and T.U.; supervision, M.N.C., E.S.R., R.Y., T.U.; project administration, T.U.; funding acquisition, T.U. All authors have read and agreed to the published version of the manuscript.

Funding: This research was funded by the Indonesia Ministry of Research and Higher Education (RISTEKDIKTI) through a competition-based research program 2020, grant number 2937/UN1.DITLIT/DITLIT/PT/2020. The APC was funded by Universitas Gadjah Mada.

Institutional Review Board Statement: Not applicable.

Informed Consent Statement: Not applicable.

Data Availability Statement: The data presented in this study are available on request from the corresponding author.

Acknowledgments: The authors acknowledge Saraswanti Indo Genetech, Bogor, Indonesia, Pri Haryono from the Center for Food and Nutrition Studies, Universitas Gadjah Mada, Indonesia, and Ashari from Laboratory of Chemical and Biochemical of Food and Agricultural Products, Department of Food and Agricultural Product Technology, Faculty of Agricultural Technology, Gadjah Mada University, Indonesia for providing technical assistant.

Conflicts of Interest: The authors declare no conflict of interest.

References

1. Kim, M.Y.; Jang, G.Y.; Oh, N.S.; Baek, S.Y.; Lee, S.H.; Kim, K.M.; Kim, T.M.; Lee, J.; Jeong, H.S. Characteristics and in Vitro Anti-Inflammatory Activities of Protein Extracts from Pre-Germinated Black Soybean [*Glycine Max* (L.)] Treated with High Hydrostatic Pressure. *Innov. Food Sci. Emerg. Technol.* **2017**, *43*, 84–91. [CrossRef]
2. Lee, J.H.; Cho, K.M. Changes Occurring in Compositional Components of Black Soybeans Maintained at Room Temperature for Different Storage Periods. *Food Chem.* **2012**, *131*, 161–169. [CrossRef]
3. Kim, G.P.; Lee, J.; Ahn, K.G.; Hwang, Y.S.; Choi, Y.; Chun, J.; Chang, W.S.; Choung, M.G. Differential Responses of B Vitamins in Black Soybean Seeds. *Food Chem.* **2014**, *153*, 101–108. [CrossRef]
4. Joshi, J.; Rahal, A. Effect of Processing Method on Nutritional Quality of Black Soybean. *Asian J. Dairy Food Res.* **2018**, *37*, 162–164. [CrossRef]
5. Widowati, W.; Prahastuti, S.; Hidayat, M.; Hasianna, S.T.; Wahyudianingsih, R.; Eltania, T.F.; Azizah, A.M.; Aviani, J.K.; Subangkit, M.; Handayani, R.A.S.; et al. Detam 1 Black Soybean against Cisplatin-Induced Acute Ren Failure on Rat Model via Antioxidant, Antiinflammatory and Antiapoptosis Potential. *J. Tradit. Complement. Med.* **2022**, *12*, 426–435. [CrossRef]
6. Huang, G.; Cai, W.; Xu, B. Improvement in Beta-Carotene, Vitamin B2, GABA, Free Amino Acids and Isoflavones in Yellow and Black Soybeans upon Germination. *LWT Food Sci. Technol.* **2017**, *75*, 488–496. [CrossRef]
7. Hati, S.; Vij, S.; Singh, B.P.; Mandal, S. β-Glucosidase Activity and Bioconversion of Isoflavones during Fermentation of Soy milk. *J. Sci. Food Agric.* **2015**, *95*, 216–220. [CrossRef] [PubMed]

8. Hsiao, Y.H.; Ho, C.T.; Pan, M.H. Bioavailability and Health Benefits of Major Isoflavone Aglycones and Their Metabolites. *J. Funct. Foods* **2020**, *74*, 104164. [CrossRef]

9. Zuo, A.; Yanying, Y.; Li, J.; Binbin, X.; Xiongying, Y.; Yan, Q.; Shuwen, C. Study on the Relation of Structure and Antioxidant Activity of Isorhamnetin, Quercetin, Phloretin, Silybin and Phloretin Isonicotinyl Hydrazone. *Free Radic. Antioxid.* **2011**, *1*, 39–47. [CrossRef]

10. Ding, W.K.; Shah, N.P. Enhancing the Biotransformation of Isoflavones in Soy milk Supplemented with Lactose Using Probiotic Bacteria during Extended Fermentation. *J. Food Sci.* **2010**, *75*, 140–149. [CrossRef]

11. Zhao, D.; Shah, N.P. Changes in Antioxidant Capacity, Isoflavone Profile, Phenolic and Vitamin Contents in Soy milk during Extended Fermentation. *LWT Food Sci. Technol.* **2014**, *58*, 454–462. [CrossRef]

12. Lee, M.; Hong, G.E.; Zhang, H.; Yang, C.Y.; Han, K.H.; Mandal, P.K.; Lee, C.H. Production of the Isoflavone Aglycone and Antioxidant Activities in Black soy milk Using Fermentation with *Streptococcus thermophilus* S10. *Food Sci. Biotechnol.* **2015**, *24*, 537–544. [CrossRef]

13. Fitrotin, U.; Utami, T.; Hastuti, P.; Santoso, U. Antioxidant Properties of Fermented Sesame Milk Using *Lactobacillus plantarum* Dad 13. *Int. Res. J. Biol.* **2015**, *4*, 56–61.

14. Utami, T.; Cindarbhumi, A.; Khuangga, M.C.; Rahayu, E.S.; Cahyanto, M.N.; Nurfiyani, S.; Zulaichah, E. Preparation of Indigenous Lactic Acid Bacteria Starter Cultures for Large Scale Production of Fermented Milk. In Proceedings of the 10th Asian Conference of Lactic Acid Bacteria, Yogyakarta, Indonesia, 10 February 2020.

15. Yudianti, N.F.; Yanti, R.; Cahyanto, M.N.; Rahayu, E.S.; Utami, T. Isolation and Characterization of Lactic Acid Bacteria from Legume Soaking Water of Tempeh Productions. In Proceedings of the 10th Asian Conference of Lactic Acid Bacteria, Yogyakarta, Indonesia, 10 February 2020.

16. Wardani, S.K.; Cahyanto, M.N.; Rahayu, E.S.; Utami, T. The Effect of Inoculum Size and Incubation Temperature on Cell Growth, Acid Production and Curd Formation during Milk Fermentation by *Lactobacillus plantarum Dad 13*. *Int. Food Res. J.* **2017**, *24*, 921–926.

17. AOAC International. *Official Methods of Analysis*, 21st ed.; AOAC International: Rockville, MD, USA, 2019.

18. Wasik, A.; McCourt, J.; Buchgraber, M. Simultaneous Determination of Nine Intense Sweeteners in Foodstuffs by High Performance Liquid Chromatography and Evaporative Light Scattering Detection-Development and Single-Laboratory Validation. *J. Chromatogr. A* **2007**, *1157*, 187–196. [CrossRef]

19. Ulyatu, F.; Pudji, H.; Tyas, U.; Umar, S. The Changes of Sesaminol Triglucoside and Antioxidant Properties during Fermentation of Sesame Milk by *Lactobacillus plantarum* Dad 13. *Int. Food Res. J.* **2015**, *22*, 1945–1952.

20. Sulistyowati, E.; Martono, S.; Riyanto, S.; Lukitaningsih, E. Development and Validation for Free Aglycones Daidzein and Genistein in Soybeans (*Glycine Max* (l.) Merr.) Using RP HPLC Method. *Int. J. Appl. Pharm.* **2019**, *11*, 138–142. [CrossRef]

21. Djafaar, T.F.; Santoso, U.; Cahyanto, M.N.; Takuya, S.; Rahayu, E.S.; Kosuke, N. Effect of Indigenous Lactic Acid Bacteria Fermentation on Enrichment of Isoflavone and Antioxidant Properties of Kerandang *(Canavalia Virosa)* Extract. *Int. Food Res. J.* **2013**, *20*, 2945–2950.

22. IBM SPSS Statistic 20. Available online: https://www.ibm.com/support/pages/downloading-ibm-spss-statistics-20 (accessed on 10 March 2021).

23. Berisvil, A.P.; Astesana, D.M.; Zimmermann, J.A.; Frizzo, L.S.; Rossler, E.; Romero-Scharpen, A.; Olivero, C.R.; Zbrun, M.V.; Signorini, M.L.; Sequeira, G.J.; et al. Low-Cost Culture Medium for Biomass Production of Lactic Acid Bacteria with Probiotic Potential Destined to Broilers. *FAVE Secc. Cienc. Vet.* **2020**, *20*, 1–9. [CrossRef]

24. Zheng, Y.; Li, L.; Jin, Z.; An, P.; Yang, S.T.; Fei, Y.; Liu, G. Characterization of Fermented Soy milk by Schleiferilactobacillus Harbinensis M1, Based on the Whole-Genome Sequence and Corresponding Phenotypes. *LWT Food Sci. Technol.* **2021**, *144*, 111237. [CrossRef]

25. Baú, T.R.; Garcia, S.; Ida, E.I. Changes in Soy milk during Fermentation with Kefir Culture: Oligosaccharides Hydrolysis and Isoflavone Aglycone Production. *Int. J. Food Sci. Nutr.* **2015**, *66*, 845–850. [CrossRef] [PubMed]

26. Bernat, N.; Cháfer, M.; Chiralt, A.; González-Martínez, C. Hazelnut Milk Fermentation Using Probiotic Lactobacillus Rhamnosus GG and Inulin. *Int. J. Food Sci. Technol.* **2014**, *49*, 2553–2562. [CrossRef]

27. Djaafar, T.F.; Cahyanto, M.N.; Santoso, U.; Rahayu, E.S. Growth of Indigenous Lactic Acid Bacteria *Lactobacillus plantarum-pentosus* T14 and *Lactobacillus plantarum-pentosus* T35 in Kerandang (*Canavalia Virosa*) Milk and Changes of Raffinose. *Malays. J. Microbiol.* **2013**, *9*, 213–218.

28. Singh, B.P.; Vij, S. α-Galactosidase Activity and Oligosaccharides Reduction Pattern of Indigenous Lactobacilli during Fermentation of Soy Milk. *Food Biosci.* **2018**, *22*, 32–37. [CrossRef]

29. Chalupa-Krebzdak, S.; Long, C.J.; Bohrer, B.M. Nutrient Density and Nutritional Value of Milk and Plant-Based Milk Alternatives. *Int. Dairy J.* **2018**, *87*, 84–92. [CrossRef]

30. Kundu, P.; Dhankhar, J.; Sharma, A. Development of Non Dairy Milk Alternative Using Soy milk and Almond Milk. *Curr. Res. Nutr. Food Sci.* **2018**, *6*, 203–210. [CrossRef]

31. Boulay, M.; al Haddad, M.; Rul, F. *Streptococcus thermophilus* Growth in Soya Milk: Sucrose Consumption, Nitrogen Metabolism, Soya Protein Hydrolysis and Role of the Cell-Wall Protease PrtS. *Int. J. Food Microbiol.* **2020**, *335*, 108903. [CrossRef]

32. Teusink, B.; Molenaar, D. Systems Biology of Lactic Acid Bacteria: For Food and Thought. *Curr. Opin. Syst. Biol.* **2017**, *6*, 7–13. [CrossRef]

33. Endo, A.; Dicks, L.M.T. Physiology of the LAB. In *Lactic Acid Bacteria: Biodiversity and Taxonomy*; Holzapfel, W.H., Wood, B.J.B., Eds.; John Wiley & Sons, Ltd.: Chichester, UK, 2014; pp. 13–20.
34. Liu, W.S.; Yang, C.Y.; Fang, T.J. Strategic Ultrasound-Induced Stress Response of Lactic Acid Bacteria on Enhancement of β-Glucosidase Activity for Bioconversion of Isoflavones in Soy milk. *J. Microbiol. Methods* **2018**, *148*, 145–150. [CrossRef]
35. Li, C.; Li, W.; Chen, X.; Feng, M.; Rui, X.; Jiang, M.; Dong, M. Microbiological, Physicochemical and Rheological Properties of Fermented Soy milk Produced with Exopolysaccharide (EPS) Producing Lactic Acid Bacteria Strains. *LWT Food Sci. Technol.* **2014**, *57*, 477–485.
36. Hong LE, P.; Parmentier, N.; Trung LE, T.; Raes, K. Evaluation of Using a Combination of Enzymatic Hydrolysis and Lactic Acid Fermentation for γ-Aminobutyric Acid Production from Soy milk. *LWT Food Sci. Technol.* **2021**, *142*, 111044. [CrossRef]
37. Kuda, T.; Kataoka, M.; Nemoto, M.; Kawahara, M.; Takahashi, H.; Kimura, B. Isolation of Lactic Acid Bacteria from Plants of the Coastal Satoumi Regions for Use as Starter Cultures in Fermented Milk and Soy milk Production. *LWT Food Sci. Technol..* **2016**, *68*, 202–207. [CrossRef]
38. Kim, H.J.; Han, M.J. The Fermentation Characteristics of Soy Yogurt with Different Content of D-Allulose and Sucrose Fermented by Lactic Acid Bacteria from Kimchi. *Food Sci. Biotechnol.* **2019**, *28*, 1155–1161. [CrossRef]
39. Salar, R.K.; Certik, M.; Brezova, V. Modulation of Phenolic Content and Antioxidant Activity of Maize by Solid State Fermentation with Thamnidium Elegans CCF 1456. *Biotechnol. Bioprocess Eng.* **2012**, *17*, 109–116. [CrossRef]
40. Rodríguez, H.; Curiel, J.A.; Landete, J.M.; de las Rivas, B.; de Felipe, F.L.; Gómez-Cordovés, C.; Mancheño, J.M.; Muñoz, R. Food Phenolics and Lactic Acid Bacteria. *Int. J. Food Microbiol.* **2009**, *132*, 79–90. [CrossRef] [PubMed]
41. Yusuf, D.; Nuraida, L.; Dewanti-Hariyadi, R.; Hunaef, D. In Vitro Antioxidant and α-Glucosidase Inhibitory Activities of Lactobacillus Spp. Isolated from Indonesian Kefir Grains. *Appl. Food Biotechnol.* **2021**, *8*, 39–46.
42. Lee, J.H.; Kim, B.; Hwang, C.E.; Haque, M.A.; Kim, S.C.; Lee, C.S.; Kang, S.S.; Cho, K.M.; Lee, D.H. Changes in Conjugated Linoleic Acid and Isoflavone Contents from Fermented Soy milks Using Lactobacillus Plantarum P1201 and Screening for Their Digestive Enzyme Inhibition and Antioxidant Properties. *J. Funct. Foods* **2018**, *43*, 17–28. [CrossRef]
43. Cairns, J.R.K.; Esen, A. β-Glucosidases. *Cell. Mol. Life Sci.* **2010**, *67*, 3389–3405. [CrossRef]

fermentation 

MDPI

Article

Probiotic and Antioxidant Potential of the *Lactobacillus* Spp. Isolated from Artisanal Fermented Pickles

Urva Akmal [1], Ifra Ghori [1,*], Abdelbaset Mohamed Elasbali [2,*], Bandar Alharbi [3], Arshad Farid [4], Abdulhakeem S. Alamri [5,6], Muhammad Muzammal [4], Syed Mohammed Basheeruddin Asdaq [7], Mohammed A. E. Naiel [8] and Shakira Ghazanfar [9,*]

1. Department of Biotechnology, Fatima Jinnah Women University Rawalpindi, Rawalpindi 46000, Pakistan; urvaakmal4@gmail.com
2. Department of Clinical Laboratory Science, College of Applied Sciences-Qurayyat, Jouf University, Sakaka 72388, Saudi Arabia
3. Department of Medical Laboratory, College of Applied Medical Science, University of Hail, Hail 55476, Saudi Arabia; b.alharbi@uoh.edu.sa
4. Gomal Center of Biochemistry and Biotechnology, Gomal University, Dera Ismail Khan 29050, Pakistan; arshadfarid@gu.edu.pk (A.F.); mustafamuzammal1@yahoo.com (M.M.)
5. Department of Clinical Laboratory Sciences, The Faculty of Applied Medical Sciences, Taif University, Taif 26521, Saudi Arabia; a.alamri@tu.edu.sa
6. Deanship of Scientific Research, Centre of Biomedical Sciences Research (CBSR), Taif University, Taif 26521, Saudi Arabia
7. Department of Pharmacy Practice, College of Pharmacy, AlMaarefa University, Dariyah, Riyadh 13713, Saudi Arabia; sasdag@mcst.edu.sa
8. Department of Animal Production, Faculty of Agriculture, Zagazig University, Zagazig 44511, Egypt; mohammednaiel.1984@gmail.com
9. National Institute for Genomics and Advanced Biotechnology, National Agricultural Research Centre, Islamabad 45500, Pakistan
* Correspondence: ifraghori@fjwu.edu.pk (I.G.); aeelasbali@ju.edu.sa (A.M.E.); shakira_akmal@yahoo.com (S.G.)

Citation: Akmal, U.; Ghori, I.; Elasbali, A.M.; Alharbi, B.; Farid, A.; Alamri, A.S.; Muzammal, M.; Asdaq, S.M.B.; Naiel, M.A.E.; Ghazanfar, S. Probiotic and Antioxidant Potential of the *Lactobacillus* Spp. Isolated from Artisanal Fermented Pickles. *Fermentation* **2022**, *8*, 328. https://doi.org/10.3390/fermentation8070328

Academic Editors: Michela Verni and Carlo Giuseppe Rizzello

Received: 6 May 2022
Accepted: 5 July 2022
Published: 13 July 2022

Publisher's Note: MDPI stays neutral with regard to jurisdictional claims in published maps and institutional affiliations.

Abstract: The present study was based on bacterial isolation with probiotic potential from artisanal fermented pickles. A total of 36 bacterial strains were isolated from 50 different artisanal fermented pickle samples. Nine isolates with promising probiotic potential (PCR99, PCR100, PCR118, PCR119, PCR121, PCR125, PCR137, PCR140 and PCR141) were selected. The strains showed varied protease, amylase, lipase and cellulase patterns. The isolated strains displayed varied responses towards various antibiotic classes, i.e., PCR140 showed resistance to penicillin G, polymyxin B, Metronidazole and Streptomycin. PCR140 showed highest resistance to bile salt concentrations (0.3% and 0.5%) and acidic conditions (pH 3 and pH 4) when exposed to mimicked gastrointestinal conditions. The cell viability against enzymes produced in stomach and intestines showed different patterns as pepsin was in the range of 94.32–91.22%, pancreatic resistance 97.32–93.11% and lysozyme resistance was detected at 99.12–92.55%. Furthermore, the auto-aggregation capability of isolated strains was in the range of 46.11–33.33% and cell surface hydrophobicity was in the range of 36.55–31.33%. PCR 140 showed maximum antioxidant activity in lyophilized cells as well as probiotic potential. A phylogenetic analysis based on 16S rRNA gene sequencing confirmed that PCR140 (NMCC91) with higher in vitro probiotic and antioxidant potential belongs to the genus *Lactobacillus* with 97% similarity with *Lacticaseibacillus paracasei*. This work demonstrated that the isolate PCR 140 (NMCC91) is suitable for use in food and medical industries.

Keywords: pickles; probiotics; lactic acid bacteria; *Lactobacillus*; probiotic property; enzymatic potentials

1. Introduction

Food fermentation has traditionally been associated with a myriad of cultures since antiquity. The outcomes of this centuries-old practice are better food storage, improved quality, and an enhanced consistency of final products through pickling or fermentation processes. Since the beginnings of human civilization, fermentation has been one of most important components in our diet in the form of wine, beer, yogurt, pickles, cheese, etc. Pickling can be defined as "a procedure of converting sugar into acids by lactic acid bacteria" [1]. These acids inhibit pathogens and spoilage-causing non-acid tolerant microorganisms [1]. Pickling is a common practice for food preservation based on the fermentation of lactic acid, alcohol and acetic acid [2]. Globally, pickles are consumed as appetizers that improve the digestion of grains and vegetables. Furthermore, pickles are stored at room temperature [3,4].

Lactic Acid bacteria involved in the pickling process can be defined as "Gram positive, organotrophic, non-sporulating, acid and air tolerant, catalase-negative, non-motile cocci or rod and carbohydrate fermentative and lactic acid forming bacteria" [5,6]. These pickle-associated bacteria produce unique flavors and also promote overall health and fitness [7]. Pickles improve human health by providing vitamins, carbohydrates, minerals, and certain pigments such as anthocyanin, glucosinolates, lycopene, β-carotene and flavonoids [8,9]. Pickling serves to preserve the nutritional value of fruits and vegetables, prevents food spoilage, and increases food safety. Thus, fermentation acts as an alternative for food additives [1,10,11].

Lactic Acid Bacteria [LAB] can be used as a starter culture for the processing and preservation of meats, fruits, vegetables and dairy items [12]. The final products must meet the standards of consumer acceptability and consistent quality [13]. *Lactiplantibacillus plantarum* and *Limosilactobacillus fermentum* are well known microorganisms that are suitable for use as starter cultures [12,13]. The safety and production of antimicrobial compounds makes LAB a befitting choice for the bio-preservation of food commodities. The antimicrobials produced by LAB are di-acetyl, ethanol, organic acid, hydrogen peroxide and bactericidal peptides [14–16].

Probiotics are known to combat the putrefactive impacts of the gut metabolism which could cause aging and illness. Probiotics were discovered by Elie Metchnikoff, who is known as the "father of probiotics". Fermented foods prepared using probiotics are functional foods and have beneficial impacts on the human gut system [17].

Food and agricultural industries are continuously improving through innovational strategies, leading to the generation of constantly emerging research technologies. The change in the acceptance, needs and preferences of consumers is a dynamic process, but food quality maintenance through technology innovations is obvious. The habits of consumers, sustainability factors and their cultural heritage affect technological innovations which are applied in our food industry. Recently, consumers became increasingly health conscious and now prefer products with higher beneficial values. Manufacturers are incentivized to produce functional foods due to these demands. Hence, the acceptance of novel products and successful marketing requires added food value functionalities. Novel food products can be processed or naturally enriched with active compounds secreted by biological agents. These compounds provide health benefits apart from those provided by nutrients when administered in a required quantity.

In the context of previous studies, LABs are recognized as the most prominent probiotic strains as few of the species inhabit the small intestine by resisting harsh environmental conditions, i.e., a low pH, various bile salts and interact with other pathogenic strains and natural inhibitors through their antagonistic activities and antibiotic susceptibility [18]. Besides, their probiotic capabilities such as production of enzymes and hydrophobicity, auto-aggregation and antioxidative activity against DPPH are well demonstrated by *L. plantarum, P. pentosaceus, P. acidilactici, E. lactis* and *E. hira* which were isolated from fermented fish *Shidal* [18], whereas Lactiplantibacillus plantarum F1 and Levilactobacillus brevis OG1 showed antagonistic activity against pathogens [19], and *E. faecalis* and *E. faecium* demonstrated high responses towards cell-surface auto-aggregation characteristics [18].

This is the first study based on microflora of traditional homemade pickles in this region of Pakistan. Therefore, this research was performed with the aim to isolate and identify the potential probiotic LAB from traditional homemade pickles. The strains were evaluated based on FAO/WHO parameters on the classification of probiotics. These parameters include the ability to survive simulated GIT conditions, bile salt tolerance, assessment of antibiotic susceptibility, antagonistic activity, screening of bile salt hydrolase (BSH) and lipase activity, antioxidative activity against DPPH free radicals and cell surface characteristics, so that they can be employed for the formation of other beneficial products for living beings.

2. Materials and Methods

2.1. Isolation and Characterization of Lactic Acid Bacteria

Fifty samples of homemade pickles made in brine were collected randomly from Islamabad, Pakistan. Carrot pickles [*Daucus carota* subsp. *Sativus*], garlic pickles (*Allium sativumi*), radish pickles (*Raphanus raphanistrum* subsp. *Sativus*) and green chili pickles (*Capsicum annuum*) were collected in sterile sample bottles. All the samples were transferred to the Probiotic Laboratory in the National Institute for Genomics and Advanced Biotechnology (NIGAB) NARC and were stored at 4 °C. Furthermore, samples were serially diluted and 100 μL of a 100-10-5 serial dilution was poured on MRS agar plates and incubated at 37 °C for 48 h. Colonies were randomly selected and purified on MRS agar by repeated streaking about 4–5 times [4,20,21]. Purified cultures were stored in 30% glycerol at −80 °C. The preliminary identification of isolates included the study of cell morphology, Gram staining, as well as oxidase and catalase testing [22].

2.2. Molecular Identification of Bacterial Isolates

The molecular identification of selected purified strains was performed by adopting the procedure described in [23]. The bacterial DNA was isolated through a DNA extraction process in which colonies were suspended in $1\times$ TE buffer and placed in a PCR machine for 2–3 min at 6000 rpm after heating at 95 °C for 10 min. The 16S rRNA gene of 9 isolated strains was amplified by PCR [Polymerase Chain Reaction] using the reaction mixture containing PCR water, Taq buffer, DNTPs, MgCl2, Universal primers, =, and Taq polymerase. After the formation of the reaction mixture, a second set of tubes was placed in PCR for 2 h and 10 min with the following PCR conditions: initial denaturation at 94 °C for 2 min then 30 cycles of denaturation at 94 °C for 1 min. Initial denaturation was followed by an annealing process for 1 min at 50 °C. After the annealing process, extension was performed at 72 °C for 1.5 min. Lastly, the final extension of the PCR product was performed at 72 °C for almost 5 min. The PCR product was electrophorized by using 2% agarose gel and visualized with ethidium bromide staining. Selected strains were sent for 16S rRNA sequencing at the commercial sequencing facility of Macrogen Inc. (Seoul, South Korea) Sequences were identified by using the BLAST system (https://blast.ncbi.nlm.nih.gov/Blast.cgi, accessed on 10 June 2019) and were submitted to NCBI on 26 June 2019 (http://www.blast.ncbi.nlm.nih.gov). Phylogenetic trees were developed by aligning the selected sequences into MEGA version X with the Neighbor-Joining method.

2.3. Probiotic Characterization

2.3.1. Acid Tolerance

Bacterial cultures were grown in MRS broth with different pH levels of 2, 3, 4 and 5 at 37 °C for 24 h in a shaking incubator followed by a measurement of Optical density at 600 nm. The experimentation was performed in triplicate for validity [24].

2.3.2. Bile Tolerance

Bile tolerance was evaluated with a previously used method as described by Kılıç and Karahan (2010) [25]. Fresh cultures of lactic acid bacterial strains were incubated in MRS broth of 0.3% and 0.5% bile acids for 24 h followed by the monitoring of bacterial growth at 600 nm with a spectrophotometer. All the experiments were performed in triplicate.

2.3.3. Cell Auto-Aggregation

Cell auto-aggregation of bacterial isolates was tested using the procedure outlined by Yadav et al. (2016) [26]. Pure bacterial cultures grown in MRS broth for 18 h were centrifuged at 6000 rpm for 5 min followed by washing and re-suspension in PBS. Absorbance was measured at 600 nm. This suspension was then incubated for 2 h at 37 °C. After 2 h, 1 mL of upper surface supernatant was removed from this suspension and absorbance was measured at 600 nm. Cell auto-aggregation was determined by the decline in absorbance capability and measured with the given formula. All the experiments were performed in triplicate.

$$\text{Auto-aggregation} = (\text{Initial OD} - \text{final OD}/\text{Initial OD}) \times 100 \tag{1}$$

2.4. Cell Surface Hydrophobicity

Cell surface hydrophobicity was carried out following the previously described protocol by Gharbi et al. (2019) [27] with slight modifications. Bacterial cultures were incubated in MRS broth for 24 h at 37 °C followed by centrifugation at 6000 rpm for 5 min. Pellets were washed with normal saline, Phosphate Saline Buffer (PSB), suspended in autoclaved distilled water and incubated for 20–30 min after adding 600 µL of Xylene. The aqueous phase was separated carefully, and its OD was recorded at 600 nm. Calculations of percentage of hydrophobicity were conducted according to the following formula:

$$\text{Hydrophobicity Percentage (\%)} = [(\text{Ao} - \text{A1})/\text{Ao}] \times 100 \tag{2}$$

where Ao = OD before Xylene addition and A1 = OD of aqueous layer.

2.5. Determination of Pepsin and Pancreatin Resistance

Lactobacillus strains were tested for their resistance to pepsin and pancreatin by following the method given by Jamaly (2011) [28]. Bacterial cultures were centrifuged at $10,000 \times g$ for 5 min followed by washing twice with Phosphate-buffered saline of neutral pH. Afterwards these cultures were resuspended in PBS solution of pH 2.0 containing 3 mg/mL pepsin and in PBS solution of pH 8 containing 1 mg/mL pancreatin for 24 h. Resistance to pepsin and pancreatin was determined by calculating viable colony counts.

2.6. Lysozyme Resistance

The lysozyme resistance of selected bacterial isolates was observed using the procedure suggested by Yadav et al. (2016) [26]. Pure Lactobacilli strains were centrifuged for 10 min at 7000 rpm. The isolates were washed with PBS twice and suspended in Ringer's solution. A total of 10 µL from this suspension was incubated at 37 °C in a sterilized electrolyte solution containing 100 mg/L lysozyme, $NaHCO_3$ 1.2 g/L, NaCl 6.2 g/L, $CaCl_2$ 0.22 g/L and KCl 2.2 g/L. A control sample was also prepared by the inoculation of strains in an electrolyte solution without lysozyme. After 2 h, the plate count method was used to determine the cell count viability.

2.7. Amylolytic Activity

Amylolytic activity was assessed using a previously recommended procedure described by Ghazanfar (2016) [29]. Media was prepared by adding nutrients agar and starch in distilled water. Media was inoculated with these strains and incubated at 37 °C for 24 h. After that, iodine was sprinkled over it and the presence or absence of luminous zones was observed.

2.8. Cellulolytic Activity

Cellulolytic activity was checked using the method described by Ghazanfar (2016) with slight modifications applied [29]. CMC media was prepared followed by inoculation and incubation for 37 °C for 24 h. The plates were first stained with Congo red dye for 15 min followed by staining with NaCl for 15 min. The presence or absence of a luminous zone was recorded.

2.9. Proteolytic Activity

Proteolytic activity of LAB strains was measured by streaking these strains on skim milk agar plates using previously described method as per Monika et al. (2017) [4]. Luminous zones were observed.

2.10. Lipolytic Activity

For the determination of the lipolytic activity of concerned strains, TSA was prepared with 1 mL Tween 80 in 100 mL distilled water and phenol red. Colonies were spread on the solid surface of agar and incubated for 37 °C for 24 h. Change in color was recorded.

2.11. Assessment of Antibiotic Susceptibility

Antibiotic susceptibility was assessed using the disc diffusion method described by Monika et al. (2017) [4]. Bacterial cultures which were 24 h old and grown at 37 °C were tested against 15 antibiotics including Penicillin G (10 iu), Polymyxin B (300 iu), Chloramphenicol (30 mcg), Ampicillin (10 mcg), Bacitracin (10 iu), Kanamycin (30 mcg), Cephalexin (30 mcg), Tetracycline (30 mcg), Amoxycillin (30 mcg), Metronidazole (5 mcg), Vancomycin (30 mcg), Streptomycin (10 mcg), Gentamycin (10 mcg) and Nalidixic Acid (30 mcg). Bacterial suspensions were swabbed on Mueller Hinton Agar (MHA) agar plates followed by the placement of antibiotics discs on the MHA agar plates, which were then incubated for 24 h at 37 °C. Results were recorded afterwards. The measurement of inhibition zones was evaluated on the basis of VETLAB standards. All the experiments were performed in triplicate.

2.12. Antagonistic Activity

The antimicrobial activity test of concerned strains was carried out using the agar-well diffusion method [4]. The pathogenic strains were *Escherichia coli* (ATCC8739), *Listeria monocytogenes* (ATCC13932), *Staphylococcus aureus* (ATCC6538), and *Bacillus cereus* (ATCC11778). Swabbed plates with a pathogen suspension were used to make wells filled by 50 µL suspension Lactobacillus strains and zones were measured after incubation at 37 °C for 24 h. The experiments were performed in triplicate.

2.13. Antioxidant Activity

The antioxidant activity of bacterial isolates was monitored against (2, 2-diphenylpicrylhydrazyl) DPPH free radicals using a protocol described by Nadri et al. (2014) [30]. To perform this assay, 0.03 g of DPPH was added to 100 mL of methanol. An 18 h fresh bacterial culture was prepared and added (200 mL) to 800 mL of DPPH and kept for 30 min at 37 °C. Ascorbic acid was used as the standard. Scavenging or the inhibition of free radicals was measured at 517 nm with a UV–visible spectrophotometer. The value was calculated using the following formula:

$$\text{Inhibition (\%)} = [(\text{Absorbance of control} - \text{Absorbance of tests}) / \text{Absorbance of control}] \times 100 \tag{3}$$

2.14. Statistical Analysis

The data were expressed as mean ± standard error mean (SEM) calculated over individual experiments which were performed in triplicate. For the inference statistics, a one-way ANOVA was used to determine the significant differences among the isolated strains on quantitative parameters ($p < 0.05$).

3. Results

3.1. Purification and Biochemical Characterization of Bacterial Isolates

In this experimental study, nine bacterial isolates were isolated as LAB strains based on their morphological, biochemical, and molecular characteristics. All the isolated strains were Gram positive rods. All bacterial isolates were catalase and oxidase negative.

3.2. Enzymatic Potential of Lactobacillus Strains

Enzymatic activity was evaluated to observe the capability of *Lactobacillus* strains for producing industrially important enzymes. Bacterial strains did not show any amylolytic activity except strain PCR125. Clear zones were formed only in the "125" strain's plate which indicated that this bacterial strain has potential to hydrolyze the starch. Five strains, i.e., PCR99, PCR100, PCR118, PCR119 and PCR141 including *L. plantarum* and *L. paracasei* were able to show positive results and to utilize Tween 80 as a lipid source. Two strains PCR137 and PCR119) showed proteolytic activity (Table 1).

Table 1. Determination of Enzymatic Potential of Selected *Lactobacillus* Strains.

Bacterial Isolates	Amylase Activity	Lipase Activity	Cellulolytic Activity	Proteolysis Activity
PCR99	− −	+	−	−
PCR100	−	+	−	−
PCR118	−	+	−	−
PCR119	−	+	−	+
PCR121	−	−	−	−
PCR125	+	−	−	−
PCR137	−	−	−	−
PCR140	−	−	−	−
PCR141	−	+	−	−

Where: + ve indicates 'positive' results as strains shows specified activity. − ve indicates 'negative' results as strain did not show that activity.

3.3. Antibiotic Suscenptibility

The antibiotic susceptibility results of nine bacterial strains against 15 different antimicrobial agents are presented in Table 2. Results demonstrated that almost all isolated strains are sensitive to protein synthesis inhibitors, i.e., Chloramohenicol, Gentamycin and Streptomycin. Strains showed varied susceptibility responses towards cell wall and DNA synthesis inhibitors, i.e., Penicilin G, Plymyxin B, Ampicillin, Bacitracin, Vancomycin, etc. Furthermore, strains were observed to be resistant to Niroimidazoles, i.e., Metronidazoles.

3.4. Determination of Cell Viability under Different Probiotic Parameters

3.4.1. pH Tolerance Assay

The isolated bacteria demonstrated diverse results when exposed to different acidic conditions, i.e., pH, as mentioned in Figure 1. The graph depicts that isolated strains are able to survive all pH conditions, specifically pH 5 and pH 7 while at pH 2, PCR118 displayed lowest and PCR140 showed highest tolerance, and PCR100 showed highest tolerance and PCR137 showed the lowest tolerance at pH 3. At pH 4, a tolerance range was observed between PCR121 and PCR100. On the whole, significant results were observed.

3.4.2. Bile Salt Tolerance Assay

While observing bacterial tolerance at different bile salt concentrations, bacterial isolates demonstrated that at a 0.3% concentration, strains were highly tolerant as compared to 0.5% concentration (Figure 2). It was observed that PCR99, PCR100, PCR140 showed the highest growth at a concentration of 0.3% and PCR121, PCR118 and PCR119 showed highest growth among other strains at 0.5%.

Table 2. Antibiotic Susceptibility Determination of Isolated *Lactobacillus* strains according to VET-LAB standards.

Antibiotics [Anti-Microbial Agents]	PCR99	PCR100	PCR118	PCR119	PCR121	PCR125	PCR137	PCR140	PCR141
Penicillin G	S	I	R	R	R	R	R	R	I
Polymyxin B	R	R	R	S	S	S	S	R	I
Chloramphenicol	S	S	S	S	S	S	S	S	S
Ampicillin	I	R	R	S	S	S	S	S	R
Bacitracin	S	S	R	S	S	S	S	S	R
Kanamycin	R	R	S	S	S	S	S	S	S
Cephalexin	I	S	S	S	S	S	S	S	R
Tetracycline	I	R	S	S	S	S	S	S	S
Amoxycillin	S	R	R	S	S	S	S	S	S
Metronidazole	R	R	R	R	R	R	R	R	R
Vancomycin	S	R	S	S	S	S	R	S	S
Cefuroxime	I	R	R	S	S	S	S	S	S
Streptomycin	R	I	S	S	R	S	S	R	S
Gentamycin	S	I	S	S	S	S	S	S	I
Nalidixic Acid	S	R	S	S	I	S	R	S	R

Where R = RESISTANT; S = SENSITIVE; I = INTERMEDIATE.

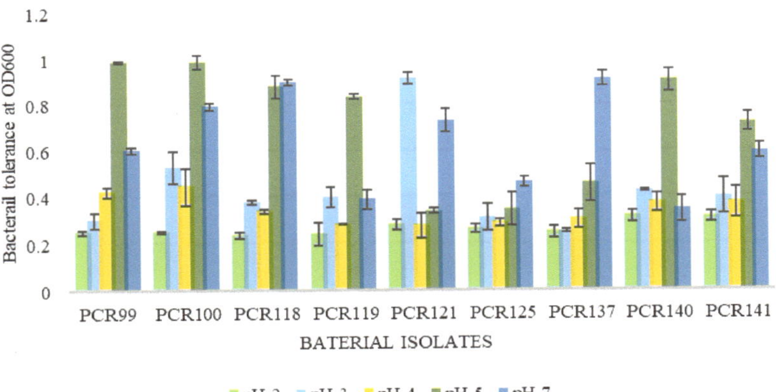

Figure 1. Detection of pH tolerance in bacterial strains. Values are means of replicate experiments; $\pm$ indicates standard deviation from the mean. Values differs significantly ($p < 0.05$).

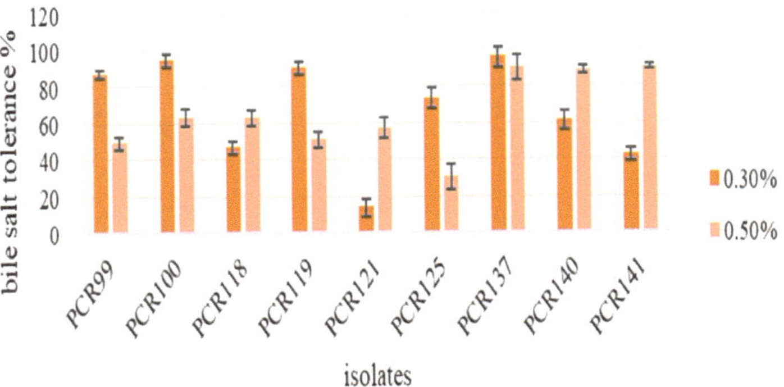

Figure 2. Detection of Bile Salt tolerance in bacterial strains. Values are means of replicate experiments; $\pm$indicates standard deviation from the mean. Values differs significantly ($p < 0.05$).

3.4.3. Auto-Aggregation and Cell Surface Hydrophobicity Determination

Among the tested strains, all bacterial isolates showed good auto-aggregation in the range of 46.11–33.33%. Among them, PCR141 was found to be the best auto-aggregating strain and showed 46.11% aggregation over the 24 h incubation period which means this strain is capable of aggregating on epithelial layers of intestine (Figure 3). In the present study, six strains showed highest hydrophobicity, i.e., PCR99, PCR100, PCR118, PCR121 and PCR140, in the range of 36.65–31.33% while the capacity of other strains was also satisfactory, thereby depicting that the strains are able to survive and adhere themselves with the epithelial intestinal layers (Figure 4).

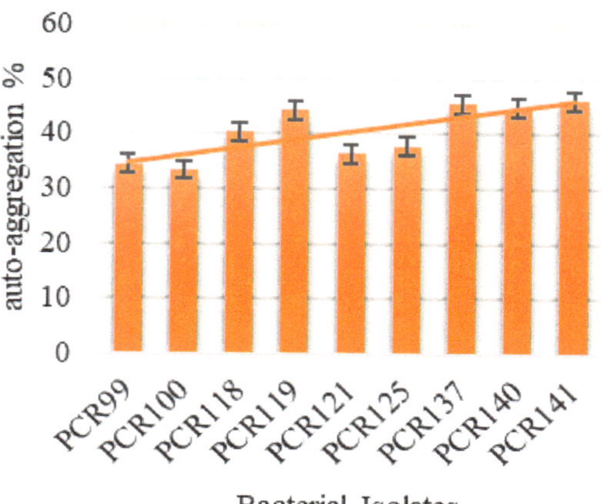

Figure 3. Determination of Auto-aggregation % of isolated bacterial strains.

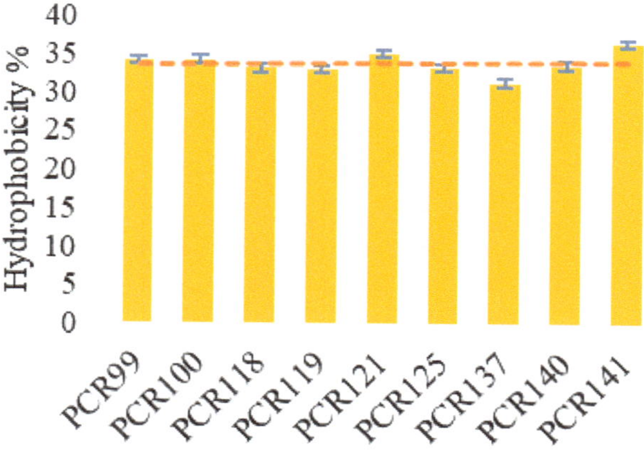

Figure 4. Determination of Hydrophobicity % of bacterial isolates. Values are means of replicate experiments; ±indicates standard deviation from the mean. Values differs significantly ($p < 0.05$).

3.4.4. Effect of Lysozyme on Cell Viability

Among the tested bacterial isolates against the bactericidal activity of lysozyme, all strains showed resistance in the range of 99.12–92.55%. PCR141 was observed to be highly resistant to lysozyme, showing viability of up to 99.12%, while the trend showed that the property is strain-specific as all isolates displayed different patterns (Figure 5).

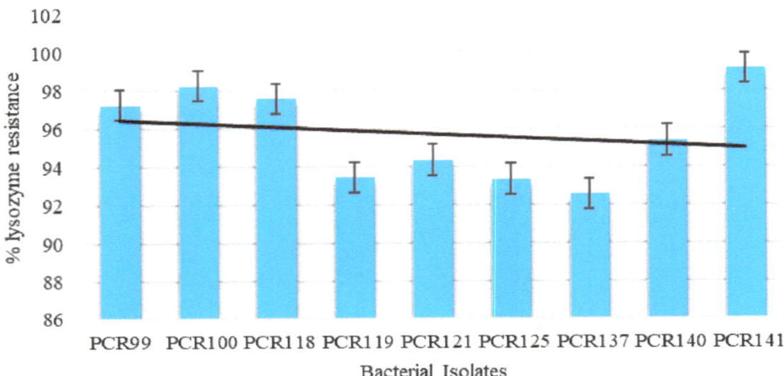

Figure 5. Determination of cell viability of bacterial isolates against Lysozyme resistance. Values are means of replicate experiments; ± indicates standard deviation from the mean. Values differ significantly ($p < 0.05$).

3.4.5. Pepsin and Pancreatin Resistance Assay

The pepsin and pancreatin resistance test revealed that the selected bacterial isolates are capable of surviving under intestinal and gastric enzymes as their viability rates were in the range of 94.34–90.34% for pepsin resistance (Figure 6a) and 97.32–93.11% for pancreatin resistance (Figure 6b). The results show that PCR140 and PCR137 demonstrated the highest pancreatin resistance while PCR125 showed lowest resistance. Furthermore, PCR118 and PCR140 possessed the highest pepsin resistance in our study and PCR119 showed lowest resistance.

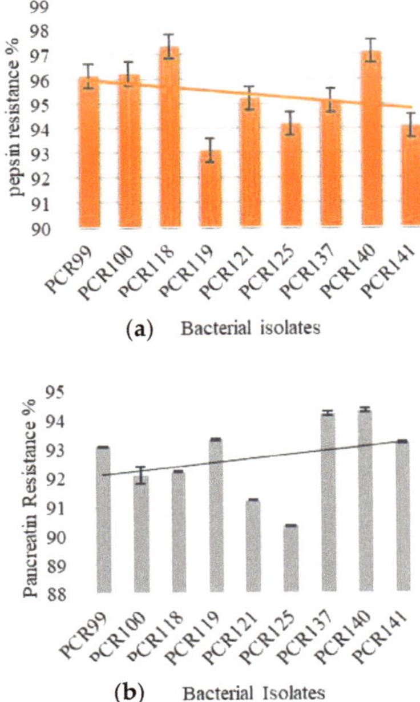

Figure 6. Determination of bacterial viability in the presence of; (**a**) pepsin conditions and (**b**) pancreatin conditions.

3.4.6. Antioxidant Activity against DPPH

Table 3 displays the antioxidant capability of PCR140. The results show that PCR140 displayed the highest DPPH free radical scavenging property. The probiotic strains were observed with almost the same inhibition ability as displayed with ascorbic acid.

Table 3. In vitro antioxidant of the PCR140.

ID	Treatments	Inhibition %
PCR140	Supernatant	67.000 ± 0.82
	Ascorbic acid [drug]	71.1 ± 0.83
	Control	0 ± 0.81

3.5. Phylogenetic Analysis

The bacterial isolates were identified on a molecular level by using the BLAST tool provided on the National Center for Biotechnology Information [NCBI] website. Four of the sequences were submitted to the NCBI database and the taxonomy of these bacterial isolates is illustrated in Table 4.

Table 4. Provisionally Taxonomy of Bacterial Isolates.

Strain ID	Strain Name/Genus	Length of 16s RNA	Bacterial Taxonomy	Similarity % of 16s RNA Gene Sequencing	Accession Numbers
PCR118	*Lactobacillus*	354	*Lacticaseibacillus paracasei*	99.08%	MN088781.1
PCR119	*Lactobacillus*	417	*Lacticaseibacillus paracasei*	99.53%	MN088853.1
PCR121	*Lactobacillus*	334	*Lacticaseibacillus paracasei*	99.30%	MN088850.1
PCR125	*Lactobacillus*	338	*Lactiplantibacillus plantarum*	99.53%	MN089481.1

Provisionally molecular identification of strain PCR140 was followed by a phylogenetic analysis of *Lacticaseibacillus paracasei* to study their relationship to the nearest species, as mentioned in Figure 7.

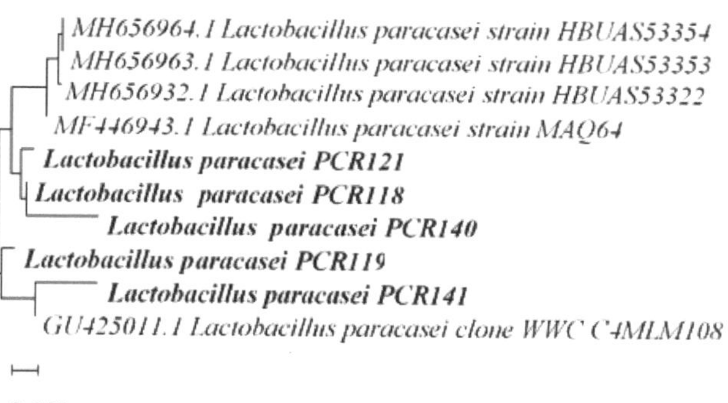

Figure 7. Phylogenetic Analysis of *Lacticaseibacillus paracasei* PCR121, PCR118, PCR140, PCR141, PCR119.

4. Discussion

In the present study, nine bacterial isolates were isolated and characterized from fifty different pickled samples, which displayed potent probiotic potential. Despite the fact that the isolates of this genus are already being consumed in the market for the production of many nutraceutical products, no reports have been published reported about their isolation from pickles made in brine solution. Furthermore, the present research is novel due to its sampling area, i.e., Islamabad, where no reports have as of yet been published. The samples selected for the isolation of LAB strains form an indigenous part of the Pakistani diet and are consumed unknowingly as probiotic drinks/foods by people. The presence of LAB strains with potential probiotic characteristics has been reported in many studies. In a study by Kumar et al. (2011), many Lactic Acid Bacteria were isolated from Persian traditional vegetable pickles in which Lactobacillus was the predominant species along with *L. Paracasei, L. casei, L. Pentosus, L. brevis and L. mesenteroides* [31]. Monika et al. (2017) isolated *Lactiplantibacillus plantarum* from pickles made in Himachal Pradesh, India [4].

In terms of the enzymatic activity of functional foods in industries, *Lactiplantibacillus plantarum* [PCR125] strains showed amylase activity while the other eight strains did not exhibit this activity, as they are not involved in α-amylase production, which is in agreement with the results of our research [32]. The literature shows that some strains of *Lactiplantibacillus plantarum* are capable of α-amylase production [33,34]. In contrast to the literature, *Lacticaseibacillus paracasei* did not show amylolytic activity [35]. The amylase enzyme has its applications in food, textile, paper, pharmaceutical and detergent manufacturing industries as it can be used for corn syrup, glucose syrups, alcohol fermentation, maltose syrups and in detergents [36,37]. In the present study, the absence of cellulolytic activity revealed that these strains are incapable of metabolizing carboxymethyl cellulose. Therefore, it was concluded that cellulase activity is strain dependent. Cellulolytic activity was observed in the present study, contrary to the study by Singhvi et al. (2010), which declared mutant LAB strains (*Lactobacillus lactis* mutant RM2-24) were used for cellulose production [38]. In accordance with a study conducted by Dinçer et al. (2018), *L. plantarum* has the highest lipolysis activity as this activity has been observed in strains isolated from Pastirma in Turkey [39]. The development and production of cheese flavors such as cheddar cheese can be achieved with lipase enzymes [40], which reveals that they are able to make peptide populations of a medium size as a result of proteolytic processes [41]. Protease enzymes can be used for bread quality improvement, brewing, meat tenderization and for the coagulation of milk [40,42].

In this study, antibacterial tests were performed in which bacterial strains were tested against pathogenic strains of *E. coli* (ATCC8739), *Listeria monocytogenes* (ATCC13932), *Staphylococcus aureus* (ATCC6538), and *Bacillus cereus*. No bacterial strain was detected to be resistant against pathogenic strains. The results revealed that most strains are resistant to Penicillin G and metronidazole, which is in accordance with previous research [43–45]. Multiple drug resistance was observed in strains PCR100 and PCR141 while most strains were observed to be sensitive to these antimicrobial agents. The resistance observed against some antibiotics tested suggests that our strains would not be affected by therapies using these antibiotics and might help maintain the natural balance of intestinal microflora during antibiotic treatments.

Microorganisms must be able to resist uncongenial Gastrointestinal Tract (GIT) conditions to be classified as probiotics and to exert beneficial effects. The stomach pH can be reduced to 1.0 in the presence of pepsin but in vitro experimentations were performed with a pH of 1.5–4.5 due to the buffering mechanisms of the food matrix, which produces shielding effects on gut microbiota [16,23]. Resistance to low pH is essential not only for use as a probiotic for humans, but also to produce various food products because low pH resistance helps them to survive in acidic conditions such as in yogurt, etc. Moreover, researchers also discussed that low acidic conditions affect the survival of lactic acid bacterial strains, especially at pH 2 [23,46,47]. The ability to resist bile salts is crucial for colonization and metabolic activities of bacterial isolates in intestines. In intestines, bile salt, at a concentration of 0.3%, is secreted; therefore, bacterial tolerance was observed against bile salt concentrations of 0.3% and 0.5%. In the present study, the results demonstrated that isolated strains are significantly tolerant to pH 1–pH 4 and 0.3% bile salt concentrations. BSH activity is a relevant property for probiotic strains to survive the toxicity of conjugated bile salts in the duodenum [25]. Pancreatin enzymes help in the digestion of fats, proteins and carbohydrate and are released in intestines through the pancreatin duct, while pepsin is secreted by gastric chief cells to digest proteins in an inactive form, i.e., zymogens, which become active when HCl lowers the stomach pH. The tolerance of bacterial strains against pancreatin and pepsin is considered as another criterion to predict their survival in harsh GIT conditions [48]. The nine isolated strains displayed 90–95% resistance against pancreatin.

The auto-aggregation property of bacterial isolates allows them to adhere to the epithelial lining of intestines, which is considered as a beneficial property because it prevents the flush out of bacterial strains from the body through peristalsis. The auto-aggregation percentages of isolated strains increased with time which is consistent with previous findings [49,50]. In a study conducted by Abid et al., it was revealed that Lactobacilli strains, i.e., NMCC-14, exhibit the highest auto-aggregation properties, i.e., 47.55 ± 0.08.

A hydrophobicity test was conducted to detect bacterial attachment capability to intestinal epithelial cells in the presence of hydrocarbons. Bacterial potential colonization is considered an important property for the selection of strains in probiotics. Researchers have discussed hydrophobicity values of some *Lactobacillus* strains in the range of 82.41–97.96% but values of 15–60% were also observed [51,52]. The interactive forces involved in the adhesion process of *Lactobacilli* strains include electrostatic interaction, passive forces, and steric and hydrophobic forces. The adherence property of probiotic strains inhibits or prevents the attachment of pathogenic strains to the epithelial layer of the gut which, in turn, prevents pathogenic activity in the GIT tract of organisms [53,54]. Our results are in accordance with the literature [53,55] which demonstrated that *Lactobacilli* strains isolated in Islamabad are capable of growing under lysozyme conditions in the range of 96.69–53.45%. The results of gastric conditions are in accordance with the literature; however, according to Tokatlı et al., pancreatin has a negative impact on strain-survival rates which is in contrast to our results [54]. Furthermore, these results are in accordance with the data published by Adnan et al. (2017) [56].

In this study, the antioxidant activity of strain PCR140 was observed using the DPPH radical method. The intracellular LAB extracts are reported to have a chelating ability due to metal ions. Besides, intact LAB cells are known to have antioxidant capabilities; therefore, they are present in the GIT tract. [55,57,58]. While passing through GIT tract, LAB released an antioxidant constituent which was reported as a healthy mechanism. Our experimental results showed higher antioxidant capability which is in accordance with the previous literature [59–61]. Therefore, strain PCR 140 can be classified as probiotic. The phylogenetic analysis of isolated strains demonstrated the evolutionary relationship and showed that isolates possess > 95% similarity with *Lacticaseibacillus paracasei* strains.

5. Conclusions

The need to isolate *Lactobacillus* strains with potent probiotic potential was satisfied with the isolation and characterization of nine *Lactobacillus* strains from fifty traditional pickles from Islamabad, Pakistan. This was the first study on *Lactobacillus* isolates conducted in Islamabad on fermented products prepared in brine solution. All the nine isolates were examined on the basis of FAO/WHO guidelines and possessed the minimum criteria to be classified as probiotics. *Lactobacillus* strains demonstrated satisfactory resistance against gastrointestinal conditions such as an acidic environment, bile salt, bile salt hydrolase, pepsin and pancreatin conditions. Isolated bacterial strains displayed remarkable cell surface characteristics, i.e., hydrophobicity and auto-aggregation properties. The antibiotic susceptibility assay concluded that isolated *Lactobacilli* strains can be classified as safe. Moreover, *Lactobacilli* strains with an enzymatic potential of varied significant performances were observed. *Lacticaseibacillus paracasei* (PCR140) displayed higher antioxidant potential as well. The molecular identification of isolated strains showed that three strains were of *Lactiplantibacillus plantarum,* five of *Lacticaseibacillus paracasei* and one was identified as Levilactobacillus brevis. The phylogenetic analysis of these bacterial isolates was performed to analyze the evolutionary relationship. After all these experiments, their significance results demonstrated that these isolates are suitable for use in food and medical industries for the benefits of humans and animals. Furthermore, assays such as the hemolysis assay and gelatinase assay, and the co-aggregation of bacterial isolates with pathogenic strains need to be studied for a further safety evaluation of isolated strains. Moreover, in vivo studies evaluating their safety need to be conducted.

Author Contributions: Conceptualization, U.A. and I.G.; methodology, S.G.; software, A.F. and M.M.; validation, S.G., I.G. and U.A.; formal analysis, A.M.E.; investigation, S.G. and M.A.E.N.; resources, S.M.B.A.; data curation, S.G.; writing—original draft preparation, S.G. and A.S.A.; writing—review and editing, S.G., A.S.A. and S.M.B.A.; visualization, M.A.E.N.; supervision, I.G.; project administration, S.G.; funding acquisition, B.A., A.S.A. and S.M.B.A. All authors have read and agreed to the published version of the manuscript.

Funding: Abdulhakeem S. Alamri would like to acknowledge Taif University for support No. TURSP (2020/288). Syed Mohammed Basheeruddin Asdaq wishes to express his gratitude to AlMaarefa University in Riyadh, Saudi Arabia, for providing support (TUMA-2021-1) to do this study. The authors also acknowledge PARC laboratory for providing financial support for this research project.

Institutional Review Board Statement: Not applicable.

Informed Consent Statement: Not applicable.

Data Availability Statement: All data generated and analyzed during this study are included in this published article.

Acknowledgments: The authors extend their appreciation to the Deputyship for Research and Innovation, Ministry of Education in Saudi Arabia for funding this work through grant no. 375213500. Abdulhakeem S. Alamri would like to acknowledge Taif university for support No. TURSP (2020/288).

Conflicts of Interest: The authors declare no conflict of interest.

References

1. Sayın, F.K. The effect of pickling on total phenolic contents and antioxidant activity of 10 vegetables. *J. Food Sci.* **2015**, *1*, 135–141. [CrossRef]
2. Lee, E.; Jung, S.-R.; Lee, S.-Y.; Lee, N.-K.; Paik, H.-D.; Lim, S.-I. *Lactiplantibacillus plantarum* strain LN4 attenuates diet-induced obesity, insulin resistance, and changes in hepatic mrna levels associated with glucose and lipid metabolism. *Nutrients* **2018**, *10*, 643. [CrossRef] [PubMed]
3. Khaskheli, G.B.; Zuo, F.L.; Yu, R.; Chen, S.W. Overexpression of small heat shock protein enhances heat- and salt-stress tolerance of Bifidobacterium Longum NCC2705. *Curr. Microbiol.* **2015**, *71*, 8–15. [CrossRef] [PubMed]
4. Monika, S.; Kumar, V.; Kumari, A.; Angmo, K.; Bhalla, T.C. Isolation and characterization of lactic acid bacteria from traditional pickles of Himachal Pradesh, India. *J. Food Sci. Technol.* **2017**, *54*, 1945–1952. [CrossRef]
5. Perczak, A.; Goliński, P.; Bryła, M.; Waśkiewicz, A. The efficiency of lactic acid bacteria against pathogenic fungi and mycotoxins. *Arch. Ind. Hyg. Toxicol.* **2018**, *69*, 32–45. [CrossRef]
6. Alhashem, Y.N.; Farid, A.; Al Mohaini, M.; Muzammal, M.; Khan, M.H.; Dadrasnia, A.; Alsalman, A.J.; Al Hawaj, M.A.; Ghazanfar, S.; Almusalami, E.M.; et al. Protein Isolation and Separation Techniques of Pasteurella multocidavia One-and Two-Dimensional Gel Electrophoresis. *Int. J. Cur. Res. Rev.* **2022**, *14*, 1–8. [CrossRef]
7. Choi, I.H.; Noh, J.S.; Han, J.-S.; Kim, H.J.; Han, E.-S.; Song, Y.O. Kimchi, a fermented vegetable, improves serum lipid profiles in healthy young adults: Randomized clinical trial. *J. Med. Food* **2013**, *16*, 223–229. [CrossRef]
8. Isabelle, M.; Lee, B.L.; Lim, M.T.; Koh, W.-P.; Huang, D.; Ong, C.N. Antioxidant activity and profiles of common vegetables in Singapore. *Food Chem.* **2010**, *120*, 993–1003. [CrossRef]
9. Takebayashi, J.; Oki, T.; Watanabe, J.; Yamasaki, K.; Chen, J.; Sato-Furukawa, M.; Tsubota-Utsugi, M.; Taku, K.; Goto, K.; Matsumoto, T.; et al. Hydrophilic antioxidant capacities of vegetables and fruits commonly consumed in Japan and estimated average daily intake of hydrophilic antioxidants from these foods. *J. Food Compos. Anal.* **2013**, *29*, 25–31. [CrossRef]
10. Klein, M.; Sanders, M.E.; Duong, T.; Young, H.A. Probiotics: From bench to market. *Ann. N. Y. Acad. Sci.* **2010**, *1212*, E1–E14. [CrossRef]
11. Zuo, F.; Yu, R.; Feng, X.; Chen, L.; Zeng, Z.; Khaskheli, G.B.; Ma, H.; Chen, S. Characterization and in vitro properties of potential probiotic bifidobacterium strains isolated from breast-fed infant feces. *Ann. Microbiol.* **2015**, *66*, 1027–1037. [CrossRef]
12. Olaoye, O.A.; Ndife, J.; Raymond, V.I. Use of *Lactiplantibacillus plantarum* as starter culture and its influence on physicochemical, microbiological, and sensory characteristics of *kunnu-aya* produced from Sorghum and tigernut. *J. Food Qual.* **2017**, *2017*, 6738137. [CrossRef]

13. Akabanda, F.; Owusu-Kwarteng, J.; Tano-Debrah, K.; Parkouda, C.; Jespersen, L. The use of lactic acid bacteria starter culture in the production of *Nunu*, a spontaneously fermented milk product in Ghana. *Int. J. Food Sci.* **2014**, *2014*, 721067. [CrossRef] [PubMed]

14. Yaseen, M.; Kamran, M.; Farid, A.; Ismail, S.; Muzammal, M.; Amir, K.A.; Rashid, S.A. Antibacterial, Hemagglutination, and Insecticidal Activity Studies on the Solvent Extracts of the Roots of Olea ferruginea. *Makara J. Sci.* **2022**, *26*, 8.

15. Salomskiene, J.; Jonkuviene, D.; Macioniene, I.; Abraitiene, A.; Zeime, J.; Repeckiene, J.; Vaiciulyte-Funk, L. Differences in the occurence and efficiency of antimicrobial compounds produced by lactic acid bacteria. *Eur. Food Res. Technol.* **2019**, *245*, 569–579. [CrossRef]

16. Kanwal, H.; Di Cerbo, A.; Zulfiqar, F.; Sabia, C.; Nawaz, A.; Siddiqui, F.M.; Aqeel, M.; Ghazanfar, S. Probiotic characterization and population diversity analysis of gut-associated Pediococcus acidilactici for its potential use in the dairy industry. *Appl. Sci.* **2021**, *11*, 9586. [CrossRef]

17. Daliri, E.B.-M.; Lee, B.H. New Perspectives on probiotics in health and disease. *Food Sci. Hum. Wellness* **2015**, *4*, 56–65. [CrossRef]

18. Gupta, S.; Mohanty, U.; Majumdar, R.K. Isolation and characterization of lactic acid bacteria from traditional fermented fish product *Shidal* of India with reference to their probiotic potential. *LWT* **2021**, *146*, 111641. [CrossRef]

19. Ogunbanwo, S.T.; Sanni, A.I.; Onilude, A.A. Characterization of bacteriocin produced by *Lactiplantibacillus plantarum* F1 and *Lactobacillus brevis* OG1. *Afr. J. Biotechnol.* **2003**, *2*, 219–227.

20. Mohammed, S.; Çon, A.H. Isolation and characterization of potential probiotic lactic acid bacteria from traditional cheese. *LWT* **2021**, *152*, 112319. [CrossRef]

21. Prawan, K.; Bhima, B. Isolation and characterization of lactic acid bacteria for probiotic application from plant sources. *Int. J. Adv. Res.* **2017**, *5*, 869–876. [CrossRef]

22. Breed, R.S.; Bergey, D.H. *Bergey's Manual of Determinative Bacteriology*, 7th ed.; Williams & Wilkins Co.: Baltimore, MD, USA, 1957.

23. Guo, X.-H.; Kim, J.-M.; Nam, H.-M.; Park, S.-Y.; Kim, J.-M. Screening lactic acid bacteria from swine origins for multistrain probiotics based on in vitro functional properties. *Anaerobe* **2010**, *16*, 321–326. [CrossRef] [PubMed]

24. Singhal, N.; Singh, N.S.; Mohanty, S.; Kumar, M.; Virdi, J.S. Rhizospheric lactobacillus plantarum (*Lactiplantibacillus plantarum*) strains exhibit bile salt hydrolysis, hypocholestrolemic and probiotic capabilities in vitro. *Sci. Rep.* **2021**, *11*, 15288. [CrossRef] [PubMed]

25. Kılıç, G.B.; Karahan, A.G. Identification of lactic acid bacteria isolated from the fecal samples of healthy humans and patients with dyspepsia, and determination of their ph, bile, and antibiotic tolerance properties. *J. Mol. Microbiol. Biotechnol.* **2010**, *18*, 220–229. [CrossRef]

26. Yadav, R.; Puniya, A.K.; Shukla, P. Probiotic properties of lactobacillus plantarum rypr1 from an indigenous fermented beverage Raabadi. *Front. Microbiol.* **2016**, *7*, 1683. [CrossRef]

27. Gharbi, Y.; Fhoula, I.; Ruas-Madiedo, P.; Afef, N.; Boudabous, A.; Gueimonde, M.; Ouzari, H.-I. In-vitro characterization of potentially probiotic lactobacillus strains isolated from human microbiota: Interaction with pathogenic bacteria and the enteric cell line HT29. *Ann. Microbiol.* **2018**, *69*, 61–72. [CrossRef]

28. Jamaly, N. Probiotic potential of lactobacillus strains isolated from known popular traditional Moroccan dairy products. *Br. Microbiol. Res. J.* **2011**, *1*, 79–94. [CrossRef]

29. Ghazanfar, S. Study on the Effect of Dietary Supplementation of Saccharomyces cerevisiae on Performance of Dairy Cattle and Heifers. Ph.D. Thesis, Department of Microbiology, Quaid-i-Azam University, Islamabad, Pakistan, 2016.

30. Nadri, M.H.; Salim, Y.; Basar, N.; Yahya, A.; Zulkifli, R.M. Antioxidant activities and tyrosinase inhibition effects of *phaleria macrocarpa* extracts. *Afr. J. Tradit. Complement. Altern. Med.* **2014**, *11*, 107–111. [CrossRef]

31. Al Mohaini, M.; Farid, A.; Muzammal, M.; Dadrasnia, A.; Alsalman, A.J.; Al Hawaj, M.A.; Alhashem, Y.N.; Ismail, S. Pathological study of Pasteurella Multocida Recombinant Clone ABA392. *Pak. J. Med. Health Sci.* **2022**, *16*, 1112–1116. [CrossRef]

32. Adesulu-Dahunsi, A.T.; Jeyaram, K.; Sanni, A.I. Probiotic and technological properties of exopolysaccharide producing lactic acid bacteria isolated from cereal-based Nigerian fermented food products. *Food Control* **2018**, *92*, 225–231. [CrossRef]

33. Hattingh, M.; Alexander, A.; Meijering, I.; Van, R.C.; Dicks, L.M. Amylolytic strains of lactobacillus plantarum isolated from Barley. *Afr. J. Biotechnol.* **2015**, *14*, 310–318. [CrossRef]

34. Khalil, T.; Okla, M.K.; Al-Qahtani, W.H.; Ali, F.; Zahra, M.; Shakeela, Q.; Ahmed, S.; Akhtar, N.; AbdElgawad, H.; Asif, R.; et al. Tracing probiotic producing bacterial species from gut of buffalo (bubalus bubalis), south-east-Asia. *Braz. J. Biol.* **2022**, *84*, e259094. [CrossRef]

35. Atanassova, M. Isolation and partial biochemical characterization of a proteinaceous anti-bacteria and anti-yeast compound produced by *Lacticaseibacillus paracasei* subsp. paracasei strain M3. *Int. J. Food Microbiol.* **2003**, *87*, 63–73. [CrossRef]

36. Khan, A.N.; Yasmin, H.; Ghazanfar, S.; Hassan, M.N.; Keyani, R.; Khan, I.; Gohar, M.; Shahzad, A.; Hashim, M.J.; Ahmad, A. Antagonistic, anti-oxidant, anti-inflammatory and anti-diabetic probiotic potential of lactobacillus agilis isolated from the rhizosphere of the medicinal plants. *Saudi J. Biol. Sci.* **2021**, *28*, 6069–6076. [CrossRef] [PubMed]

37. Singhvi, M.; Joshi, D.; Adsul, M.; Varma, A.; Gokhale, D. D-(−)-lactic acid production from cellobiose and cellulose by lactobacillus lactis mutant RM2-24. *Curr. Green Chem.* **2010**, *12*, 1106–1109. [CrossRef]

38. Dinçer, E.; Kıvanç, M. Lipolytic activity of lactic acid bacteria isolated from Turkish pastırma. *Anadolu Univ. J. Sci. Technol.–C Life Sci. Biotechnol.* **2018**, *7*, 12–19. [CrossRef]

39. Farid, A.; Shah, A.H.; Ayaz, M.; Amin, A.; Yaseen, M.; Ullah, H.; Haq, F. Comparative study of biological activity of glutathione, sodium tungstate and glutathione-tungstate mixture. *Afr. J. Biotechnol.* **2012**, *11*, 10431–10437.

40. Abid, R.; Ghazanfar, S.; Farid, A.; Sulaman, S.M.; Idrees, M.; Amen, R.A.; Muzammal, M.; Shahzad, M.K.; Mohamed, M.O.; Khaled, A.A.; et al. Pharmacological Properties of 4′, 5,7-Trihydroxyflavone (Apigenin) and Its Impact on Cell Signaling Pathways. *Molecules* **2022**, *27*, 4304. [CrossRef]

41. Patel, M.; Siddiqui, A.J.; Hamadou, W.S.; Surti, M.; Awadelkareem, A.M.; Ashraf, S.A.; Alreshidi, M.; Snoussi, M.; Rizvi, S.M.; Bardakci, F.; et al. Inhibition of bacterial adhesion and antibiofilm activities of a glycolipid biosurfactant from lactobacillus rhamnosus with its physicochemical and functional properties. *Antibiotics* **2021**, *10*, 1546. [CrossRef]

42. Abriouel, H.; Casado Muñoz, M.d.C.; Lavilla Lerma, L.; Pérez Montoro, B.; Bockelmann, W.; Pichner, R.; Kabisch, J.; Cho, G.-S.; Franz, C.M.A.P.; Gálvez, A.; et al. New insights in antibiotic resistance of lactobacillus species from fermented foods. *Int. Food Res. J.* **2015**, *78*, 465–481. [CrossRef]

43. Alsalman, A.J.; Farid, A.; Al Mohaini, M.; Al Hawaj, M.A.; Muzammal, M.; Khan, M.H.; Dadrasnia, A.; Alhashem, Y.N.; Ghazanfar, S.; Almusalami, E.M.; et al. Analysis and Characterization of Chitinase in Bacillus salmalaya Strain 139SI. *Int. J. Curr. Res. Rev.* **2022**, *14*, 31–36. [CrossRef]

44. Al Hawaj, M.A.; Farid, A.; Al Mohaini, M.; Alsalman, A.J.; Muzammal, M.; Khan, M.H.; Dadrasnia, A.; Alhashem, Y.N.; Ghazanfar, S.; Almusalami, E.M.; et al. Biosurfactant Screening and Antibiotic Analysis of Bacillus salmalaya. *Int. J. Cur. Res. Rev.* **2022**, *14*, 56–64. [CrossRef]

45. Al Mohaini, M.; Farid, A.; Alsalman, A.J.; Al Hawaj, M.A.; Alhashem, Y.N.; Ghazanfar, S.; Muzammal, M.; Khan, M.H.; Dadrasnia, A.; Ismail, S. Screening of Anticancer and Immunomodulatory Properties of Recombinant pQE-HAS113 Clone Derived from Streptococcus Equi. *Pak. J. Med. Health Sci.* **2022**, *16*, 1100. [CrossRef]

46. Angmo, K.; Kumari, A.; Savitri; Bhalla, T.C. Probiotic characterization of lactic acid bacteria isolated from fermented foods and beverage of Ladakh. *LWT—Food Sci. Tech.* **2016**, *66*, 428–435. [CrossRef]

47. FAO/WHO. Probiotics in food. In *Health and Nutritional Properties and Guidelines for Evaluation*; FAO Food and Nutrition Paper 85; FAO: Rome, Italy, 2006.

48. Saadullah, M.; Asif, M.; Farid, A.; Naseem, F.; Rashid, S.A.; Ghazanfar, S.; Muzammal, M.; Ahmad, S.; Bin Jardan, Y.A.; Alshaya, H.; et al. A Novel Distachionate from Breynia distachia Treats Inflammations by Modulating COX-2 and Inflammatory Cytokines in Rat Liver Tissue. *Molecules* **2022**, *27*, 2596. [CrossRef]

49. Borah, D.; Gogoi, O.; Adhikari, C.; Kakoti, B.B. Isolation and characterization of the new indigenous *Staphylococcus* sp. DBOCP06 as a probiotic bacterium from traditionally fermented fish and meat products of Assam State. *Egypt. J. Basic Appl. Sci.* **2016**, *3*, 232–240. [CrossRef]

50. Al Mohaini, M.; Farid, A.; Muzammal, M.; Gazanffar, S.; Dadrasnia, A.; Alsalman, A.J.; Al Hawaj, M.A.; Alhashem, Y.N.; Ismail, S. Enhancing Lipase Production of Bacillus salmalay Strain 139SI Using Different Carbon Sources and Surfactants. *Appl. Microbiol.* **2022**, *2*, 237–247. [CrossRef]

51. Ghalouni, E.; Hassaine, O.; Karam, N.-E. Phenotypic identification and technological characterization of lactic acid bacteria isolated from l'ben, an Algerian traditional fermented cow milk. *J. Pure Appl. Microbiol.* **2018**, *12*, 521–532. [CrossRef]

52. Alsalman, A.J.; Farid, A.; Al Mohaini, M.; Muzammal, M.; Khan, M.H.; Dadrasnia, A.; Ismail, S. Chitinase Activity by Chitin Degrading Strain (Bacillus Salmalaya) in Shrimp Waste. *Int. J. Curr. Res. Rev.* **2022**, *14*, 11–17. [CrossRef]

53. Tokatlı, M.; Gülgör, G.; Bağder Elmacı, S.; Arslankoz İşleyen, N.; Özçelik, F. In vitro properties of potential probiotic indigenous lactic acid bacteria originating from traditional pickles. *Biomed Res. Int.* **2015**, *2015*, 315819. [CrossRef]

54. Adnan, M.; Patel, M.; Hadi, S. Functional and health promoting inherent attributes ofenterococcus hiraeF2 as a novel probiotic isolated from the digestive tract of the freshwater fishcatla catla. *PeerJ* **2017**, *5*, e3085. [CrossRef] [PubMed]

55. Abid, S.; Farid, A.; Abid, R.; Rehman, M.U.; Alsanie, W.F.; Alhomrani, M.; Alamri, A.S.; Asdaq, S.M.B.; Hefft, D.I.; Saqib, S.; et al. Identification, Biochemical Characterization, and Safety Attributes of Locally Isolated Lactobacillus fermentum from Bubalus bubalis (Buffalo) Milk as a Probiotic. *Microorganisms* **2022**, *10*, 954. [CrossRef]

56. Shivangi, S.; Devi, P.B.; Ragul, K.; Shetty, P.H. Probiotic Potential of Bacillus Strains Isolated from an Acidic Fermented Food Idli. *Probiotics Antimicrob. Proteins* **2020**, *12*, 1502–1513.

57. Fatima, S.; Muzammal, M.; Khan, M.A.; Farid, A.; Kamran, M.; Qayum, J.; Qureshi, M.; Khan, M.N.; Khan, M.A. CRISPR/Cas9 endonucleases: A new era of genetic engineering. *Abasyn J. life Sci.* **2021**, *4*, 29–39. [CrossRef]

58. Farid, A.; Javed, R.; Hayat, M.; Muzammal, M.; Khan, M.H.; Ismail, S.; Rashid, S.A. Screening of Strobilanthes urticifolia wall.ex kuntze for Antitermite and insecticidal activities. *Abasyn J. life Sci.* **2021**, *4*, 40–45. [CrossRef]

59. Amaretti, A.; Di Nunzio, M.; Pompei, A.; Raimondi, S.; Rossi, M.; Bordoni, A. Antioxidant properties of potentially probiotic bacteria: In vitro and in vivo activities. *Appl. Microbiol. Biotechnol.* **2013**, *97*, 809–817. [CrossRef]

60. Hameed, A.; Condò, C.; Tauseef, I.; Idrees, M.; Ghazanfar, S.; Farid, A.; Muzammal, M.; Al Mohaini, M.; Alsalman, A.J.; Al Hawaj, M.A.; et al. Isolation and Characterization of a Cholesterol-Lowering Bacteria from Bubalus bubalis Raw Milk. *Fermentation* **2022**, *8*, 163. [CrossRef]

61. Ghazanfar, S. Understanding the Mechanism of Action of Indigenous Target Probiotic Yeast: Linking the Manipulation of Gut Microbiota and Performance in Animals. In *Saccharomyces*; Intech Open: London, UK, 2021.

 *fermentation*

Article

Fermentation Enhances the Anti-Inflammatory and Anti-Platelet Properties of Both Bovine Dairy and Plant-Derived Dairy Alternatives

Kyeesha Glenn-Davi [1], Alison Hurley [1], Eireann Brennan [1], Jack Coughlan [1], Katie Shiels [2], Donal Moran [1], Sushanta Kumar Saha [2], Ioannis Zabetakis [1,3,4] and Alexandros Tsoupras [1,3,4,*]

[1] Department of Biological Sciences, University of Limerick, V94 T9PX Limerick, Ireland; 17079705@studentmail.ul.ie (K.G.-D.); alisonhurley97@gmail.com (A.H.); brennan.eireann@gmail.com (E.B.); coughlanjack3@gmail.com (J.C.); donal.moran@ul.ie (D.M.); ioannis.zabetakis@ul.ie (I.Z.)
[2] Shannon Applied Biotechnology Centre, Technological University of Shannon, Moylish Park, V94 E8YF Limerick, Ireland; katie.shiels@lit.ie (K.S.); sushanta.saha@lit.ie (S.K.S.)
[3] Bernal Institute, University of Limerick, V94 T9PX Limerick, Ireland
[4] Health Research Institute, University of Limerick, V94 T9PX Limerick, Ireland
* Correspondence: alexandros.tsoupras@ul.ie

Abstract: Within the present study, the effects of fermentation on the anti-inflammatory and anti-platelet properties of both homemade and commercially purchased bovine dairy and almond, coconut, and rice-based dairy alternatives were evaluated. The extracted total lipids (TL) from homemade and commercially purchased fermented and unfermented bovine, almond, coconut, and rice-based products were further separated into their neutral lipids (NL) and polar lipids (PL) fractions by counter current distribution. The TL, PL, and NL of each sample were assessed in human platelets against the inflammatory and thrombotic mediator, platelet-activating factor (PAF), and the well-established platelet agonist, adenosine $5'$ diphosphate (ADP). In all samples, the PL fractions showed significantly stronger inhibitory effects against human platelet aggregation induced by PAF or ADP, in comparison to the TL and NL, with higher specificity against PAF. PL of all fermented products (bovine yogurt and fermented dairy alternatives from almond, rice, and coconut), exhibited the strongest anti-inflammatory and anti-platelet potency, in comparison to PL from their initial pasteurized materials (bovine milk and rice, almond, and coconut-based dairy alternative drinks). PL of the pasteurized rice-based drink and, especially PL from the novel homemade rice-based fermented product (HMFRD), showed the strongest anti-PAF and anti-ADP potency compared to all samples, with anti-PAF activity being most potent overall. The unfermented pasteurized coconut-based drink showed the lowest anti-inflammatory and anti-platelet potency, and the bovine and almond-based fermented products showed an intermediate effect. Further lipidomics with LC-MS analysis of all these PL fractions revealed that fermentation altered their fatty acid content in a way that decreased their degree of saturation and increased the content of unsaturated fatty acids, thus providing a rationale for the stronger anti-inflammatory and anti-platelet potency of the more unsaturated PL fractions of the fermented products. This study has shown that fermentation alters the fatty acid content and the bio-functionality of the PL bioactives in both fermented bovine dairy and plant-based dairy alternatives, and subsequently improved their anti-inflammatory and anti-platelet functional properties.

Keywords: fermentation; bovine; dairy alternatives; yogurt; anti-inflammatory; anti-platelet; PAF; ADP

 check for updates

Citation: Glenn-Davi, K.; Hurley, A.; Brennan, E.; Coughlan, J.; Shiels, K.; Moran, D.; Saha, S.K.; Zabetakis, I.; Tsoupras, A. Fermentation Enhances the Anti-Inflammatory and Anti-Platelet Properties of Both Bovine Dairy and Plant-Derived Dairy Alternatives. *Fermentation* **2022**, *8*, 292. https://doi.org/10.3390/fermentation8070292

Academic Editors: Carlo Giuseppe Rizzello and Michela Verni

Received: 5 May 2022
Accepted: 17 June 2022
Published: 21 June 2022

Publisher's Note: MDPI stays neutral with regard to jurisdictional claims in published maps and institutional affiliations.

1. Introduction

Chronic disorders, including cardiovascular diseases (CVD), respiratory disease, diabetes, and cancer account for over 70% of global deaths [1]. Chronic inflammation has been implicated as the cause of such diseases. More specifically, inflammatory and thrombotic mediators like platelet-activating factor (PAF) and thrombin, as well as platelet agonists like adenosine 5′ diphosphate (ADP) and collagen, play a crucial role in the inflammatory and thrombotic activation and response of key cells that are implicated in the onset and development of such chronic disorders [2–4]. Tackling inflammation through pharmaceutical therapies to treat inflammation-related chronic disorders can have unwanted side effects so dietary intervention and especially novel functional foods with anti-inflammatory properties are of particular interest due to the absence of side effects [2,3,5,6]. Bioactive compounds from natural sources with strong anti-inflammatory potency, such as the polar lipid bioactives found in several foods, such as dairy and fermented products, have been found to inhibit the action of PAF and the activities of other inflammatory mediators showing the promise for natural bioactive compounds for the treatment of disease [2,3,5–10].

Bovine dairy and, especially dairy lipids, are sometimes tied to negative perceptions regarding their effects on human health [11]. This is due to the increased degree of saturation and the higher levels of saturated fatty acids in such dairy, the consumption of which may increase serum levels of triglycerides and cholesterol and thus increase the risk for the development of CVD [12,13]. Evidence suggests, however, that the link between dairy fats and negative health outcomes is not well supported, while fermented dairy products like yogurt and kefir have exhibited positive effects on cardiovascular health [14–16]. Nevertheless, the consumption of plant-based alternative drinks has risen due to concerns about dairy consumption on human health, animal welfare, and environmental sustainability [17]. The economic demand for dairy alternatives is expected to reach \$14.36 billion [18]. However, there are several limitations to the replacement of dairy products with plant-based alternatives, such as their low protein content and poorer protein quality [19] and differences in nutritional quality, and the presence of anti-nutritional factors [20–22]. Plant-based dairy alternative drinks are produced from legumes, cereals, nuts, or seeds by either pulverizing plant material with water to extract the water-soluble nutrients or by forming an oil-in-water emulsion using plant oils [23]. Many of these products are fortified with vitamins and minerals, including vitamin B12, calcium, and vitamin D, to bring their nutritional composition closer to that of dairy milk before being bottled and subjected to ultra-high temperature treatment (UHT). Additionally, sugars, flavors, and other vegetable oils may be added to improve the sensory properties.

Fermented plant-based dairy alternatives to cheese and yogurt are also widespread. Fermentation of plant material has been shown to improve flavor, texture, and nutritional profile [24,25]. Some common substrates for commercially purchased plant-based fermented products are soy, oat, almond, and coconut, while traditional recipes also exist for rice-based fermented products. Classic fermented dairy products are mainly produced by incubation with monocultures of lactic acid bacteria (LAB) strains [15], while other approaches involving mixed-culture fermentation and utilizing two or more microbial species are gaining ground. Independent of the method and starter culture, fermentation of both bovine dairy and dairy alternatives have been shown to beneficially modulate their lipid content and functionality [6–8,15,16,24,25]. Nevertheless, the research on the anti-inflammatory and anti-platelet potency of such fermented products and especially their lipid content is limited.

The aim of this study was to evaluate the anti-platelet and anti-inflammatory properties of both commercially purchased and homemade, fermented and unfermented, bovine dairy and plant-based dairy alternative products of a high-fat plant source, coconut, an intermediate-fat plant source, almond, and of a plant source, rice, for which fermented products as dairy alternatives do not exist in the market. For the first time, these products were evaluated against PAF-associated inflammatory and thrombotic activation, as well as against the platelet activation effects of a well-established platelet agonist, ADP, in human platelets. Additionally, the most bioactive lipid fractions found in each product were further analyzed for their fatty acid content by LC-MS analysis, in order to evaluate structure–activity relationships. This study will provide valuable insights into the effect of fermentation in dairy and dairy alternatives, as well as the potential use of plant-based fermented products, as novel functional foods against the development of CVD and other inflammation-related chronic disorders.

2. Materials and Methods

2.1. Materials, Reagents, and Instrumentation

Analytical glass/plastic consumables, reagents, and solvents were purchased from Fisher Scientific Ltd. (Dublin, Ireland). Solvent evaporation was carried out by flash rotatory evaporation (Buchi Rotavapor, Mason Technology, Dublin, Ireland) and nitrogen stream using nitrogen cylinders (BOC, Dublin, Ireland). Platelet aggregation bioassay materials were purchased from Labmedics LLP (Abingdon on the Thames, UK). Blood sampling was carried out using 20G safety needles and evacuated sodium citrate S-monovettes from Sarstedt Ltd. (Wexford, Ireland). Standard PAF and bovine serum albumin (BSA) were purchased from Merk (Wicklow, Ireland) while standard ADP was purchased from CHRONOLOG (Havertown, PA, USA). Platelet aggregation bioassays of human platelet-rich plasma (hPRP) were carried out using a Chronolog-490 two-channel turbidimetric platelet aggregometer (Havertown, PA, USA) coupled to the accompanying AGGRO/LINK software. Centrifugation was performed using Eppendorf 5702R centrifuge (Eppendorf Ltd., Stevenage, UK) and Heraeus Biofuge Stratos centrifuge (Fisher Scientific Ltd., Dublin, Ireland). Spectrophotometric analysis was conducted using the Shimadzu UV-1800 spectrophotometer (Tokyo, Japan) with 1 cm quartz cuvettes. LC-MS analysis was carried out using HPLC (Agilent 1260 series, Agilent Technologies Ireland Ltd., Little Island, Co. Cork, Ireland), Q-TOF mass spectrometer (Agilent 6520), and an Agilent C18 Poroshell 120 column (Agilent Technologies Ireland Ltd., Little Island, Co. Cork, Ireland).

2.2. Sample Preparation

Home-made plant-based dairy alternative drinks were prepared using commercially available rice, desiccated coconut, and almonds. Homemade rice-based dairy alternative drink was prepared by homogenizing 100 g cooked rice with 500 mL water and straining through a muslin cloth. Homemade coconut-based dairy alternative drink was prepared by homogenizing 200 g shredded coconut with 1 L boiling water and straining through a muslin cloth. Homemade almond-based dairy alternative drink was prepared by soaking 100 g almonds overnight to remove the skin, homogenizing with 500 mL water then straining through a muslin cloth. Alternative versions of the almond and rice drinks were prepared using long grain rice and basmati rice and raw and pasteurized almonds.

The fermented bovine dairy (bovine yogurt) and the fermented plant-based products of all these dairy alternative sources were prepared by inoculation of 200 mL pasteurized bovine milk or each one of the aforementioned home-made purchased plant-based drinks with commercially available yogurt starter culture containing *Streptococcus thermophilus* and *Lactobacillus casei* obtained from bio yogurt. The viability of the starter bacteria was assessed using flow cytometry. All samples were incubated at 37 °C for 24 h. the initial microbial load of all samples was 10^4 CFU/mL. The microbial load was determined using flow cytometry and pH was measured before and after fermentation. Samples were stored at 4 °C before testing.

Commercially purchased pasteurized bovine milk, rice-based drink ("Rice Dream Original Organic", Hain Daniels Group, Leeds, UK), coconut-based drink ("Alpro Coconut No Sugars", Alpro Ltd., Northamptonshire, UK), and almond-based drink ("Alpro Almond No Sugars", Alpro Ltd., UK) were obtained from local grocery stores and stored at 4 °C before analysis. Commercially purchased bovine yogurt and fermented coconut product ("Coconut Collaborative Dairy-Free Natural Coconut Yog", The Coconut Collaborative Ltd., London, UK) and commercially fermented almond products ("Nush Dairy Free Almond M*lk Yog", Nush Foods, London, UK) were also obtained from local grocery stores and stored at 4 °C before analysis. Commercially purchased products were selected based on the purity of ingredients, the most minimal ingredients, and for being produced in similar ways as the homemade products were produced in the lab. Especially for the commercially purchased bovine yogurt and the fermented plant-based dairy alternative products, these were selected on the basis of purity of ingredients and microbial species present in the final product, as being similar to the microbial species used for the production of the homemade fermented products (*Streptococcus thermophilus* and *Lactobacillus casei*). No fermented rice dairy alternative product was found in the market to be purchased and tested. All other products were analyzed on the same day of their production or their purchase.

2.3. Extraction of Total Lipids and Separation into Polar Lipids and Neutral Lipids

The total lipids (TL) from each sample type were extracted in triplicate ($n = 3$), as previously described [9,26], using the Bligh and Dyer extraction method [27]. Briefly, each sample was homogenized using a Waring blender (Fisher Scientific Ltd., Dublin, Ireland) with 1:2:0.8 ($v/v/v$) chloroform/methanol/water then filtered using Whatman Grade 1 filter paper (Whatman, Maidstone, UK) with a Büchner vacuum filtration device. Phase separation was induced by transferring the filtrate into a separatory funnel and adding enough chloroform/methanol/water to achieve a 1:1:0.9 ($v/v/v$) ratio. Ten percent of the resulting TL fraction was retained for further analysis, while the remainder was further separated into the neutral lipids (NL) and the polar lipids (PL) fractions as previously described [9,26], using the Galanos and Kapoulas counter current distribution method [28]. Solvents of the obtained lipid fractions were evaporated using a flash rotary evaporator at a maximum of 40 °C and then the lipid samples were re-dissolved in 1:1 (v/v) chloroform/methanol and transferred to pre-weighted small glass vials, where the remaining solvent was further evaporated using nitrogen stream. The remaining evaporated lipid samples were weighed and stored in glass vials at -20 °C for a maximum of 8 weeks before further analysis.

2.4. Platelet Aggregometry Biological Assays

The anti-inflammatory and anti-platelet properties of all TL extracts and PL and NL fractions from the fermented and unfermented homemade and commercially purchased plant-based dairy alternative products from coconut, almond, and rice, as well as those of bovine milk and yogurt, were evaluated in human platelets for their ability to inhibit the aggregation of human platelet-rich plasma (hPRP) induced by the inflammatory and thrombotic mediator, PAF, and the well-established platelet agonist, ADP, using platelet aggregometry biological assays, as previously described [9,26,29].

All platelet aggregation bioassays were carried out in a Chronolog-490 two-channel turbidimetric platelet aggregometer (Havertown, PA, USA) coupled to the accompanying AGGRO/LINK® software (Version Opti8 for performing Aggregation, CHRONOLOG, Havertown, PA, USA) package, for analyses. For the sampling of blood, 20 G safety needles and evacuated sodium citrate S-monovettes were used, which were purchased from Sarstedt Ltd. (Wexford, Ireland), while the isolation of hPRP was performed by centrifugation using an Eppendorf 5702R centrifuge (Eppendorf Ltd., Stevenage, UK), as previously described [9,26,29]. A Shimadzu UV-1800 spectrophotometer (Kyoto, Japan) was used with a quartz 1 cm cuvette for the spectrophotometric analysis. All the other materials, consumables, and reagents for platelet aggregation were purchased from Labmedics LLP (Abingdon, UK) and Chronolog (Havertown, PA, USA), apart from standard PAF and BSA that were purchased from Sigma Aldrich (Wicklow, Ireland).

The anti-inflammatory and antithrombotic potency of all lipid extracts and fractions were expressed as means of their IC_{50} (half-maximal inhibitory concentrations) values $\pm$ standard deviation (SD), presented in mass (µg) of the bioactive lipid compounds in the aggregometer cuvette of 0.25 mL that causes 50% inhibition of PAF/ADP-induced platelet aggregation. Briefly, a range of concentrations for each lipid sample was assessed and a linear and dose-dependent relationship of the inhibitory effects of the lipid bioactives with the concentrations of each lipid sample tested was observed, within the 20–80% range of the percentage of inhibition against PAF/ADP induced aggregation of hPRP. In order to derive such inhibitory curves, a range of 10 to 200 µg amount of lipids was assessed for the most bioactive lipid extracts (PL), while a range of 50 to 800 µg was also utilized for the less bioactive ones (TL and NL). From this derived curve, for each lipid sample assessed, the concentration (µg) of the lipid sample that led to 50% of PAF/ADP induced aggregation of hPRP was calculated as the 50% inhibitory concentration value (IC_{50} value) for each lipid sample. Using blood samples from different donors, all experiments for evaluating the bioactivities of each lipid extract from all dairy and dairy alternative products. were performed several times ($n = 6$), for each replicate, in order to ensure reproducibility.

2.5. Fatty Acid Composition by LC-MS Analysis

The bioactive PL fractions of all fermented and unfermented dairy and dairy alternative samples against the PAF and ADP pathways of human platelet aggregation were analyzed by LC-MS as previously described [10], in order to elucidate their saponified fatty acid composition (free fatty acids; FFA). Briefly, each of these PL fractions was saponified by adding 1.5 mL of saponification reagent (2.5 M KOH: methanol (1:4, v/v)) and by a gentle vortex. Then the tubes were incubated at 72 °C for 15 min before the addition of 225 µL of formic acid. Then, 1725 µL of chloroform and 375 µL of ultrapure water were added to the tube, and vortexed to separate the content into two layers. The lower chloroform layer containing FFA was carefully transferred to amber gas chromatography vials and evaporated to dryness and stored at 20 °C until used for LC-MS analysis.

For LC-MS analysis, all dried lipids were re-constituted in 500 µL of methanol: dichloromethane (2:1, v/v), centrifuged at 13,000 rpm for 6 min (Heraeus Biofuge Stratos, Fisher Scientific Ltd., Dublin, Ireland) and the content was filtered through 3 kDa ultracentrifuge filters (Amicon Ultra 3k, Merck Millipore Ltd., Carrigtwohill, Co. Cork, Ireland). Then, 10 µL of the filtrate was injected and the fatty acid profiles were obtained from an HPLC (Agilent 1260 series, Agilent Technologies Ireland Ltd., Little Island, Co., Cork, Ireland) equipped with a Q-TOF mass spectrometer (Agilent 6520) and the source type was electrospray ionization (ESI). The column used for the resolution of fatty acids was an Agilent C18 Poroshell 120 column (2.7 µm, 3.0–150 mm). Mobile phase A consisted of 2 mM ammonium acetate in water and mobile phase B consisted of 2 mM ammonium acetate in 95% acetonitrile. Chromatographic separation was performed by gradient elution starting with 60% B for 1 min, then increasing to 90% B over 2.5 min. Subsequently, 90% B was held for 1.5 min and increased afterward to 100% over 5 min. Then, 100% B was held for 4 min, reducing afterward to 60% B over 0.5 min, and held for 1 min until the next run. The mobile phase flow rate was 0.3 mL/min until 5 min elapsed, increasing up to 0.6 mL/min after 10 min, and held at this flow rate until the end of the run. The mass spectrometer was operated in negative ionization mode, scanning from m/z 50–1100. Drying gas flow rate, nebulizer pressure, and temperature were 5 L min^{-1}, 30 psi, and 325 °C, respectively. The fragmentor and skimmer voltages were maintained respectively at 175 V and 65 V, and the capillary voltage was 3500 V. The monitoring reference masses used were 1033.988 and 112.9855 in the negative ion mode. Standard fatty acids such as palmitic (C16:0), oleic (C18:1n-9 cis), and eicosapentaenoic (EPA, 20:5n-3) acids were used to validate the LC-MS protocol by comparing their specific accurate mass and their retention time (RT) (Sigma Aldrich, Wicklow, Ireland). Each FFA from the corresponding lipid fractions was then identified based on their known accurate mass. The peak area of each identified fatty acid was the average of triplicate samples, which was used to determine the relative % of each FFA of the total fatty acids identified in each sample.

2.6. Statistical Analysis

Kolmogorov–Smirnov criterion was used to test the normality of the yield of extraction, the IC_{50} values, and fatty acid composition obtained for each lipid sample. Subsequently, for comparisons of the lipid content and FA composition of all PL fractions, acquired from the LC–MS analysis, the Kruskal–Wallis nonparametric multiple comparison test was used, while one-way analysis of variance (ANOVA) was used for all comparisons of the IC_{50} values of these lipid bioactives against the ADP-/PAF-induced platelet aggregation. The differences were statistically significant when the *p*-values were less than 0.05 ($p < 0.05$). The resulting IC_{50} values were expressed as a mean value of the mass of lipid (μg) in the aggregometer cuvette $\pm$ standard deviation (SD), while fatty acids content was expressed as the mean % percentage of total fatty acids of each sample (mean $\pm$ SD). Analysis of the data was carried out using a statistical software package (IBM-SPSS statistics 26 for Windows, SPSS Inc., Chicago, IL, USA).

3. Results and Discussion

3.1. Lipid Yield

The mass of recovered TL, PL, and NL of all homemade and commercial samples (expressed as g of lipids/100 g of sample) are given in Table 1. The combination of methods applied for this extraction and separation of lipids into PL and NL fractions in all samples, was the Bligh and Dyer [27] extraction method in conjunction with the Galanos and Kapoulas [28] counter-current distribution technique, as previously described, [9,26], which ensured a minimal loss of yield and retention of integrity of bioactive PL compounds in comparison to traditional methods like Soxhlet extraction that uses heat which threatens to degrade double bonds and other functional groups in the fatty acids and other lipid molecules of the lipid extracts. The methodology in which the lipid extracts and fractions were obtained and stored was thus suitable for the minimum loss of their functionality and their use in assessing their anti-inflammatory and anti-platelet potency using the well-established platelet aggregometry.

Table 1. Lipid extraction yield for homemade plant drinks and fermented plant products.

Samples	TL [1]	NL [1]	PL [1]
BM	2.87 $\pm$ 0.25 *	2.01 $\pm$ 0.22 *	0.80 $\pm$ 0.12 *
HMFBM	2.36 $\pm$ 0.41 *	1.55 $\pm$ 0.45 *	0.85 $\pm$ 0.17 *
HMAD	0.43 $\pm$ 0.29	0.32 $\pm$ 0.27	0.11 $\pm$ 0.07
HMFAD	0.25 $\pm$ 0.27	0.19 $\pm$ 0.20	0.06 $\pm$ 0.07
HMRD	0.20 $\pm$ 0.10	0.14 $\pm$ 0.04 #	0.06 $\pm$ 0.02
HMFRD	0.15 $\pm$ 0.05 #	0.09 $\pm$ 0.06 #	0.06 $\pm$ 0.05
HMCD	2.52 $\pm$ 0.81 *	2.47 $\pm$ 0.80 *	0.05 $\pm$ 0.01
HMFCD	1.88 $\pm$ 0.64 *	1.83 $\pm$ 0.64 *	0.05 $\pm$ 0.01
CPFBM	2.80 $\pm$ 0.25 *	2.41 $\pm$ 0.25 *	0.38 $\pm$ 0.03 *
CPAD	0.62 $\pm$ 0.06	0.59 $\pm$ 0.07	0.04 $\pm$ 0.01
CPFAD	3.42 $\pm$ 0.52 *	3.26 $\pm$ 0.51 *	0.16 $\pm$ 0.01
CPCD	0.77 $\pm$ 0.03	0.75 $\pm$ 0.03	0.02 $\pm$ 0.01 #
CPFCD	7.52 $\pm$ 0.47 **	7.39 $\pm$ 0.46 **	0.12 $\pm$ 0.01
CPRD	0.70 $\pm$ 0.08	0.56 $\pm$ 0.12	0.14 $\pm$ 0.05

[1] Expressed as mean $\pm$ SD ($n = 3$) of g of lipids per 100 g of source. ** Denotes the highest yield with a statistically significant difference ($p < 0.01$), while * denotes high yield in comparison to the other samples with lower yield ($p < 0.05$). # Denotes the lowest yields within all the TL, NL, and PL extracts respectively, with statistically significant difference ($p < 0.05$) Abbreviations: TL, total lipid; NL, neutral lipid; PL, polar lipid; BM, bovine milk; HMFBM, homemade fermented bovine milk; HMAD, homemade almond drink; HMFAD, homemade fermented almond drink; HMRD, homemade rice drink; HMFRD, homemade fermented rice drink; HMCD, homemade coconut drink; HMFCD, homemade fermented coconut drink; CPAD, commercially purchased almond drink; CPFAD, commercially purchased fermented almond drink; CPRD, commercially purchased rice drink; CPCD, commercially purchased coconut drink; CPFCD, commercially purchased fermented coconut drink.

As shown in Table 1, the higher TL content was observed in the bovine dairy products (the pasteurized bovine milk BM, and the homemade and commercially purchased bovine yogurts, HMFBM and CPFBM), but mostly in the unfermented and fermented coconut-based homemade and commercially purchased dairy alternative drinks HMCD and CPCD, and their fermented products, HMFCD and CPFCD, with the commercially purchased fermented coconut-based CPFCD product containing the highest TL content (approx 7.5 g of TL per 100 g of source). Apart from the commercially purchased fermented almond-based CPFAD product, all the other homemade and commercially purchased plant-based dairy alternatives, almond-based and rice-based fermented and unfermented HMAD, HMFAD, HMRD, HMFRD, CPAD, and CPRD products showed much lower TL content, approximately one-fifth of the TL content of the BM, yogurt, and coconut-based products' overall lipid content.

Independently of the overall TL content, all samples contained higher amounts of NL, with their PL comprising a smaller proportion of their TL content, while in some cases, the homemade products' fermentation increased the Pl content in the final fermented product in comparison to the unfermented raw material. For example, BM samples contained ~75% NL and ~25% PL of their TL, while fermentation resulted in a relative reduction of the NL content (65% of TL) and an increase of the PL content (35% of TL) of the homemade HMFBM yogurt. Similarly, the homemade fermented rice-based HMFRD product showed much higher PL content (40% of the TL) in comparison to either the homemade or the commercially purchased unfermented raw rice-based HMRD (PL was 30% of the TL) and CPRD (PL was 20% of the TL) dairy alternative drinks.

However, this was not the case for the commercially purchased CPFBM yogurt, where the NL content was increased and the Pl content was decreased in comparison to the initial raw BM. Fermentation did not alter the % PL content of the homemade HMAD, since this dairy alternative drink and its fermented HMFAD contained a similar proportion of NL (75% of TL) and PL (25% of TL), which were similar also to the proportions of NL and PL observed in the unfermented BM, while both commercially purchased CPAD and CPFAD showed lower content of PL (4–6% of TL) and higher NL content (94–96% of TL), suggesting that fermentation did not seem to influence the ratio of the PL and NL proportions in the almond-based products. Both homemade and commercially purchased coconut-based fermented and unfermented dairy alternatives, HMCD, HMFCD, CPCD, and CPFCD had a significantly higher proportion of NL to PL (97.5% of NL and 2.5% of PL to their TL content, respectively), while it seems that in these coconut-based dairy alternatives, fermentation again did not seem to influence their ratio of the PL and NL proportions. A commercially fermented rice-based dairy alternative drink was not available in the market to be purchased, for any analysis.

Overall, homemade versions had higher PL proportions than commercially available counterparts except for coconut products which had very similar proportions to homemade. Variability between homemade and commercial plant-based beverages and fermented products may be explained by the difference in the plant substrate used. Variability in lipid content is known to occur between varieties of plant substrates [30–32]. Fermentation seemed to only affect the lipid yield of homemade bovine and rice-based dairy alternatives. In terms of lipid content, fermentation did not have a significant effect on commercially purchased samples.

The current study demonstrated that plant-based dairy alternatives were found to contain substantial amounts of PL that differ in yield based on the substrate used. In comparison to dairy, rice-derived products were found to have the most similar or, in some cases (HMFRD), better polar lipid proportion. The PL content of the plant-based dairy alternatives assessed in this study was much lower than other plant substrates like apple pomace [26] and tea [33], but similar to other plant-based fermented products/beverages, such as beer and cider [9,10], indicating that the plant-based alternatives may be good candidates for functional foods containing considerable amounts of PL bioactives per 100 g of food source, with enhanced health-promoting effects.

3.2. Anti-Inflammatory and Anti-Platelet Properties of Fermented and Unfermented Plant Drinks

The anti-inflammatory and anti-platelet bioactivities of the TL extracts and NL and PL fractions from all fermented and unfermented plant-based dairy alternative drinks and bovine dairy were evaluated by assessing their inhibitory effects against aggregation of human platelets induced by the inflammatory pathways of PAF, as well as against platelet aggregation induced by the well-established platelet agonist, ADP (Figure 1). Results depicted in Figure 1A–C are expressed in IC_{50} values (half-maximal inhibitory concentrations; μg of lipid samples needed for 50% inhibition of platelet aggregation). Low IC_{50} values represent a more beneficial inhibitory effect against PAF pathways of inflammation and

ADP-induced platelet aggregation. In all samples assessed in this study (bovine dairy and plant-based alternatives), PL displayed the strongest anti-inflammatory potency against PAF and anti-platelet effect against ADP (Figure 1A) while TL extracts displayed an intermediate potency against both PAF and ADP (Figure 1B) with NL having the lowest potency against both mediators (Figure 1C). The finding that the PL bioactives from both bovine dairy and plant-based dairy alternatives showed the strongest anti-PAF and anti-ADP activities than the NL and TL comes in accordance with similar results observed for the anti-inflammatory properties of PL, TL, and NL in other dairy sources (ovine and caprine), as well as in other plant-based fermented and non-fermented foods, beverages, and their by-products [7–10,26,33]. Both homemade and commercially purchased fermented bovine yogurts (HMFBM and CPFBM) had stronger anti-PAF and anti-ADP effects compared to their unfermented raw material, pasteurized commercially purchased bovine milk (BM). This suggests that fermentation of bovine milk by *Streptococcus thermophilus*, and *Lactobacillus casei* increased the anti-inflammatory and anti-platelet bio-functionality of their PL present.

A similar effect was observed in homemade almond, coconut, and rice-based fermented dairy alternatives, whereby the PL-bioactives of these plant-based fermented products showed much stronger anti-PAF and anti-ADP potency in comparison to their unfermented raw materials (almond, coconut, and rice dairy alternative drinks). This relationship was also seen in commercially purchased fermented almond and coconut products (CPFAD and CPFCD), the PL of which had higher bioactivities than the commercially purchased unfermented almond and coconut drinks. Since this effect can be seen in both bovine and plant-based products, it can be reasonably suggested that fermentation by species of *Lactobacillus casei* and *Streptococcus thermophilus* led to the increase in bio-functionality of PL fractions. Although it should be noted that neither plant-based fermented product contained the specific species *Lactobacillus casei* and the CPFAD stated the use of "live vegan cultures" including *L. acidiphilus* and *Bifidobacterium*, therefore, the complete microbial composition is unknown and may differ from other samples. Additionally, CPFCD listed *L. bulgaricus* and *L. acidiphilus* in addition to *S. thermophilus* and *Bifidobacterium* being present in CPFBM.

It should be stressed that the strongest anti-PAF and anti-ADP effects of PL fraction from all products were observed in the novel homemade fermented rice-based product (HMFRD), which surprisingly had stronger anti-inflammatory and anti-platelet effects than the ones observed in this study for the PL of the commercially purchased bovine yogurt and previously studied fermented dairy [7,8]. Such a rice-based fermented product was not available in the market to purchase and assess as a base for a comparison with the observed activities of the homemade novel rice product. Nevertheless, the very potent anti-inflammatory and anti-platelet potency of the PL from this novel homemade fermented rice-based dairy alternative product further suggest new perspectives for novel fermented plant-based foods and dairy alternatives as functional foods with anti-inflammatory and antithrombotic potential. Next to the HMFRD, the PL from the commercially purchased CPFCD and CPFAD also had strong anti-inflammatory and anti-platelet properties, with a potency that was similar to that of the PL from the CPFBM.

Interestingly, the PL bioactives in all these rice, almond, and coconut-based fermented dairy alternative products, as well as the PL from the bovine fermented dairy yogurt, showed strong inhibition of the PAF-induced and inflammatory activation and ADP-induced platelet aggregation, but with significantly higher anti-inflammatory specificity against the PAF-pathway than their anti-platelet potency against the ADP-induced platelet aggregation. This result also follows previously reported results for PL bioactives from several sources, since these PL have higher specificity as direct antagonists and agonists for the G-coupled protein PAF-receptor in cell membranes, due to structural homology to PAF, they inhibit strongly the binding of PAF to its receptor and subsequently the PAF-induced inflammatory activation [2,3,34]. Such PL bioactives can indirectly affect other pathways of platelet activation, like that of the platelet agonist ADP [34], which provides a rationale for the higher specificity of the PL bioactives from both dairy and dairy alternative-fermented products against the PAF-pathway.

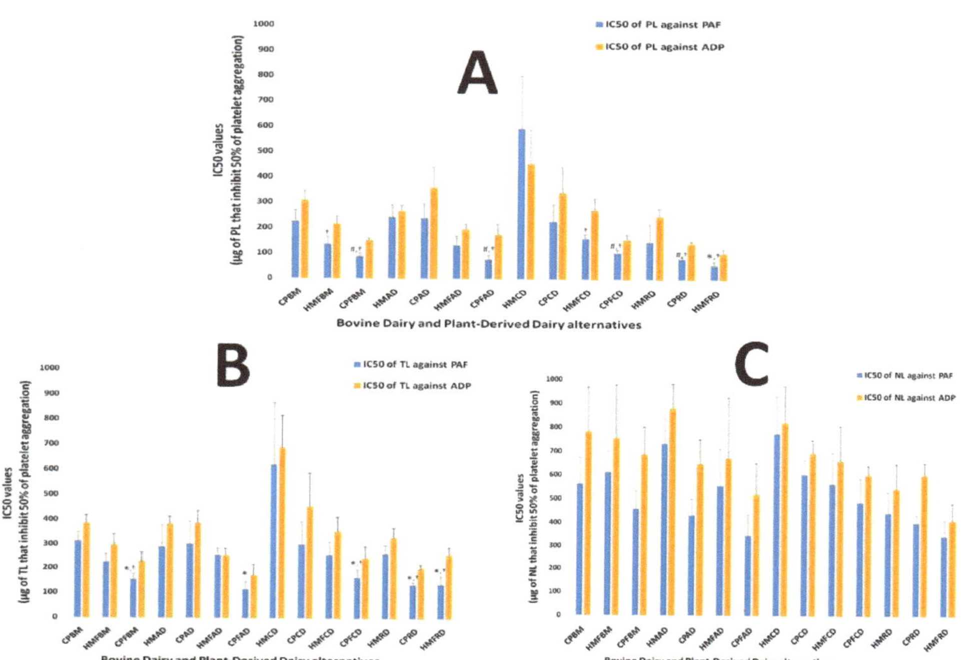

Figure 1. The anti-inflammatory and anti-platelet potency of bioactive PL (**A**), TL (**B**), and NL (**C**) from fermented and non-fermented plant-derived dairy alternatives, versus bovine milk and yogurt. The anti-inflammatory and anti-platelet potency of lipid bioactives from all samples were assessed against human platelet aggregation induced by the inflammatory and thrombotic mediator PAF (anti-PAF effects in blue bars) or by a classic platelet agonist, ADP (anti-ADP effects in yellow bars). Results are expressed as means of the IC50 values (half-maximal inhibitory concentrations) in µg of lipid extract in the aggregometer cuvette that causes 50% inhibition of PAF/ADP-induced platelet aggregation (the lower the IC50 value for a lipid extract the higher its inhibitory effect against the specific agonist of platelet aggregation). * Denotes statistically significant difference ($p < 0.05$) when the strongest anti-PAF potency (IC50 value) of the most bioactive lipid extracts were compared with those of all the other samples. # Denotes statistically significant difference ($p < 0.05$) when the intermediate anti-PAF effects of the PL bioactives from the samples assessed were compared with those showing the lowest anti-PAF effects. † Denotes statistically significant difference ($p < 0.05$) when the anti-PAF potency (IC50 values against PAF) of the lipid bioactives from a sample was compared to its anti-ADP potency (IC50 values against ADP) of the same sample. Abbreviations: PL: polar lipids; TL: total lipids; NL: neutral lipids; PAF: platelet-activating factor; ADP: adenosine 5′ diphosphate; CPBM: Commercially Purchased Bovine Milk; HMFBM: Homemade Fermented Bovine Milk (Homemade Yogurt); CPFBM: Commercially Purchased Fermented Bovine Milk (Commercially Purchased Yogurt); HMAD: Homemade Almond Drink; CPAD: Commercially Purchased Almond Drink; HMFAD: Homemade Fermented Almond Drink; CPFAD: Commercially Purchased Fermented Almond Drink; HMCD: Homemade Coconut Drink; CPCD: Commercially Purchased Coconut Drink; HMFCD: Homemade Fermented Coconut Drink; CPFCD: Commercially Purchased Fermented Coconut Drink; HMRD: Homemade Rice Drink; CPRD: Commercially Purchased Rice Drink; HMFRD: Homemade Fermented Rice Drink.

Lower anti-inflammatory and anti-platelet properties were observed in unfermented plant-based dairy alternatives drinks which were in line with anti-PAF and anti-ADP effects of PL from CPBM apart from CPRD and HMCD. Interestingly, the unfermented rice-based dairy alternative drinks, and especially the commercially purchased ones (CPRD) showed the strongest anti-PAF and anti-ADP effects within the unfermented products, which was comparable to that of the fermented products and higher than that of the PL from bovine milk. In contrast, the homemade coconut drink (HMCD) displayed the lowest bioactivity against both PAF and ADP of all samples assessed. Both commercially purchased and homemade almond-based drinks (HMAD and CPAD) showed an intermediate potency, similar to that of the BM.

Moreover, there did not appear to be an overall difference in the bioactivity of commercially purchased samples versus the homemade samples of the same source, therefore, it can be suggested that UHT treatment (above 135 °C for 2–5 s) did not result in the reduction of the bioactivity of their PL against the PAF and ADP-induced platelet aggregation.

While TL extracts had intermediate potency, they followed a similar trend relative to their PL bioactivities. More specifically, TL from fermented bovine products (CPFBM and HMFBM) had greater anti-PAF and anti-ADP than the unfermented CPBM. The same effect of fermentation can be seen with CPFAD and CPFCD which had greater bio-functionality compared to unfermented CPAD and CPCD. In homemade products, only HMFRD and HMFCD had stronger anti-PAF and anti-ADP than their unfermented drinks (HMRD and HMCD). Again, a higher specificity for PAF-pathway over ADP-pathway of TL fractions was seen with fermented bovine yogurts (AMFBM, CPFBM), some fermented plant-based products based in almond and coconut (CPFAD and CPFCD but, especially in the rice-based fermented product (HMFRD) and the unfermented rice-based dairy alternative drink (CPRD). Lastly, all NL extracts showed very low bioactivity against both PAF and ADP. A similar potency was observed for all NL samples assessed.

There are previous studies showing anti-platelet effects of coconut, almond, and rice-based products, however, these studies were mostly based on oils and/or extracts from these plant-derived sources and against the activity of classic platelet agonists, such as ADP and collagen, and not against prothrombotic inflammatory mediators like the PAF. For example, in a study conducted in 2010, it was found that rats fed a diet of rice bran oil, sesame oil, or a blend of either with coconut oil reduced platelet aggregation induced by ADP and collagen compared to coconut oil alone [35]. Moreover, a sweet almond (*Prunus amygdalus*) extract with beneficial effects against blood lipid biomarkers, such as lowering the total cholesterol, triacylglycerol, LDL-C, and VLDL-C while increasing HDL-C levels, also showed anti-coagulant properties, by increasing the prothrombin time, partial thromboplastin time, and clotting time [36]. The effects of rice bran (*Oryza sativa*) extract on platelet aggregation and adhesion induced by ADP and collagen were investigated by oral administration to male rats. Significant inhibition of aggregation and adhesion was observed [37]. Another study found that rats fed a hyperlipidemic diet supplemented with anthocyanin extracts from black rice (AEBR) had reduced serum TAG levels and improved platelet function [38].

Within the present study, the anti-inflammatory and anti-platelet potency of lipid bioactives from plant-based dairy alternatives and dairy products, fermented and unfermented, commercially purchased and homemade products, were studied against the inflammatory mediator for the first time, along with their anti-platelet potency against the well established classic platelet agonist ADP. The initial results of the present study further support the potential anti-inflammatory and anti-platelet functional properties of plant-based dairy alternatives, especially those from rice and almond, with similar or even better potency to that of the bovine dairy, which further suggests their use as novel functional foods with protective properties against inflammation, platelet aggregation, and associated disorders. Nevertheless, more research is needed to support this notion.

3.3. Fatty Acid Composition of the Fermented and Unfermented Dairy and Plant-Based Dairy Alternatives

The alteration of the FA profile of plant substrates by fermentation is particularly interesting in developing novel products, functional foods, and nutraceuticals. Fermented foods are also of increasing interest due to their probiotic and prebiotic properties, specifically the production of short-chain fatty acids (SCFAs), which are beneficial to gut health [39]. In addition, the alteration of FAs by fermentation was found to influence bio-functionality in dairy and non-dairy fermented products and, especially increased the anti-inflammatory potency of specific PL bioactives of these products [6–10,40]. This is because FAs and other bioactive lipids survive the digestion process, enter the blood through the intestinal epithelium, and participate in metabolic pathways. Specific FAs, including the monounsaturated fatty acid (MUFA) oleic acid, and the polyunsaturated fatty acids (PUFA) linoleic, linolenic eicosapentaenoic (EPA), and docosahexaenoic (DHA) fatty acids, have been found to possess anti-inflammatory properties and cardioprotective effects, while foods and diets rich in them have been associated with lower disease incidence and mortality [34,41]. A similar effect has been seen in legume-based dairy alternatives subjected to fermentation by lactic acid bacteria. For example, lactic acid fermentation of white beans resulted in increased oleic acid and decreased palmitic acid, indicating an increased health benefit from fermentation [30].

Even though there are studies on the FA composition of plant-based alternative dairy drinks [42], limited research compares the FA composition of several plant-based substrates before and after fermentation. The FA composition of the PL bioactives from the fermented and unfermented bovine dairy and alternative dairy products, either homemade or commercially purchased, were identified within the present study using LC-MS analysis as previously described [10]. The results of this analysis on homemade samples are shown in Table 2, while the results for commercially purchased samples are shown in Table 3. BM has been included in both tables for comparison purposes.

Concerning the dairy samples assessed in the present study, in BM, saturated fatty acids (SFA) were the dominant class of FAs in BM PL, with the most abundant being medium- and long-chain SFAs like lauric (12:0), myristic (14:0), and palmitic (16:0) acids, followed by unsaturated fatty acids (UFA), mainly MUFA, especially the omega-9 (n-9) oleic acid (19:1c9), and lower amounts of PUFA, such as linoleic (18:2) and linolenic (18:3) acids. Compared to the FA content of PL from BM, PL from HMFBM yogurt had decreased oleic acid content and increased lauric acid content. This difference explains the increase in the SFA content and a decrease in the UFA content, even though the PUFA content and its representatives linoleic, linolenic, docosapentaenoic acid (DPA), and EPA increased. Similar outcomes were observed for the PL from the HMFBM, where SFA was again the dominant class of FAs.

In contrast, PL from the CPFBM increased PUFA, especially the linoleic, linolenic, DPA, DHA, and EPA, increased the overall UFA content, while the MUFA content remained stable. The SFA content was substantially reduced, which resulted in UFA content equalizing the SFA content, increasing the UFA/SFA ratio, compared to the value for the same ratio of the BM. It should also be stressed that the PL of the CPFMB had the highest omega-3 (n-3) PUFA content of all the other bovine dairy and plant-based dairy alternative samples tested.

It has been suggested that substitution of SFA by UFA and thus an increase in the UFA/SFA ratio in food, either by specific processing or by fortification of the initial food matrix with bioactive UFA (MUFA and especially PUFA) from sidestreams, to design functional foods that prevent specific nutrient deficiencies and promote population health [43,44], by reducing the risk of chronic disorders. It seems that the benefits of the increase of the UFA content in the novel products seem to be associated mainly with their anti-inflammatory potency [34] rather than any other proposed effects on dyslipidemia and body composition since there is no substantial evidence that the replacement of SFA with UFA may benefit lipid profiles in metabolically healthy adults with overweight and obesity on markers of dyslipidemia and body composition [44], due to null results and a small number of studies.

Table 2. Fatty acid composition of home-made dairy and dairy alternative products.

Fatty Acid	Lipid Number	BM	HMFBM	HMRD	HMFRD	HMAD	HMFAD	HMCD	HMFCD
Caprylic	8:0	0.04 ± 0.011	0.04 ± 0.001	0.26 ± 0.163	0.15 ± 0.035	ND	ND	0.02 ± 0.005	0.13 ± 0.004
Pelargonic	9:0	0.02 ± 0.001	ND	ND	ND	ND	ND	ND	ND
Capric	10:0	0.69 ± 0.010	0.41 ± 0.010	0.09 ± 0.038	0.19 ± 0.043	ND	0.11 ± 0.010	0.36 ± 0.018	1.01 ± 0.026
Undecylic	11:0	0.16 ± 0.001	ND	ND	ND	ND	ND	ND	0.04 ± 0.000
Lauric	12:0	7.33 ± 0.086	35.11 ± 0.835	1.97 ± 0.110	9.82 ± 0.037	0.67 ± 0.023	9.12 ± 0.602	20.35 ± 0.096	46.95 ± 0.206
Tridecylic	13:0	0.38 ± 0.004	0.08 ± 0.004	0.06 ± 0.000	0.05 ± 0.001	0.04 ± 0.001	0.03 ± 0.002	0.06 ± 0.000	0.07 ± 0.001
Myristic	14:0	16.78 ± 0.113	15.92 ± 0.230	11.54 ± 0.305	11.02 ± 0.195	1.71 ± 0.034	4.78 ± 0.281	9.35 ± 0.084	21.82 ± 0.022
Pentadecylic	15:0	4.18 ± 0.015	0.65 ± 0.007	0.44 ± 0.007	0.33 ± 0.003	0.84 ± 0.670	0.22 ± 0.012	0.23 ± 0.003	0.07 ± 0.000
Palmitic	16:0	27.26 ± 0.442	18.53 ± 0.085	47.73 ± 0.142	34.18 ± 0.294	31.71 ± 0.280	15.57 ± 1.076	14.42 ± 0.412	9.68 ± 0.078
Palmitoleic	16:1(9)	4.02 ± 0.020	1.10 ± 0.010	0.64 ± 0.017	0.64 ± 0.008	0.74 ± 0.011	0.97 ± 0.059	1.31 ± 0.012	0.08 ± 0.000
Palmitelaidic	16:1(9t)	ND	ND	ND	ND	ND	ND	ND	ND
Margaric	17:0	3.18 ± 0.115	0.6 ± 0.005	0.35 ± 0.019	0.27 ± 0.011	0.39 ± 0.019	0.32 ± 0.016	0.19 ± 0.012	0.08 ± 0.003
Stearic	18:0	0.65 ± 0.046	0.44 ± 0.009	0.62 ± 0.012	0.38 ± 0.018	1.37 ± 0.019	4.80 ± 4.954	1.12 ± 0.135	0.38 ± 0.034
Oleic	18:1(9)	33.09 ± 0.270	21.77 ± 1.049	29.28 ± 0.260	18.52 ± 0.766	61.79 ± 0.269	49.07 ± 2.335	47.48 ± 0.507	17.11 ± 0.113
Elaidic	18:1(9t)	ND	ND	ND	ND	ND	ND	ND	ND
Linoleic	18:2(9,12) n-6	1.18 ± 0.004	3.88 ± 0.038	6.58 ± 0.119	23.41 ± 0.316	0.36 ± 0.004	14.06 ± 0.768	4.50 ± 0.050	2.40 ± 0.019
Linolenic (α + γ)	18:3(9,12,15) n-3/ 18:3(6,9,12) n-6	ND	0.39 ± 0.004	0.02 ± 0.001	0.86 ± 0.009	ND	0.19 ± 0.011	0.10 ± 0.000	0.08 ± 0.000
Stearidonic	18:4(6,9,12,15) n-3	ND	ND	ND	ND	ND	ND	ND	ND
Nonadecylic	19:0	0.02 ± 0.001	ND	ND	ND	ND	ND	ND	ND
Arachidic	20:0	0.48 ± 0.128	0.23 ± 0.108	0.23 ± 0.162	ND	0.38 ± 0.231	0.21 ± 0.150	0.21 ± 0.150	ND
Gadoleic	20:1(9)	0.26 ± 0.047	0.14 ± 0.015	0.20 ± 0.007	ND	ND	0.24 ± 0.018	0.31 ± 0.011	0.08 ± 0.010
Gondoic	20:1(11)	ND	ND	ND	ND	ND	ND	ND	ND
DihomoLinoleic	18:2(10,12) n-6	0.04 ± 0.002	ND	ND	ND	ND	0.05 ± 0.001	ND	ND
Dihomolinolenic	18:3(8,11,14) n-6	0.01 ± 0.000	0.16 ± 0.002	ND	ND	ND	0.01 ± 0.001	0.01 ± 0.000	ND
Mead Acid	20:3(5,8,11)	ND	ND	ND	ND	ND	ND	ND	ND
Arachidonic	20:4(5,8,11,14) n-6	ND	0.24 ± 0.005	ND	ND	ND	0.03 ± 0.006	ND	ND
Eicosatetraenoic	20:4	ND	ND	ND	ND	ND	ND	ND	ND
EPA	20:5(5,8,11,14,17) n-3	ND	0.12 ± 0.002	ND	ND	ND	0.05 ± 0.004	ND	ND
Heneicosylic	21:0	ND	ND	ND	ND	ND	ND	ND	ND
Behenic	22:0	ND	ND	ND	ND	ND	ND	ND	ND
Erucic	22:1(13)	0.16 ± 0.111	ND	ND	ND	ND	0.1 ± 0.074	ND	ND
Docosadienoic	22:2(13,16) n-6	ND	ND	ND	ND	ND	ND	ND	ND
Eranthic	22:3(5,13,16) n-6	ND	ND	ND	ND	ND	ND	ND	ND
Ardenic	22:4(7,10,13,16) n-6	ND	0.03 ± 0.003	ND	ND	ND	ND	ND	ND
DPA	22:5(4,7,10,13,16) n-3	ND	0.16 ± 0.001	ND	ND	ND	ND	ND	ND
DHA	22:6 (4,7,10,13,16,19) n-3	0.01 ± 0.002	ND	ND	ND	0.02 ± 0.008	0.05 ± 0.007	ND	ND
Tricosylic	23:0	ND	ND	ND	0.19 ± 0.264	ND	ND	ND	ND
Lignoceric	24:0	0.06 ± 0.080	ND	ND	ND	ND	ND	ND	ND
SFA		61.23	72.02	63.29	56.57	37.10	35.16	46.30	80.24
MUFA		37.53	23.01	30.11	19.17	62.53	50.39	49.10	17.27
PUFA		1.24	4.97	6.60	24.26	0.37	14.45	4.60	2.49
UFA		38.77	27.98	36.71	43.43	62.90	64.84	53.70	19.76
UFA/SFA		0.63	0.39	0.58	0.77	1.70	1.84	1.16	0.25

Values are expressed as the mean % percentage of total fatty acids of each sample (mean ± standard deviation (SD), *n* = 3). Abbreviations: TL, total lipid; NL, neutral lipid; PL, polar lipid; BM, bovine milk; HMFBM, homemade fermented bovine milk; HMAD, homemade almond drink; HMFAD, homemade fermented almond drink; HMRD, homemade rice drink; HMFRD, homemade fermented rice drink; HMCD, homemade coconut drink; HMFCD, homemade fermented coconut drink; SFA, saturated fatty acids; MUFA, monounsaturated fatty acids; PUFA, polyunsaturated fatty acids; UFA, unsaturated fatty acids; ND, not detected.

Table 3. Fatty acid composition of commercially purchased dairy and dairy alternative products.

Fatty Acid	Lipid Numbers	BM	CPFBM	CPRD	CPAD	CPFAD	CPCD	CPFCD
Caprylic	8:0	0.04 ± 0.011	0.05 ± 0.006	0.02 ± 0.008	0.01 ± 0.001	ND	0.06 ± 0.005	0.31 ± 0.031
Pelargonic	9:0	0.02 ± 0.001	ND	ND	ND	ND	0.01 ± 0.001	ND
Capric	10:0	0.69 ± 0.010	0.49 ± 0.063	0.12 ± 0.006	0.29 ± 0.027	0.01 ± 0.003	0.75 ± 0.021	0.81 ± 0.030
Undecylic	11:0	0.16 ± 0.001	0.18 ± 0.015	ND	ND	ND	0.17 ± 0.009	0.04 ± 0.001
Lauric	12:0	7.33 ± 0.086	8.21 ± 0.342	7.92 ± 0.102	5.24 ± 0.057	1.91 ± 0.011	16.5 ± 1.20	23.76 ± 0.310
Tridecylic	13:0	0.38 ± 0.004	0.32 ± 0.001	0.02 ± 0.000	0.24 ± 0.002	ND	0.21 ± 0.014	0.03 ± 0.001
Myristic	14:0	16.78 ± 0.113	11.72 ± 0.050	5.59 ± 0.031	10.31 ± 0.100	0.96 ± 0.016	12.48 ± 1.010	10.54 ± 0.038
Pentadecylic	15:0	4.18 ± 0.015	2.47 ± 0.031	0.13 ± 0.004	2.68 ± 0.043	0.06 ± 0.000	1.91 ± 0.201	0.06 ± 0.000
Palmitic	16:0	27.26 ± 0.442	20.1 ± 0.466	35.05 ± 0.087	27.49 ± 0.323	13.81 ± 0.248	19.73 ± 1.445	14.41 ± 0.073
Palmitoleic	16:1(9)	4.02 ± 0.020	3.44 ± 0.038	0.70 ± 0.005	2.48 ± 0.031	1.23 ± 0.019	7.26 ± 7.150	0.47 ± 0.004
Palmitelaidic	16:1(9t)	ND	ND	ND	ND	ND	ND	ND
Margaric	17:0	3.18 ± 0.115	4.78 ± 0.111	0.31 ± 0.010	2.66 ± 0.063	0.53 ± 0.036	1.86 ± 0.142	0.12 ± 0.003
Stearic	18:0	0.65 ± 0.046	1.05 ± 0.018	0.78 ± 0.004	1.4 ± 0.369	2.16 ± 0.066	0.97 ± 0.066	1.43 ± 0.172
Oleic	18:1(9)	33.09 ± 0.270	33.7 ± 0.241	24.25 ± 0.132	44.23 ± 0.315	55.41 ± 0.257	29.45 ± 1.915	39.66 ± 0.131
Elaidic	18:1(9t)	ND	ND	ND	ND	ND	ND	ND
Linoleic	18:2(9,12) n-6	1.18 ± 0.004	6.99 ± 0.075	22.97 ± 0.187	1.60 ± 0.028	21.94 ± 0.086	5.13 ± 0.509	7.29 ± 0.051
Linolenic (α + γ)	18:3(9,12,15) n-3/18:3(6,9,12) n-6	ND	2.37 ± 0.143	1.59 ± 0.025	0.09 ± 0.002	0.54 ± 0.005	1.04 ± 0.069	0.13 ± 0.003
Stearidonic	18:4(6,9,12,15) n-3	ND	0.06 ± 0.002	ND	ND	ND	ND	ND
Nonadecylic	19:0	0.02 ± 0.001	0.03 ± 0.015	ND	0.06 ± 0.002	ND	0.04 ± 0.003	ND
Arachidic	20:0	0.48 ± 0.128	0.37 ± 0.107	0.19 ± 0.138	0.71 ± 0.281	0.65 ± 0.460	1.2 ± 0.930	0.26 ± 0.220
Gadoleic	20:1(9)	0.26 ± 0.047	0.84 ± 0.144	0.30 ± 0.047	0.37 ± 0.009	0.71 ± 0.034	0.48 ± 0.074	0.63 ± 0.077
Gondoic	20:1(11)	ND	ND	ND	ND	ND	ND	ND
DihomoLinoleic	18:2(10,12) n-6	0.04 ± 0.002	0.39 ± 0.015	0.06 ± 0.002	0.04 ± 0.002	0.08 ± 0.003	0.08 ± 0.006	0.04 ± 0.001
Dihomolinolenic	18:3(8,11,14) n-6	0.01 ± 0.000	0.49 ± 0.004	ND	ND	ND	0.1 ± 0.009	ND
Mead Acid	20:3(5,8,11)	ND	ND	ND	ND	ND	ND	ND
Arachidonic	20:4(5,8,11,14) n-6	ND	0.77 ± 0.009	ND	ND	ND	0.17 ± 0.018	ND
Eicosatetraenoic	20:4	ND	ND	ND	ND	ND	ND	ND
EPA	20:5(5,8,11,14,17) n-3	ND	0.41 ± 0.006	ND	ND	ND	0.13 ± 0.008	ND
Heneicosylic	21:0	ND	ND	ND	ND	ND	ND	ND
Behenic	22:0	ND	ND	ND	ND	ND	ND	ND
Erucic	22:1(13)	0.16 ± 0.111	0.15 ± 0.108	ND	0.07 ± 0.049	ND	0.14 ± 0.100	0.02 ± 0.012
Docosadienoic	22:2(13,16) n-6	ND	ND	ND	ND	ND	ND	ND
Eranthic	22:3(5,13,16) n-6	ND	0.02 ± 0.001	ND	ND	ND	ND	ND
Ardenic	22:4(7,10,13,16) n-6	ND	0.08 ± 0.001	ND	ND	ND	0.02 ± 0.001	ND
DPA	22:5(4,7,10,13,16) n-3	ND	0.48 ± 0.004	ND	ND	ND	0.11 ± 0.010	ND
DHA	22:6(4,7,10,13,16,19) n-3	0.01 ± 0.002	0.04 ± 0.028	ND	0.02 ± 0.014	0.01 ± 0.001	ND	ND
Tricosylic	23:0	ND	ND	ND	ND	ND	ND	ND
Lignoceric	24:0	0.06 ± 0.080	ND	ND	ND	ND	ND	ND
SFA		61.23	49.76	50.13	51.10	20.08	55.89	51.77
MUFA		37.53	38.14	25.25	47.15	57.35	37.32	40.78
PUFA		1.24	12.10	24.62	1.75	22.57	6.79	7.46
UFA		38.77	50.24	49.87	48.9	79.92	44.11	48.23
UFA/SFA		0.63	1.01	0.99	0.66	3.98	0.79	0.93

Values are expressed as the mean % percentage of total fatty acids of each sample (mean ± standard deviation (SD), $n = 3$). Abbreviations: BM, bovine milk; CPAD, commercially purchased almond drink; CPFAD, commercial purchased fermented almond drink; CPRD, commercially purchased rice drink; CPCD, commercially purchased coconut drink; CPFCD, commercially purchased fermented coconut drink; SFA, saturated fatty acids; MUFA, monounsaturated fatty acids; PUFA, polyunsaturated fatty acids; UFA, unsaturated fatty acids; ND, not detected.

In yogurt's PL, the substitution of the SFA by UFA takes place through the natural process of fermentation of the BM during yogurt production. At the same time, such changes in the FAs' content have been found to increase the anti-inflammatory and anti-thrombotic potency of the PL of yogurt in comparison to the bioactivities of the milk PL [10], suggesting a structure-activity relationship between the FA content of the PL bioactives in fermented dairy and their anti-inflammatory bioactivities. The increase of UFA content and, especially that of the oleic acid and the n-3 PUFA content in the PL of the CPFBM may also explain why these PL bioactives had much stronger anti-inflammatory and anti-platelet potency, especially against PAF, in comparison to the HMFBM. Nevertheless, further research is needed to fully evaluate the fermentation process and the strains needed for increasing the UFA content and anti-inflammatory bio-functionality of yogurt's PL as a natural way of dairy fortification with UFA.

In rice-based samples, in the PL of both the unfermented and fermented homemade products, the SFA was the main class of FAs, with the most dominant being the palmitic acid, followed by the myristic and the lauric acids, with lower but considerable amounts of UFA, mainly by MUFA, especially the n-9 oleic acid, and lower PUFA levels, mainly linoleic acid followed by low but detectable levels of linolenic acids. These results agree with the previously reported FA profile for rice-based products, where oleic acid and linoleic acid were the dominant UFA [42]. The overall UFA content and the UFA/SFA ratio of the rice-based drink were higher than the BM. Similar to the dairy products, differences were observed in the FA content of the rice-based plant-derived dairy alternatives' PL due to fermentation.

Specifically, in the PL of HMFRD, an increase in lauric acid was observed, accompanied by a decrease in palmitic acid and oleic acid. In contrast, a substantial increase in the linolenic acid was observed, accompanied by an increase in the linolenic acids, compared to the content of these FAs in the PL HMRD. Such changes in specific FAs resulted in decreased SFA, increased UFA content, and increased UFA/SFA ratio in PL of HMFRD. This is due to a substantial increase in their PUFA content, even though MUFA content decreased. Overall, PL from HMFRD and UFA/SFA ratio were increased in comparison to the values of the same ratio for HMRD. This difference seems rational for the stronger anti-inflammatory and anti-platelet properties of HMFRD against PAF, compared to the strong but lower anti-PAF properties of HMRD.

On the other hand, CPRD showed equal levels of SFA with UFA, with dominant SFA being again the palmitic acid followed this time by the lauric acid and with lower amounts of the myristic acid this time. In the CPRD, the UFA content and the UFA/SFA ratio were much higher than the BM, with the n-9 oleic acid being the main MUFA and linoleic acid the main PUFA, followed by lower but considerable amounts of linolenic acids. The high UFA content and UFA/SFA ratio of the CPRD were higher even than that of the homemade one due to the higher PUFA content observed in CPRD. Nevertheless, there was no availability in the market of a fermented rice-based dairy alternative product to be commercially purchased. Thus similar comparisons made for the homemade products could not occur in the present study.

Overall, the higher values of the UFA/SFA ratio of the PL bioactives in the rice-based dairy alternatives, especially the fermented HMFRD, in comparison to the BM, also support the stronger anti-PAF anti-inflammatory potency of the PL bioactives of the rice-based dairy alternatives. Nevertheless, more studies are needed to fully evaluate the FA modifications of the PL bioactives in HMFRD and whether these changes are associated with structure-activity relationships of the anti-inflammatory bio-functionality of these PL from rice-based products.

In contrast to the bovine dairy and rice-based products, in almond-based products, the UFA was dominant in the PL of both the unfermented and fermented homemade and commercially purchased products, with the MUFA being the main class of FAs due to the dominance of the n-9 oleic acid. More specifically, in HMAD, the main SFA was palmitic acid, lower than those of the oleic MUFA. In contrast, only the linoleic acid was detected in very low amounts in this almond-based product from the PUFA. Similar results were also observed in CPAD, where again, the dominant FA was the n-9 oleic acid, followed by the SFA palmitic acid. In contrast, higher amounts were observed for the myristic and lauric acid and the PUFA linoleic and linolenic acids than those observed in HMAD. These results agree with the previously reported FA profile for almond-based products, where oleic acid was the dominant FA [42]. Subsequently, in both these almond-based dairy alternative drinks, their overall UFA content levels were higher than the levels of their SFA content. Thus the ratio of UFA/SFA was higher than that of the relevant unfermented rice-based drink and BM.

Again, differences were observed in the FA content of the almond-based plant-derived dairy alternatives' PL due to fermentation. More specifically, in the PL of both HMFAD and CPFAD, a substantial decrease of the SFA palmitic, myristic, and lauric acids with a subsequent reduction in their SFA content was observed. More importantly, their PUFA content substantially increased due to an increase in the levels of the linoleic acid and an increase in other PUFA like the linoleic acids compared to these FAs in the PL HMAD, and CPAD. There was a difference in the MUFA content changes between the HMFAD and the CPFAD, especially in the oleic acid content, which was decreased during the homemade fermentation process and increased in the commercially purchased fermented products.

However, the substantial increase in the PUFA content of the PL of both almond-based fermented products resulted in a substantial increase in their overall UFA content and their UFA/SFA ratio levels. For example, the high UFA content and the UFA/SFA increased ratio of the PL from the CPFAD was the highest observed within all the samples assessed, followed by that of the PL from the HMFAD. These results seem to provide a rationale for the strong anti-inflammatory and anti-platelet properties of the fermented almond-based dairy alternative products against the inflammatory PAF-associated pathways compared to the lower anti-PAF properties of the non-fermented almond-based dairy alternatives and BM.

Apart from the direct anti-PAF inhibitory effect of these dietary PL bioactives on the PAF-receptor, the high UFA content observed in the PL of the fermented almond-based products seems to provide additional bio-functionality. After digestion in the intestine, they travel on the surface of lipoproteins in the bloodstream. These dietary PL bioactives with UFA in their structure are usually also transferred in cell membranes of target cells due to their amphiphilic properties, where specific phospholipases A2 (PLA_2) exist. From the activities of the UFA, the dietary PL is released intracellularly, affecting several intracellular inflammatory processes [34]. For example, the released free forms of several UFA, such as the n-9 MUFA oleic acid and the PUFA linoleic acid, especially the n-3 PUFA, DHA, EPA, and linolenic acids, have been found to possess strong anti-platelet effects against several mediators, with specific intracellular signaling [34,45,46]. Thus, the increase in the degree of unsaturation, the levels of UFA, and the UFA/SFA content of the fermented products in both the dairy and plant-based dairy alternatives, due to the fermentation process, further enhances the anti-inflammatory and anti-platelet bio-functionality of their PL bioactives

In contrast to all aforementioned samples, the PL from all coconut samples had high SFA and MUFA but very low PUFA levels. The dominant FA was the MUFA oleic acid, in agreement with previously reported results in coconut products [42]. In the SFA of the PL of the HMCD, the more abundant were those of medium-chain SFA and especially lauric acid, followed by palmitic and myristic acid. In contrast, from PUFA, only linoleic acid was present. The high levels of oleic acid present resulted in UFA having similar levels with the SFA content in the PL of the HMCD.

In contrast to all the other samples assessed, a substantial increase in the SFA content was observed in the HMFCD PL after fermentation. Increased SFA content is due to increased lauric acid content, which is the dominant FA in the PL of this product, followed by palmitic and myristic acids that were reduced after fermentation. In addition, both oleic acid and linoleic acids were substantially reduced. Thus, the MUFA, PUFA, and overall UFA content were reduced, resulting in the lowest UFA/SFA ratio levels for the PL of this fermented coconut-based dairy alternative compared to the PL of the other bovine dairy and almond and rice-based dairy alternative samples.

In the PL of CPCD, the oleic acid was the dominant FA. However, this time it was followed by high levels of the SFA palmitic, lauric, and myristic acids. In contrast, linoleic acid was the main PUFA, followed by low but considerable amounts of linolenic acids and very low but detectable amounts of other PUFA, such as the EPA, DHA, and arachidonic acid. Subsequently, the SFA content was higher than the UFA content in the PL of this coconut-based drink. Unlike HMFCD, after fermentation, the oleic acid was substantially increased in the PL of the CPFCD. Thus, it was retained as the dominant FA in this fermented product, followed by lauric acid as the main SFA and palmitic and myristic SFA. In the PUFA, the linoleic acid was increased, and the linolenic acid content was reduced to very low but detectable levels, while no other PUFA was detected. Thus, differently than the PL of the HMFCD, the UFA content of the PL from the CPFCD was higher and at similar levels to their SFA content and subsequently a higher level of the UFA/SFA ratio.

Overall, the increase of the UFA content and the UFA/SFA ratio of the PL of the CPFCD, in contrast to the decreased levels of the UFA content and the UFA/SFA ratio in the PL of HMFCD, seems to be associated with the stronger anti-PAF anti-inflammatory potency observed in the PL of CPFCD, in comparison to the less bioactive PL from HMFCD.

It should also be stressed that higher SFA and lower UFA content of the PL from some coconut-based products resulted in the lowest anti-inflammatory and anti-platelet potency of these coconut-based PL against both the PAF and ADP.

Even though coconut oil has a relatively high medium-chain FA concentration, the clinical benefits of commercial oils based on such FAs cannot be generalized to coconut oil and coconut products. Nonetheless, apart from the potential anti-inflammatory functional properties of the coconut-based products, the abundance of medium-chain SFA lauric acid in this product has been proposed to modify the blood lipid profile by increasing the low-density lipoprotein cholesterol (LDL-C) and high-density lipoprotein cholesterol (HDL-C) concentrations. Specifically, it plays a main role as a substrate for apolipoprotein (apo)A1 and apoB synthesis, the key molecules in HDL-C and LDL-C particles, respectively [47]. Several studies consistently showed that consumption of coconut products increases mostly LDL-C and could increase adverse cardiovascular health. Until the long-term effects of coconut products on cardiovascular health are established, coconut oil should be considered a saturated fat, and its consumption should not exceed the USDA's daily recommendation (less than 10% of total calorie intake) [47].

Several variations were observed in the FA composition of both the dairy and the plant-based dairy alternatives assessed in the present study. These variations in their FA composition, especially before and after fermentation, agree with other relevant studies where similar variations were detected in the FA composition of bovine dairy and rice, almond, and coconut-based products [7,8,20–22,25,30–32,42]. This variation proposes that other parameters besides FA content play essential roles in the overall functionality of the PL bioactives in such products. Thus, further studies are needed to evaluate how fermentation can increase the functional anti-inflammatory properties of PL bioactives in dairy and dairy alternatives.

4. Conclusions

The present study showed that fermentation could enhance the anti-inflammatory and anti-platelet potency in dairy and plant-based dairy alternatives. The bio-functionality of these products and, especially of their fermented representatives, in both homemade and commercially purchased ones, was mainly attributed to their dietary PL bioactives and the increase of the UFA in these PL. These bioactive PL had potent activity against platelet aggregation initiated by the inflammatory and thrombotic mediator PAF and against that of the well-established platelet agonist ADP but with higher anti-PAF specificity. Comparing plant drink samples, fermentation resulted in PL with more potent bioactivities against platelet aggregation. The dietary PL bioactives of the fermented rice-based dairy alternative had the greatest anti-inflammatory potency against PAF and anti-platelet capacity against ADP, compared to the PL from all the other samples, followed by the PL of the bovine yogurt and the fermented almond-based products. In contrast, the PL from the coconut-based products showed the lowest bioactivities.

The substitution of the SFA by UFA during the natural fermentation process seems to be associated with the increased anti-inflammatory and anti-thrombotic potency of the fermented products, especially against the PAF pathway, in comparison to the bioactivities of the PL of the non-fermented ones. This suggests a structure-activity relationship between the fatty acid content of the PL bioactives in fermented dairy and dairy alternatives, with their potent anti-inflammatory bioactivities. However, there did not appear to be a strong correlation between fermentation and the anti-inflammatory n-3 PUFA content changes. An increase in the bioactive n-3 PUFA, such as linolenic acids, EPA, DPA, and DHA, was found in low levels and very few samples, especially in the bovine CPFBM yogurt, were in smaller amounts compared to other food products. On the other hand, other UFAs like the n-9 MUFA oleic acid and the PUFA linoleic acid were dominant in all samples. At the same time, in most fermented products, their levels were increased, especially those of the linoleic acid, with a subsequent increase in the overall UFA content and the UFA/SFA ratio due to fermentation. Substitution of the SFA content by increasing the UFAs in both bovine dairy and plant-based dairy alternatives due to the natural process of fermentation is associated with several health benefits. It is also associated with enhancing the anti-inflammatory and anti-platelet bio-functionality of the dietary PL bioactives in these fermented bovine yogurt and dairy alternatives, as observed in the present study.

Nevertheless, more studies are needed to unveil the full benefits of the fermentation process and the strains needed for increasing the UFA content and the anti-inflammatory bio-functionality of both dairy and dairy alternatives, as a natural way of their fortification with UFA and PL bioactives, with a subsequent enhancement of their functional anti-inflammatory properties. These results indicate the potential for developing novel bio-functional sustainable dairy and plant-based dairy alternative products with potent anti-inflammatory, anti-platelet, and cardio-protective properties. However, additional research is needed to thoroughly assess the benefits and application of bioactive PL derived from fermented dairy and plant-based dairy alternatives.

Author Contributions: Conceptualization, A.T.; methodology, A.T., K.S. and S.K.S.; software, K.G.-D., D.M., A.T., K.S., S.K.S., A.H., E.B. and J.C.; validation, A.T.; formal analysis, K.G.-D., D.M., A.T., A.H., E.B., K.S., S.K.S. and J.C.; investigation, A.T.; resources, A.T., I.Z. and S.K.S.; data curation, K.G.-D., D.M., A.T., A.H., E.B., K.S., S.K.S. and J.C.; writing—original draft preparation, A.T. and K.G.-D.; writing—review and editing, A.T., S.K.S. and K.G.-D.; visualization, A.T.; supervision, A.T.; project administration, A.T.; funding acquisition, A.T. All authors have read and agreed to the published version of the manuscript.

Funding: This research received no external funding.

Institutional Review Board Statement: The study was conducted according to the guidelines of the Declaration of Helsinki and approved by the Ethics Committee of the Faculty of Science & Engineering of the University of Limerick.

Informed Consent Statement: Informed consent was obtained from all subjects involved in the study.

Acknowledgments: The authors are grateful to the volunteers who took part in the study, to Elaine Ahern for her phlebotomy support, and to the Department of Biological Sciences at the University of Limerick, Ireland, for their continued support.

Conflicts of Interest: The authors declare no conflict of interest.

References

1. WHO. Non-Communicable Disease. Available online: https://www.who.int/news-room/fact-sheets/detail/noncommunicable-diseases#:~{}:text=Key%20facts. (accessed on 2 June 2022).
2. Tsoupras, A.; Lordan, R.; Zabetakis, I. Inflammation, not Cholesterol, Is a Cause of Chronic Disease. *Nutrients* **2018**, *10*, 604. [CrossRef] [PubMed]
3. Tsoupras, A.B.; Iatrou, C.; Frangia, C.; Demopoulos, C.A. The implication of platelet activating factor in cancer growth and metastasis: Potent beneficial role of PAF-inhibitors and antioxidants. *Infect. Disord. Drug Targets* **2009**, *9*, 390–399. [CrossRef] [PubMed]
4. Tsoupras, A.; Lordan, R.; Zabetakis, I. Inflammation and Cardiovascular Diseases. In *The Impact of Nutrition and Statins on Cardiovascular Diseases*; Zabetakis, I., Lordan, R., Tsoupras, A., Eds.; Academic Press: Cambridge, MA, USA, 2019; pp. 53–117.
5. Tierney, A.; Lordan, R.; Tsoupas, A.; Zabetakis, I. Diet and Cardiovascular Disease: The Mediterranean Diet. In *The Impact of Nutrition and Statins on Cardiovascular Diseases*; Zabetakis, I., Lordan, R., Tsoupras, A., Eds.; Academic Press: Cambridge, MA, USA, 2019; pp. 267–288.
6. Zabetakis, I.; Lordan, R.; Tsoupras, A.; Ramji, D. *Functional Foods and Their Implications for Health Promotion*; Zabetakis, I., Lordan, R., Tsoupras, A., Ramji, D., Eds.; Academic Press: Cambridge, MA, USA, 2022; ISBN 0128238127/9780128238127.
7. Lordan, R.; Walsh, A.; Crispie, F.; Finnegan, L.; Demuru, M.; Tsoupras, A.; Cotter, P.D.; Zabetakis, I. Caprine milk fermentation enhances the antithrombotic properties of cheese polar lipids. *J. Funct. Foods* **2019**, *61*, 103507. [CrossRef]
8. Lordan, R.; Vidal, N.P.; Huong Pham, T.; Tsoupras, A.; Thomas, R.H.; Zabetakis, I. Yoghurt fermentation alters the composition and antiplatelet properties of milk polar lipids. *Food Chem.* **2020**, *332*, 127384. [CrossRef] [PubMed]
9. Tsoupras, A.; Moran, D.; Pleskach, H.; Durkin, M.; Traas, C.; Zabetakis, I. Beneficial Anti-Platelet and Anti-Inflammatory Properties of Irish Apple Juice and Cider Bioactives. *Foods* **2021**, *10*, 412. [CrossRef] [PubMed]
10. Tsoupras, A.; Lordan, R.; O'Keefe, E.; Shiels, K.; Saham, S.K.; Zabetakis, I. Structural Elucidation of Irish Ale Bioactive Polar Lipids with Antithrombotic Properties. *Biomolecules* **2020**, *10*, 1075. [CrossRef]

11. Vargas-Bello-Perez, E.; Faber, I.; Osorio, J.S.; Stergiadis, S. Consumer knowledge and perceptions of milk fat in Denmark, the United Kingdom, and the United States. *J. Dairy Sci.* **2020**, *103*, 4151–4163. [CrossRef]

12. EFSA Panel on Dietetic Products, Nutrition, and Allergies (NDA). Scientific Opinion on Dietary Reference Values for fats, including saturated fatty acids, polyunsaturated fatty acids, monounsaturated fatty acids, trans fatty acids, and cholesterol. *EFSA J.* **2010**, *8*, 1461. [CrossRef]

13. Mensink, R.P. *Effects of Saturated Fatty Acids on Serum Lipids and Lipoproteins: A Systematic Review and Regression Analysis*; World Health Organization: Geneva, Switzerland, 2016.

14. Lordan, R.; Tsoupras, A.; Mitra, B.; Zabetakis, I. Dairy Fats and Cardiovascular Disease: Do We Really Need to be Concerned? *Foods* **2018**, *7*, 29. [CrossRef] [PubMed]

15. García-Burgos, M.; Moreno-Fernández, J.; Alférez, M.J.M.; Díaz-Castro, J.; López-Aliaga, I. New perspectives in fermented dairy products and their health relevance. *J. Funct. Foods* **2020**, *72*, 104059. [CrossRef]

16. Soedamah-Muthu, S.S.; Ding, E.L.; Al-Delaimy, W.K.; Hu, F.B.; Engberink, M.F.; Willett, W.C.; Geleijnse, J.M. Milk and dairy consumption and incidence of cardiovascular diseases and all-cause mortality: Dose-response meta-analysis of prospective cohort studies. *Am. J. Clin. Nutr.* **2011**, *93*, 158–171. [CrossRef]

17. Janssen, M.; Busch, C.; Rodiger, M.; Hamm, U. Motives of consumers following a vegan diet and their attitudes towards animal agriculture. *Appetite* **2016**, *105*, 643–651. [CrossRef] [PubMed]

18. Scholz-Ahrens, K.E.; Ahrens, F.; Barth, C.A. Nutritional and health attributes of milk and milk imitations. *Eur. J. Nutr.* **2020**, *59*, 19–34. [CrossRef] [PubMed]

19. Rivera Del Rio, A.; Boom, R.M.; Janssen, A.E.M. Effect of Fractionation and Processing Conditions on the Digestibility of Plant Proteins as Food Ingredients. *Foods.* **2022**, *11*, 870. [CrossRef] [PubMed]

20. Aydar, E.F.; Tutuncu, S.; Ozcelik, B. Plant-based milk substitutes: Bioactive compounds, conventional and novel processes, bioavailability studies, and health effects. *J. Funct. Foods* **2020**, *70*, 103975. [CrossRef]

21. Clegg, M.E.; Tarrado Ribes, A.; Reynolds, R.; Kliem, K.; Stergiadis, S. A comparative assessment of the nutritional composition of dairy and plant-based dairy alternatives available for sale in the UK and the implications for consumers' dietary intakes. *Food Res. Int.* **2021**, *148*, 110586. [CrossRef]

22. Chalupa-Krebzdak, S.; Long, C.J.; Bohrer, B.M. Nutrient density and nutritional value of milk and plant-based milk alternatives. *Int. Dairy J.* **2018**, *87*, 84–92. [CrossRef]

23. McClements, D.J.; Newman, E.; McClements, I.F. Plant-based Milks: A Review of the Science Underpinning Their Design, Fabrication, and Performance. *Compr. Rev. Food Sci. Food Saf.* **2019**, *18*, 2047–2067. [CrossRef] [PubMed]

24. Drabińska, N.; Ogrodowczyk, A. Crossroad of Tradition and Innovation—The Application of Lactic Acid Fermentation to Increase the Nutritional and Health-Promoting Potential of Plant-Based Food Products—A Review. *Pol. J. Food Nutr. Sci.* **2021**, *71*, 107–134. [CrossRef]

25. Tangyu, M.; Muller, J.; Bolten, C.J.; Wittmann, C. Fermentation of plant-based milk alternatives for improved flavour and nutritional value. *Appl. Microbiol. Biotechnol.* **2019**, *103*, 9263–9275. [CrossRef]

26. Tsoupras, A.; Moran, D.; Byrne, T.; Ryan, J.; Barrett, L.; Traas, C.; Zabetakis, I. Anti-Inflammatory and Anti-Platelet Properties of Lipid Bioactives from Apple Cider By-Products. *Molecules* **2021**, *26*, 2869. [CrossRef] [PubMed]

27. Bligh, E.G.; Dyer, W.J. A Rapid Method of Total Lipid Extraction and Purification. *Can. J. Biochem. Physiol.* **1959**, *37*, 911–917. [CrossRef] [PubMed]

28. Galanos, D.S.; Kapoulas, V.M. Isolation of polar lipids from triglyceride mixtures. *J. Lipid Res.* **1962**, *3*, 134–136. [CrossRef]

29. Tsoupras, A.; Zabetakis, I.; Lordan, R. Platelet aggregometry assay for evaluating the effects of platelet agonists and antiplatelet compounds on platelet function in vitro. *MethodsX* **2018**, *6*, 63–70. [CrossRef] [PubMed]

30. Ferreira, J.A.; Santos, J.M.; Breitkreitz, M.C.; Ferreira, J.M.S.; Lins, P.M.P.; Farias, S.C.; de Morais, D.R.; Eberlin, M.N.; Bottoli, C.B.G. Characterization of the lipid profile from coconut (*Cocos nucifera* L.) oil of different varieties by electrospray ionization mass spectrometry associated with principal component analysis and independent component analysis. *Food Res. Int.* **2019**, *123*, 189–197. [CrossRef]

31. Beltran Sanahuja, A.; Maestre Perez, S.E.; Grane Teruel, N.; Valdes Garcia, A.; Prats Moya, M.S. Variability of Chemical Profile in Almonds (*Prunus dulcis*) of Different Cultivars and Origins. *Foods* **2021**, *10*, 153. [CrossRef]

32. Fernando, D.; Goffman, S.P.; Bergman, C. Genetic Diversity for Lipid Content and Fatty Acid Profile in Rice Bran. *JAOCS* **2003**, *80*, 485–490.

33. Tsoupras, A.; Lordan, R.; Harrington, J.; Pienaar, R.; Devaney, K.; Heaney, S.; Koidis, A.; Zabetakis, I. The Effects of Oxidation on the Antithrombotic Properties of Tea Lipids Against PAF, Thrombin, Collagen, and ADP. *Foods* **2020**, *9*, 385. [CrossRef]

34. Tsoupras, A.; Brummell, C.; Kealy, C.; Vitkaitis, K.; Redfern, S.; Zabetakis, I. Cardio-Protective Properties and Health Benefits of Fish Lipid Bioactives; The Effects of Thermal Processing. *Mar. Drugs* **2022**, *20*, 187. [CrossRef]

35. Reena, M.B.; Krishnakantha, T.P.; Lokesh, B.R. Lowering of platelet aggregation and serum eicosanoid levels in rats fed with a diet containing coconut oil blends with rice bran oil or sesame oil. *Prostaglandins Leukot. Essent. Fat. Acids* **2010**, *83*, 151–160. [CrossRef]

36. Tarmoos, A.A.; Kafi, L.A. Effects of sweet almond (*Prunus amygdalus*) suspension on blood biochemical parameters in experimentally induced hyperlipidemic mice. *Vet World* **2019**, *12*, 1966–1969. [CrossRef] [PubMed]

37. Wong, W.T.; Ismail, M.; Imam, M.U.; Zhang, Y.D. Modulation of platelet functions by crude rice (*Oryza sativa*) bran policosanol extract. *BMC Complement. Altern. Med.* **2016**, *16*, 252. [CrossRef] [PubMed]

38. Yang, Y.; Andrews, M.C.; Hu, Y.; Wang, D.; Qin, Y.; Zhu, Y.; Ni, H.; Ling, W. Anthocyanin extract from black rice significantly ameliorates platelet hyperactivity and hypertriglyceridemia in dyslipidemic rats induced by high fat diets. *J. Agric. Food Chem.* **2011**, *59*, 6759–6764. [CrossRef]

39. Valero-Cases, E.; Cerda-Bernad, D.; Pastor, J.J.; Frutos, M.J. Non-Dairy Fermented Beverages as Potential Carriers to Ensure Probiotics, Prebiotics, and Bioactive Compounds Arrival to the Gut and Their Health Benefits. *Nutrients* **2020**, *12*, 1666. [CrossRef]

40. Moran, D.; Fleming, M.; Daly, E.; Gaughan, N.; Zabetakis, I.; Traas, C.; Tsoupras, A. Anti-Platelet Properties of Apple Must/Skin Yeasts and of Their Fermented Apple Cider Products. *Beverages* **2021**, *7*, 54. [CrossRef]

41. Hang, Y.; Zhang, T.; Liang, Y.; Jiang, L.; Sui, X. Dietary Bioactive Lipids: A Review on Absorption, Metabolism, and Health Properties. *J. Agric. Food Chem.* **2021**, *69*, 8929–8943. [CrossRef]

42. Martinez-Padilla, E.; Li, K.; Blok Frandsen, H.; Skejovic Joehnke, M.; Vargas-Bello-Perez, E.; Lykke Petersen, I. In Vitro Protein Digestibility and Fatty Acid Profile of Commercial Plant-Based Milk Alternatives. *Foods* **2020**, *9*, 1784. [CrossRef]

43. Granato, D.; Carocho, M.; Barros, L.; Zabetakis, I.; Mocan, A.; Tsoupras, A.; Cruz, A.G.; Pimentel, T.C. Implementation of Sustainable Development Goals in the dairy sector: Perspectives on the use of agro-industrial side-streams to design functional foods. *Trends Food Sci. Technol.* **2022**, *124*, 128–139. [CrossRef]

44. Hannon, B.A.; Thompson, S.V.; An, R.; Teran-Garcia, M. Clinical Outcomes of Dietary Replacement of Saturated Fatty Acids with Unsaturated Fat Sources in Adults with Overweight and Obesity: A Systematic Review and Meta-Analysis of Randomized Control Trials. *Ann. Nutr. Metab.* **2017**, *71*, 107–117. [CrossRef]

45. Nunez, D.; Randon, J.; Gandhi, C.; Siafaka-Kapadai, A.; Olson, M.S.; Hanahan, D.J. The inhibition of platelet-activating factor-induced platelet activation by oleic acid is associated with a decrease in polyphosphoinositide metabolism. *J. Biol. Chem.* **1990**, *265*, 18330–18338. [CrossRef]

46. Bazán-Salinas, I.L.; Matías-Pérez, D.; Pérez-Campos, E.; Pérez-Campos Mayoral, L.; García-Montalvo, I.A. Reduction of Platelet Aggregation From Ingestion of Oleic and Linoleic Acids Found in *Vitis vinifera* and *Arachis hypogaea* Oils. *Am. J. Ther.* **2016**, *23*, e1315–e1319. [CrossRef] [PubMed]

47. Santos, H.O.; Howell, S.; Earnest, C.P.; Teixeira, F.J. Coconut oil intake and its effects on the cardiometabolic profile—A structured literature review. *Prog. Cardiovasc. Dis.* **2019**, *62*, 436–443. [CrossRef] [PubMed]

Article

Spray-Dried Nipa Palm Vinegar Powder: Production and Evaluation of Physicochemical, Nutritional, Sensory, and Storage Aspects

Wilawan Palachum [1,2], Wiyada Kwanhian Klangbud [1,2,*] and Yusuf Chisti [3]

1 Department of Medical Technology, School of Allied Health Sciences, Walailak University, Thasala, Nakhon Si Thammarat 80161, Thailand; pwilawan.wp@gmail.com
2 Center of Excellence Research for Melioidosis and Microorganisms, Walailak University, Thasala, Nakhon Si Thammarat 80161, Thailand
3 School of Engineering, Massey University, Private Bag 11 222, Palmerston North 4442, New Zealand; ychisti@hotmail.com
* Correspondence: wiyadakwanhian@gmail.com

Abstract: Nipa palm vinegar (NPV) is a naturally fermented vinegar derived from the nipa palm (*Nypa fruticans* Wurmb) sap. This work optimized production of spray-dried nipa palm vinegar powder. The influence of the various drier air inlet temperatures (150, 170, and 190 °C) and maltodextrin DE10 carrier concentrations (15 and 20% w/v) in the feed, on the characteristics of the product powder was investigated. Nipa palm vinegar powder (NPVp) was evaluated in terms of the following responses: physicochemical and nutritional properties, sensory acceptability, and storage stability. All processing variables affected the responses. Based on product desirability as the optimization criterion, spray-drying with a hot air inlet temperature of 170 °C with a 15% w/v maltodextrin DE10 in the feed was optimal. The nutritional characteristics of the product made under the above identified optimal conditions were (per 100 g dry product): a calorific value of 366.2 kcal; 1.3 g protein; 88.1 g carbohydrate; 0.96 g fat; 883.9 mg potassium; 12.7 mg vitamin C; and 105 mg gallic acid equivalent (GAE) phenolics content. The product, vacuum-packed and heat-sealed in aluminum laminated polyethylene bags, could be stored at 25 °C for at least 180 days without noticeable loss in quality.

Keywords: nipa palm vinegar; nipa palm vinegar powder; *Nypa fruticans* Wurmb; spray-drying of nipa palm vinegar

Citation: Palachum, W.; Klangbud, W.K.; Chisti, Y. Spray-Dried Nipa Palm Vinegar Powder: Production and Evaluation of Physicochemical, Nutritional, Sensory, and Storage Aspects. *Fermentation* **2022**, *8*, 272. https://doi.org/10.3390/fermentation8060272

Academic Editors: Michela Verni and Carlo Giuseppe Rizzello

Received: 25 May 2022
Accepted: 10 June 2022
Published: 11 June 2022

Publisher's Note: MDPI stays neutral with regard to jurisdictional claims in published maps and institutional affiliations.

1. Introduction

Nipa palm (*Nypa fruticans* Wurmb) is a high-yielding sugar palm found commonly in mangrove ecosystems in Southeast Asia and Oceania [1–3]. In Thailand, nipa palm occurs in southwestern coastal regions bordering the Gulf of Thailand [4]. Nipa palm trees are tapped for the sugar-rich nipa palm sap [1,5] that is consumed fresh, and fermented to wine. The wine may be distilled to produce other alcoholic beverages, or further fermented to produce nipa palm vinegar (NPV) [1,6]. Production of NPV from fresh sap involves a two-stage fermentation. In the first stage, yeasts ferment the carbohydrates and sugars to alcohols and carbon dioxide. In the second aerobic fermentation, acetic acid bacteria convert the ethanol to acetic acid [4,5].

NPV is used as a food condiment and also consumed locally between meals and before bedtime [4,7]. NPV is also used as a food preservative [1]. Various health benefits are ascribed to NPV based mostly on in vitro studies [4,5,7,8]. NPV is mostly a regional product. Its utility as a food condiment can be enhanced by drying it to a stable, easy to distribute, NPV powder.

Spray drying is generally the method of choice for liquid foods as it is rapid and less damaging to the sensory and nutritional characteristics. Spray drying involves atomizing the liquid into fine droplets in a hot gas stream to produce a dry powder [9]. Compared to the liquid form, a dry product has a reduced volume and weight and superior storage stability over extended periods [10]. Spray drying is also widely used for other highly heat-sensitive biological products such as pharmaceutical proteins and enzymes because the drying is rapid and, therefore, the activity loss is minimal [11]. A carrier agent such as the polysaccharide maltodextrin (MD) is often added to liquid products prior to spray drying. Maltodextrin acts as a filler, thickener, and texturizing agent. Maltodextrin is edible and generally recognized as safe (GRAS) according to the classification of the United States Federal Drug Administration (FDA) [10,12,13].

Fruit juices [12–15] and certain vinegars are commonly spray dried. The latter include the Chinese black vinegar [16,17], coconut water vinegar [18], and bamboo vinegar [19]. Unlike some other vinegar powders, nipa palm vinegar powder is not commercially available. No work has been reported on the effects of spray drying on nutritional and sensory characteristics of NPV, the organoleptic characteristics of the nipa palm vinegar powder (NPVp) in foods and beverages, and the storage stability of NPVp.

The present work aimed to identify the suitable conditions (drying air temperature, maltodextrin concentration in the NPV) for producing spray-dried NPV powder with acceptable sensory and nutritional attributes. The storage stability of the dried product was assessed as an important commercial characteristic.

2. Materials and Methods

2.1. Chemicals and Materials

The chemicals and reagents were of analytical grade and purchased from Sigma-Aldrich (St. Louis, MO, USA) unless specified otherwise. Maltodextrin with a dextrose equivalent (DE) value of 10 was obtained from Perfect Natural Food Powder and Flavor 2002 (Thailand) Co., Ltd. (Kratumban, Samutsakorn, Thailand).

All materials used in preparing consumable products were of food grade or higher quality. Hygienic practices in food preparation were followed.

2.2. Nipa Palm Vinegar

The nipa palm (*Nypa fruticans* Wurmb) vinegar product was obtained from a local market (Kanapnak sub-district, Pak Phanang district, Nakhon Si Thammarat province; 8°12′25.1″ N latitude, 100°14′51.7″ E longitude) in Thailand. The vinegar had been naturally fermented using a traditional local method, which involved incubating nipa palm sap from cut stalks in terracotta containers at room temperature (25–30 °C) for 40 days. The microorganisms found naturally in the raw sap were the fermenting agents [4]. The nipa palm vinegar lots were obtained from the same supplier. Fermented nipa palm vinegar had a pH of 2.67 ± 0.02 and an acetic acid content of $4.42 \pm 0.02\%$ w/v (g/100 mL).

The nipa palm vinegar was transferred to a clean container and filtered through a fine-mesh sifter (40 mesh, 420 µm) to remove any suspended matter and stored at 4 °C until use.

2.3. Spray Drying of Nipa Palm Vinegar

Different concentrations of maltodextrin DE10 (10, 15, and 20% w/v (g/100 mL)) were mixed with nipa palm vinegar in different experiments. The resulting solution was filtered through a fine-mesh sifter (40 mesh, 420 µm) and used as the feed for spray drying.

A pilot-scale spray drier with a rotary atomizer (model SDE-10, JCS Technic Line Co., Ltd., Samutsakorn, Thailand) was used in all work. The cylindrical drying chamber made of stainless steel had an internal diameter of 0.9 m and a total height of 2.3 m. The variables studied were the maltodextrin concentration (10, 15, and 20% w/v) in the feed and the inlet air temperatures (150, 170, and 190 °C). The outlet air temperature was held at 90 °C. The feed flow rate was controlled at 500 mL h^{-1}. The atomization air flow rate was always 5 m^3 h^{-1} and the fan blower speed was 13,000 rpm. The spray dried product (NPV powder) was sifted through a 60-mesh (250 µm) sifter, collected, weighed, and kept in sealed containers for analysis. Based on the above noted dimensions of the drying chamber and the flow rate of the atomization air, the product residence time (or contact time with the drying air) in the drying chamber was always well below 42 s.

Preliminary spray drying trials revealed no powder recovery in the cyclone when feed contained 10% w/v of maltodextrin as the solid particles stuck to the walls of the drying chamber. Therefore, the maltodextrin level in the redesigned experiments ranged from 15% w/v (low, coded value = -1) to 20% w/v (high, coded value = 1).

2.3.1. Experimental Design

Based on preliminary studies, the experiments involved two factors: the inlet temperature (X_1) and the maltodextrin concentration (X_2). The factor X_1 was studied at three levels: coded values of -1 (low), 0 (medium), and 1 (high). The factor X_2 was studied at two levels with coded values of -1 (low) and 1 (high) (Table S1). The six experimental trials involving various combinations of factors were generated using Multilevel Categoric Design (Table S2) implemented via the Design-Expert® software (trial version 13; Stat-Ease, Minneapolis, MN, USA; www.statease.com, accessed on 24 May 2022). The powder collected for each set of drying condition was used to calculate the product yield and analyzed for physicochemical properties.

2.3.2. Selection of the Optimal Point

The optimization process entailed identifying the best independent factor values (i.e., X_1 and X_2) to obtain a product with the desired characteristics. The desirability function was used to optimize the independent variables, with individual responses being combined to provide the total desirability. The Design-Expert software was used for the numeric optimization. Each variable was optimized numerically using the desirability function (range 0–1). The goal was to maximize the product output yield and bulk density while minimizing the moisture content and water activity. The constraints adopted for the estimating the overall desirability are shown in Table S3.

2.4. Physicochemical Characterization

2.4.1. Product Yield

The product yield Y (%) of spray-dried NPV powders was calculated using the following equation:

$$Y\,(\%) = \left(W_f/W_i\right) \times 100\%$$ (1)

where W_i was the mass of total solids fed (g) to the spray drier and W_f was the mass of the spray-dried NPV powder (g) recovered.

2.4.2. Bulk Density

The bulk density (BD, g mL^{-1}) of NPV powder was determined according to the method of Goula and Adamopoulos [20]. An exact mass (2 g) of NPV powder was transferred to a 10 mL graduated cylinder. The cylinder was held on the vortex vibrator for 1 min and the volume of the powder was measured. The bulk density was calculated as the known mass divided by the measured volume.

2.4.3. Total Soluble Solids, Moisture Content, and Water Activity

The total soluble solids (TSS, °Bx) was measured using a refractometer (ATAGO DR-A1, Tokyo, Japan).

The moisture content was determined using the oven drying method [21]. A precisely weighed (2 g) sample of the NPV powder was oven-dried (105 °C) to a constant weight. Triplicate samples were measured.

The water activity was measured using a water activity meter (Aqualab Series 3TE; Decagon Devices, Pullman, WA, USA). Measurements were performed at room temperature.

2.4.4. The pH and Acetic Acid Determination

The pH was determined using the Method 981.12 [21] with a pH meter (Mettler-Toledo, LLC, Columbus, OH, USA).

The acetic acid concentration of NPV powder was determined by colorimetric titration [4,22]. The titrant was 1 M sodium hydroxide, and the indicator was 1% (w/v, g/100 mL) phenolphthalein solution. NPV powder (10 g) was added to a 250 mL Erlenmeyer flask and dissolved in 20 mL distilled water. Then, three drops of 1% (w/v) phenolphthalein solution were added and mixed. The titrant volume at the endpoint when the color turned red was used to calculate the percentage of acetic acid in the sample. Each mole of sodium hydroxide consumed was equivalent to 1 mol acetic acid in the sample.

2.4.5. Color Analysis

The electronic color parameters (L^*, a^*, and b^*) of NPV powder were measured using the Hunterlab Miniscan/EX instrument ($10°$ standard observer, illuminant D65; Hunter Associates Laboratory, Inc., Reston, VA, USA). The device had been calibrated to a white and black standard. For the measurements, the NPV powder sample was weighed and placed on a transparent plastic plate of the instrument and the Hunter color parameters L^*, a^*, and b^* were measured. L^* denoted lightness in the dark–light spectrum range (black = 0, white = 100), a^* was a hue in the red–green color range where negative values indicated green and positive values indicated red, and b^* was a hue in the yellow–blue spectral range where negative values indicated blue and positive values indicated yellow [13,23].

2.5. Nutritional Characteristics

Nutritional composition of the NPVp was analyzed according to AOAC [21] method as follows: total fat (Method 948.15); cholesterol (Method 976.26); protein (Method 981.10); total sugar (Method 925.35); dietary fiber (Method 985.29); the mineral profiles of calcium (Ca), sodium (Na), and iron (Fe) (Method 984.27); and ash (Method 923.03). The amounts of vitamin A (retinol), vitamin B1 (thiamin), and vitamin C (ascorbic acid) were quantified using adaptations of published methods [24]. Carbohydrate content was determined by the method of Sullivan and Carpenter [25]. The calorific value was obtained by using a bomb calorimeter. The total phenolic content in the NPV samples was measured using the method of Chatatikun and Kwanhian [8] with gallic acid as the standard.

2.6. Sensory Evaluations of NPV and NPV Powder

The sensory analyses of the NPV and the NPV powder had been reviewed and approved by the Human Research Ethics Committee of Walailak University, Thailand (approval no. WUEC-20-048-01). A sensory evaluation involving 59 untrained panelists was carried out in a sensory laboratory (Center for Scientific and Technological Equipment 3, Walailak University, Nakhon Si Thammarat, Thailand). The freshly prepared samples of spray-dried NPV powder and the original NPV were evaluated by the panelists.

The recipe for sensory evaluation consisted of Recipe 1 (R1), Recipe 2 (R2), Recipe 3 (R3), Recipe 4 (R4), Recipe 5 (R5), and Recipe 6 (R6). The original NPV was used as R1 (control R1), and the rehydrated NPV powder was used as R2 (NPVp-3 product, control R2). For sensory evaluations, 200 g NPV or NPVp was mixed with 800 g of drinking water (sample: water mass ratio of 1:4) to produce a drink.

In the beverage recipes (R3 and R4), the original NPV was used for R3 and NPV powder was reconstituted as liquid vinegar as R4. For the preparation of the beverage, the NPV (200 g) was mixed with 200 g of honey (Thai Honey, Thailand; http://www.thaihoney.net, accessed on 24 May 2022) and 600 g of carbonated water (Singha Soda, Thailand; https://www.singhacorporation.com/singhasoda/, accessed on 24 May 2022). The mass ratio of NPV:honey:carbonated water was 1:1:3. The ingredients of R4 were the same as R3, but 200 g of NPVp was added instead of NPV.

For the food recipes (R5 and R6), vinegar was used to increase the acid taste of the food. In recipe R6, 60 g NPVp was added to 1 L of sour soup (Thai sour fish soup made with sea mullet). While the ingredients for Recipe 5 were the same as for Recipe 6, NPVp was replaced with 60 g of the original NPV.

The sensory attributes evaluated by the panelists for each sample were: color, odor, taste, and overall acceptability. Each sample (15 mL) was served in a random order in white plastic cups coded with three digits. To minimize any residual effects, the panelists were instructed to rinse their mouth with water and take all necessary time to score the samples. The participants were asked to score each sensory attribute on a 9-point scale (1 = dislike very much to 9 = like very much) [26].

2.7. Assessment of Storage Stability of NPV Powder

Spray-dried NPV powder (30 g) (product NPVp-3, Figure 1A) was packed in aluminum-laminated polyethylene bags and heat-sealed under vacuum (Figure 1B). The sealed samples were kept at room temperature (25 ± 2 °C, 75% relative humidity) for 180 days. Entire bags were sampled at 30-day intervals for measuring the pH, the moisture content, water activity, color parameters, and microbial counts. Triplicate samples were analyzed.

Figure 1. The appearance of spray-dried NPV powder product NPVp-3 (**A**); the sealed aluminum-laminated polyethylene package containing the NPV powder for stability assessments (**B**).

Microbial Counts

Each NPV powder sample (10 g) was homogenized with 90 mL of sterile 0.1% w/v (g/100 mL) peptone solution (Himedia, Mumbai, India). Serial dilutions (9 mL) were prepared using sterile peptone solutions for plating. A portion (1 mL) of the appropriate dilution was plated in duplicate on potato dextrose agar media (Merck, Germany) which had been aseptically acidified with 10% w/v (g/100 mL) tartaric acid (Ricca Chemical Co., Arlington, TX, USA). The plates were incubated at 25 °C for 5 days for the enumeration of yeasts and molds [27,28].

2.8. Statistical Analysis

All analyses were performed on triplicate samples. The results were presented as mean values (±standard deviation). The results were analyzed by one-way ANOVA (Tukey's test). The mean values at 95% significance level ($p < 0.05$) were taken as a significant difference.

3. Results and Discussion

3.1. Effects of Spray Drying Conditions

Success in spray drying depends critically on the characteristics of the feed, particularly on the concentration of the carrier [10]. The carrier in the present work was maltodextrin DE10. In addition, the inlet temperature of the drying air was important as too high a temperature could damage the product whereas an insufficiently high temperature would result in a poorly dried sticky product [10]. Preliminary studies were carried out to define the suitable ranges of inlet air temperature and the carrier concentration. The maltodextrin content had to be significantly higher than 10% w/v to produce a non-agglomerating powder with the required colloidal stability. In keeping with prior work on spray drying of blueberry juice [29], a maltodextrin content of 10% w/v in the feed resulted in no recovery of the powder possibly because the material stuck to the walls of the spray drier. Therefore, two levels of maltodextrin DE10 concentrations (15 and 20% w/v) were tested in combination with three levels of air inlet temperature (150, 170, and 190 °C) (Table S1). For the factor combinations used in the six experimental runs (Table S2), the measured responses of the physicochemical properties of the products are shown in Table 1. The data shown (Table 1) are the average values of two independent replicates. The mean squares of average experimental results were subjected to analysis of variance (Table 2).

Table 1. Physicochemical properties of the spray-dried NPV powder formulations.

Product	Condition		Y (%)	BD (g mL^{-1})	MC (% w/w)	a_w	TSS (°Bx)	Color		
	T (°C)	MD (%)						L^*	a^*	b^*
NPVp-1	150	15	53.37 ± 0.14	0.46 ± 0.00	2.56 ± 0.06	0.31 ± 0.01	19.50 ± 0.71	91.75 ± 0.06	0.18 ± 0.01	7.91 ± 0.01
NPVp-2	150	20	56.31 ± 0.44	0.47 ± 0.00	3.05 ± 0.07	0.36 ± 0.01	24.10 ± 0.14	91.95 ± 0.06	−0.27 ± 0.00	5.53 ± 0.04
NPVp-3	170	15	53.62 ± 0.11	0.47 ± 0.00	2.08 ± 0.04	0.23 ± 0.00	20.00 ± 0.00	93.57 ± 0.62	0.38 ± 0.01	10.12 ± 0.12
NPVp-4	170	20	58.29 ± 0.05	0.48 ± 0.01	3.03 ± 0.05	0.24 ± 0.00	23.70 ± 0.42	92.80 ± 0.13	0.49 ± 0.01	10.28 ± 0.02
NPVp-5	190	15	47.04 ± 0.44	0.43 ± 0.00	2.63 ± 0.04	0.23 ± 0.00	20.00 ± 0.00	91.53 ± 0.04	−0.09 ± 0.00	6.29 ± 0.03
NPVp-6	190	20	50.72 ± 0.02	0.46 ± 0.00	3.02 ± 0.03	0.24 ± 0.01	24.00 ± 0.00	92.14 ± 0.06	0.54 ± 0.00	6.21 ± 0.01

T, inlet temperature; *MD*, concentration (% w/v) of maltodextrin DE 10; *Y*, product yield; *BD*, bulk density; *MC*, moisture content, a_w, water activity; TSS, total soluble solids. Results are mean values ± standard deviations of measurements on duplicate samples of the final product.

Table 2. Analysis of variance (ANOVA) of physicochemical characteristic responses of spray-dried NPV powders.

Source of Variation	Y	BD	MC	a_w	TSS	Color		
						L^*	a^*	b^*
Model	31.94 **	0.0005 **	0.2966 **	0.0054 **	10.19 **	1.16 **	0.2132 **	8.58 **
Inlet temperature (X_1)	57.9 **	0.0008 **	0.094 **	0.0122 **	0.0433 NS	35.19 **	0.2295 **	18.59 **
Maltodextrin concentration (X_2)	42.41 **	0.0007 **	1.12 **	0.0015 **	50.43 **	0.0099 NS	0.029 **	1.76 **
$X_1 \times X_2$	0.7493 **	0.0002 **	0.0892 **	0.0006 **	0.21 NS	7.41 **	0.289 **	1.97 **
R^2	0.9974	0.9814	0.9903	0.9940	0.9864	0.9342	0.9999	0.9996

Y, product yield; *BD*, bulk density; *MC*, moisture content; a_w, water activity; *TSS*, total soluble solids.; R^2 values indicate the goodness of fit of the theoretical models (see Table S4) to the experimental data; NS, nonsignificant ($p > 0.05$); **, significant ($p < 0.05$).

3.1.1. Yield of NPV Powder

The product yield of the spray-dried NPVp samples (NPVp-1 to NPVp-6) ranged from 58.3 to 47.0% (Table 1). The inlet temperature and maltodextrin concentration both had substantial effects ($p < 0.05$) on the yield of the powdered product (Table 2). As the air inlet temperature was increased from 150 °C to 190 °C, the product yield reduced by 19.3%. This was consistent with similar findings reported for spray-drying of mountain tea extract [30], soluble sage (*Salvia fruticosa* Miller) [31], and a mandarin (*Citrus unshiu*) beverage [32]. The decreasing yield was associated with the melting of the powder at higher temperatures and deposition on the walls of the drying chamber [30–32]. In contrast to the temperature effect, a raising of the maltodextrin concentration in the feed from 15% to 20% w/v increased the product yield (Table 1). This was simply the effect of a higher concentration of soluble solids in the feed [30,31].

The interactive effect of the variables on yield was positive ($p < 0.05$; Table 2). The experimental data agreed well with the predictive model (Table S4) as evidenced by R^2 of 0.9974 (Table 2). Similar positive interactive effects of drying air temperature and maltodextrin content in the feed, on product yield were reported by others [13,14,33].

3.1.2. Bulk Density

A high bulk density reduces the volume for a given mass of product, reducing cost of packaging and shipping. In addition, powders with high bulk densities have superior flow properties compared to agglomerated powders with a low bulk density [10]. The bulk density of the NPV powders varied from 0.43 to 0.48 g mL^{-1} (Table 1). These values were higher compared to data reported for spray-dried instant soluble sage [31] and spray-dried whey protein mixes [34] made under conditions similar to those of the present work. Both the inlet temperature and maltodextrin concentration significantly ($p < 0.05$) affected the bulk density of NPVp (Table 2). The experimental data of bulk density agreed well with the predictive model ($R^2 = 0.9814$; Table 2) shown in Table S4.

3.1.3. Moisture Content and Water Activity

Dry products generally have a good storage stability if the moisture content is less than 5% [20,31,35]. In the present study, the moisture content of the NPV powders was in the range of 2.1–3.1% (Table 1). Individually, both the experimental factors (X_1, X_2) significantly ($p < 0.05$) affected the moisture contents of the NPV powders (Table 2). In addition, the moisture content was significantly affected ($p < 0.05$) by the interactive effect of the factors (Table 2). A similar behavior was reported during drying of amla juice [13]. A high drying temperature increased the rate of heat transfer into the drying particles, resulting in faster evaporation of water [36]. A higher concentration of the carrier in the feed resulted in a higher moisture content in the product (Table 2), although the moisture content was always well below 5%. A similar effect of the carrier concentration was previously reported for drying of amla juice [13].

The water activity of fruit and vegetable juice powders typically ranges from 0.2 to 0.6 [10]. The NPVp samples in the present work had water activities in the range of 0.23–0.36 (Table 1). Food with a water activity of less than 0.6 is generally regarded as microbiologically stable [37], although deterioration may occur through chemical reactions such as oxidation. The effects of the factors (X_1, X_2) on water activity were comparable to their effects on the moisture content as discussed earlier in this section. An increased concentration of the carrier increased the water activity of the product, and this effect was statistically significant ($p > 0.05$) (Table 2). Similar effects of carriers have been reported in spray drying of various fruit juices [28,32,34].

3.1.4. Total Soluble Solids

A significantly high total soluble solids were observed in the product if the feed contained 20% w/v maltodextrin instead of 15% w/v maltodextrin (Table 1). This was understandable because the total soluble solids in the feed were higher if the carrier concentration was higher. The total soluble solids were not significantly affected by the inlet temperature of the drying air as the minimum temperature used was satisfactory for drying (Table 2). Similarly, the interactive effect of factors on total soluble solids was not significant (Table 2).

3.1.5. Color

All NPV powders had an off-white general appearance (Figure 1A; Table 1). In numerical terms, the color was characterized by the parameters L^*, a^*, and b^* (Table 1). The L^* lightness scale ranged from 0 (dark) to 100 (light), with 50 as the midpoint. Thus, the NPV powders had L^* values of 91.5–93.6 (Table 1), indicating that they all were well toward the lighter end of the scale. The low positive value of a^* (-0.27 to 0.54; Table 1) revealed a hue slightly more green ($a^* = -100$) than red ($a^* = 100$). Furthermore, the low positive b^* (5.53–10.28, Table 1) indicated that the color was slightly yellowish ($b^* = -100$) rather than blue ($b^* = 100$). The product NPVp-3 is depicted in Figure 1A.

The a^* and b^* color values of the NPV powders were significantly affected ($p < 0.05$) by the inlet temperature and maltodextrin content, but maltodextrin concentration had no effect on L^* values (Table 2). Other work has shown similar trends of the drying temperature affecting the lightness (*L**) of the spray dried product [37]. The hue of the product was largely influenced by the white color of the native maltodextrin carrier.

3.2. Optimization of Design

The goal of the optimization step was to identify the optimum values of the independent variables for producing a product with the desired properties. The desirability function was created by optimizing independent variables and combining the individual responses to produce an overall desirability. The numeric optimization was implemented using the Design-Expert software. The values of the product yield, bulk density, moisture content, and water activity were set to the target values for generating the optimized product (Table S3).

The highest value of overall desirability (0.921, Figure 2) could be obtained in a product that combined a high product yield and bulk density with low moisture content and water activity. The product with a desirability of 0.921 was the NPVp-3 produced using an inlet air temperature of 170 °C and a feed with 15% *w/v* of maltodextrin. For NPVp-3, the predicted values of the product yield, bulk density, moisture content, and water activity were 53.62%, 0.47 g mL^{-1}, 2.08% *w/w*, and 0.23, respectively. The NPVp-3 had the measured attributes values that were within 5% of the predicted values, thus validating the optimized model (Table S4).

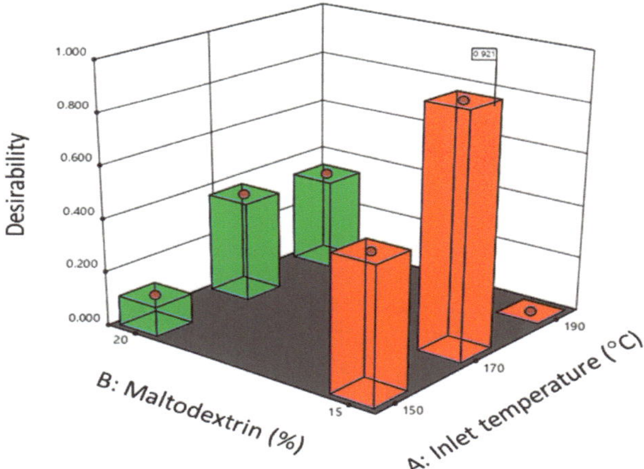

Figure 2. The overall desirability response of the products made using various combinations of the drying air inlet temperature (X_1 = A) and maltodextrin concentration (X_2 = B). The product with the highest desirability (0.921) had a drying temperature of 170 °C and 15% *w/v* of maltodextrin content. Design points are shown as red circles.

3.3. Nutritional Characteristics

A comparison of the nutritional parameters of the NPVp-3 powder with a commercial product is shown in (Table 3). The NPVp-3 powder had around 96% of the calorific content as the commercial product possibly because of its lower level (91.5% of commercial product) of total carbohydrate compared to the commercial product (Table 3). The protein content of NPVp-3 was ~27-fold greater compared to the commercial product and the total fat was also greater (~10-fold; Table 3) possibly due to natural variations of fat and protein in the vinegar, depending its original source.

Table 3. Nutritional analysis of the NPVp-3 product compared with a commercial vinegar powder sample.

Test Item	Unit/100 g [a,b]	NPVp-3 Powder	Commercial Vinegar Powder [d]
Energy	kcal	366.16 ± 3.92	380.9
Protein	g	1.33 ± 0.01	0.05
Crude fat	g	0.96 ± 0.013	0.1
Cholesterol	mg	Not detected	0
Carbohydrate	g	88.05 ± 1.88	96.2
Total sugars	g	14.16 ± 0.85	3.5
Total dietary fiber	g	0 ± 0.00	0.22
Ash	g	3.03 ± 0.01	-
Minerals			
Calcium (Ca)	mg	30.92 ± 0.71	25
Iron (Fe)	mg	0.54 ± 0.00	0.41
Sodium (Na)	mg	268.56 ± 0.12	79
Potassium (K)	mg	883.90 ± 0.35	-
Phosphorous (P)	mg	35.90 ± 0.03	-
Magnesium (Mg)	mg	21.50 ± 0.13	-
Zinc (Zn)	mg	0.26 ± 0.00	-
Vitamins			
Vitamin A	μg	0.00 ± 0.00	0.54
Vitamin B1	mg	Not detected	-
Vitamin B2	mg	Not detected	-
Vitamin C	mg	12.68 ± 0.10	0.27
Other constituents			
Total phenolics content	μg GAE [c] g^{-1}	1049.76 ± 20.03	-
Acetic acid	% w/v	5.13	12

[a] Samples were analyzed in triplicate. Data are average values ± standard deviations.; [b] Dry weight basis.; [c] GAE, gallic acid equivalent; [d] Values are based on the literature (Spice Barn® vinegar powder, Spice Barn, Inc.; https://spicebarn.com/vinegar_powder.htm (accessed on 24 May 2022)).

The NPVp-3 powder had similar levels of calcium and iron as the commercial product, but was considerably richer in the other minerals listed in Table 3. The total vitamin C in NPVp-3 powder was nearly 47-fold greater than in the commercial product, possibly because the spray drying conditions used preserved a higher proportion of this heat-labile vitamin compared to the processing conditions used in making the commercial product. The commercial product had a small amount of vitamin A (Table 3), but this was too low to be meaningful compared to ~900 μg daily requirement of this vitamin in adult males.

Gallic acid equivalent (GAE) is a commonly used measure of total phenolics in foods. Phenolics are antioxidants and also protect food against oxidative damage. The NPVp-3 powder had a high level of total phenolics (~1050 μg GAE g^{-1}; Table 3) compared to literature data (167 μg GAE mL^{-1}; [8]) for reported for NPV. The elevated level of phenolics was explained at least partly by the concentrating effect of spray drying.

Natural liquid vinegars typically have more than 4 g of acetic acid per 100 mL (FDA, https://www.fda.gov/media/71937/download (accessed on 24 May 2022)). The NPVp-3 powder had ~5% w/v acetic acid (Table 3) because much of it was lost by vaporization during drying.

3.4. Sensory Acceptance

The sensory evaluations of the beverage/food preparations containing the original NPV and the product NPVp-3 for color, odor, taste, and overall acceptability are shown in Figure 3. The six recipes (R1–R6) were scored by the 59 panelists as previously explained.

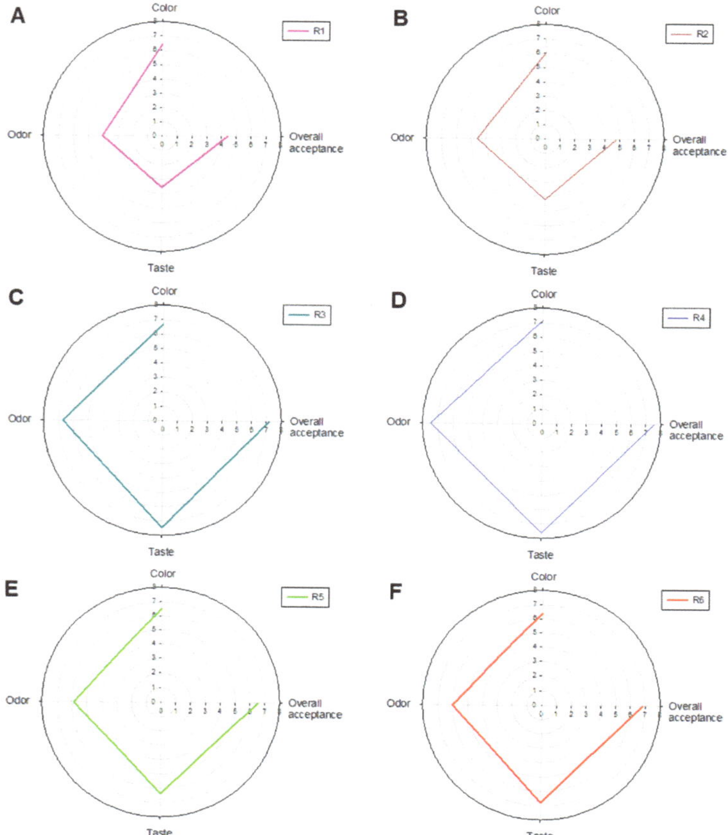

Figure 3. Spider charts of sensory evaluation scores of the original NVP and the NPVp-3 powder used as a beverage and food additive: (**A**) Recipe R1 containing the original NPV (control R1) 200 g in 800 g of drinking water; (**B**) Recipe R2 containing 200 g of rehydrated NPV powder (control R2) in 800 g of drinking water; (**C**) Recipe R3 containing the original NPV (200 g), 200 g honey, and 600 g carbonated water; (**D**) Recipe R4 containing NPVp-3 (200 g), 200 g honey, and 600 g carbonated water; (**E**) Recipe R5 containing the original NPV (60 g) in 1 L of sour soup (Thai sour sea mullet fish soup); and (**F**) Recipe R6 containing NPVp-3 (60 g) in 1 L of sour soup (as above). The data are mean values ± standard deviations (*n* = 59). Each score was generated by 59 individuals using a 9-point hedonic scale (1 = dislike very much; 9 = like very much).

The test materials containing the original NPV were used as controls (R1, Figure 3A; R3, Figure 3C; R5, Figure 3E). The various drink/food recipes are explained in the caption of Figure 3. The recipes contained either the original NPV, or NPVp-3, mixed with: drinking water (R1, R2); honey and carbonated water (R3, R4); and Thai sour fish soup (R5, R6). The recipes containing NPVp-3 (R2, Figure 3B; R4, Figure 3D; R6, Figure 3F) used the same ingredients as in the control preparations containing the original NPV (Section 2.6).

The sensory scores of color, odor, taste, and overall acceptability for the recipes R1 and R2 (Figure 3A,B) were not substantially different. R1 and R2 received overall acceptance scores of 4.41 ± 1.63 and 4.73 ± 1.72, respectively. NPVp-3 recipe R2 received a little higher rating than the recipe R1 that contained the original NPV. The difference was explained by the more intense vinegary odor and the characteristic sour flavor of NPV due to its higher content of acetic acid.

In the beverage recipes R3 and R4 (Figure 3C,D), the recipe R4 made with NPVp-3 scored somewhat higher than R3 (made with the original vinegar) for all the scored attributes. Overall, the tasters better accepted the beverages made with NPVp-3.

Original NPV (Recipe R5) and NPVp-3 powder (Recipe R6) were used to season the fish soup (Figure 3E,F). The sensory parameter ratings for both were fairly comparable, but R6 scored somewhat higher than the original NPV. Although the differences were small, the tasters preferred the beverage recipe R4 (Figure 3D) based on NPVp-3 more than the same product used as a food seasoning (R6, Figure 3F).

3.5. Product Stability during Storage

Variations in the relevant physicochemical parameters of NPVp-3 stored for various periods at 25 °C are shown in Table 4. The off-white appearance of the samples barely changed during 180 days of storage (Table 4). In terms of the electronic color parameters (L^*, a^*, b^*; Table 4), there were minor changes: a slight but fairly consistent increase in a^* value with the length of storage, and a similar increase in the b^* value. The sample lightness (L^*) was always within 94.1 and 95.1 (Table 4).

Table 4. Effect of storage duration on appearance, pH, moisture content, water activity, color, and microbial contents (molds and yeasts) of the product NPVp-3 during storage at 25 °C.

Property	Storage Duration (Days) [A]					
	0	30	60	90	120	180
Appearance						
pH	3.20 ± 0.01 [ab]	3.18 ± 0.01 [a]	3.18 ± 0.01 [a]	3.21 ± 0.01 [b]	3.21 ± 0.00 [b]	3.21 ± 0.00 [b]
Moisture (% *w/w*)	2.61 ± 0.01 [a]	2.62 ± 0.02 [a]	2.61 ± 0.01 [a]	2.61 ± 0.01 [a]	2.62 ± 0.02 [a]	2.62 ± 0.01 [a]
Water activity	0.23 ± 0.00 [a]	0.23 ± 0.00 [a]	0.23 ± 0.00 [a]	0.23 ± 0.00 [a]	0.24 ± 0.00 [a]	0.24 ± 0.00 [a]
Color						
L^*	94.13 ± 0.18 [a]	95.07 ± 0.13 [bc]	95.04 ± 0.05 [bc]	94.94 ± 0.07 [bc]	94.87 ± 0.04 [bc]	94.64 ± 0.07 [b]
a^*	0.37 ± 0.03 [a]	0.36 ± 0.01 [a]	0.42 ± 0.06 [ab]	0.49 ± 0.04 [ab]	0.54 ± 0.03 [bc]	0.58 ± 0.02 [c]
b^*	10.03 ± 0.04 [a]	10.26 ± 0.20 [ab]	10.44 ± 0.15 [abc]	10.53 ± 0.02 [bc]	10.67 ± 0.01 [bc]	10.78 ± 0.11 [c]
Molds (CFU g^{-1})	00.00 ± 0.00 [a]	00.00 ± 0.00 [a]	00.00 ± 0.00 [a]	00.00 ± 0.00 [a]	00.00 ± 0.00 [a]	00.00 ± 0.00 [a]
Yeasts (CFU g^{-1})	00.00 ± 0.00 [a]	00.00 ± 0.00 [a]	00.00 ± 0.00 [a]	00.00 ± 0.00 [a]	00.00 ± 0.00 [a]	00.00 ± 0.00 [a]

All analyses were in triplicate. Data are average values ± standard deviations.; [A] Different superscript lowercase letters within a row indicate a significant difference ($p < 0.05$).

The moisture content of the product remained stable (2.62% *w/w*; Table 4) over the full duration of storage. Similarly, the water activity remained within 0.23–0.24 throughout (Table 4). NPV powder maintained its initial water activity level throughout the study. The pH values were quite stable over the entire storage period (Table 4), although marginally higher in samples stored for 90 days and longer. A stable acidity (pH) and water activity level are particularly important in ensuring a long shelf-life of a food product. Stable values of these parameters indirectly suggest a lack of microbial growth [38].

The microbial count of the product did not change, and the product was essentially sterile (Table 4). Absence of microbial growth was consistent with the low water activity (≤ 0.24; Table 4) of the product, as microbes do not proliferate if the water activity is less than 0.5 [10].

4. Conclusions

The NPV powder with the highest desirability could be produced using a drying air inlet temperature of 170 °C in combination with a nipa palm vinegar feed containing 15% *w/v* maltodextrin DE10. This product (NPVp-3) had nearly 47-fold higher vitamin C content than a commercial vinegar powder but only ~43% of the acidity (acetic acid) compared to the commercial product. The NPVp-3 powder was stable at room temperature during 180-days of storage. The excellent stability was explained by the low water activity (a_w = 0.23) of the product, indicating highly effective drying by the method used. A relatively high content of the highly heat-sensitive vitamin C in the product indicated that the spray-drying treatment minimized damage to the product due to the short contact time (residence time in the drying chamber was less than 42 s) of the product with the drying air. The NPVp-3 was judged to be a good food seasoning and beverage ingredient.

Supplementary Materials: The following supporting information can be downloaded at: https://www.mdpi.com/article/10.3390/fermentation8060272/s1. Supplemental material for this article is available at: Table S1. The levels of the various factors in experimental design. Table S2. The factor values used in producing the various powdered NPV products. Table S3. The constraints used in estimating the overall desirability. Table S4. The statistical models and comparison of predictions with the measured values.

Author Contributions: Conceptualization, W.P. and W.K.K.; methodology, W.P. and W.K.K.; investigation, W.P. and W.K.K.; resources, W.K.K.; writing—original draft preparation, W.P. and W.K.K.; writing—review and editing, W.P., W.K.K. and Y.C.; supervision, W.K.K. and Y.C.; funding acquisition, W.K.K. and W.P. All authors have read and agreed to the published version of the manuscript.

Funding: The research was funded by Agricultural Research Development Agency (Public Organization), Thailand, under project number CRP6205031430, and partially funded by the new strategic research project of Walailak University (P2P).

Institutional Review Board Statement: This study was conducted according to the guidelines of the Declaration of Helsinki, and approved by the Human Research Ethics Committee of Walailak University (approval WUEC-20-048-01 dated 6 March 2020).

Informed Consent Statement: Informed consent was obtained from all participants of the sensory panels involved in this study.

Data Availability Statement: All data relating to this study are included within this article. Data will be made available on reasonable request.

Conflicts of Interest: The authors declare no conflict of interest.

References

1. Cheablam, O.; Chanklap, B. Sustainable nipa palm (*Nypa fruticans* Wurmb.) product utilization in Thailand. *Scientifica* **2020**, *2020*, 3856203. [CrossRef] [PubMed]
2. Hamilton, L.S.; Murphy, D.H. Use and management of Nipa palm (*Nypa fruticans*, arecaceae): A review. *Econ. Bot.* **1988**, *42*, 206–213. [CrossRef]
3. Tsuji, K.; Ghazalli, M.N.F.; Ariffin, Z.; Nordin, M.S.; Khaidizar, M.I.; Dulloo, M.E.; Sebastian, L.S. Biological and ethnobotanical characteristics of Nipa Palm (*Nypa fructicans* Wurmb.): A review. *Sains Malays.* **2011**, *40*, 1407–1412.
4. Senghoi, W.; Klangbud, W.K. Antioxidants, inhibits the growth of foodborne pathogens and reduces nitric oxide activity in LPS-stimulated RAW 264.7 cells of nipa palm vinegar. *PeerJ* **2021**, *9*, e12151. [CrossRef]
5. Laklaeng, S.-N.; Kwanhian, W. Immunomodulation effect of *Nypa fruticans* palm vinegar. *Walailak J. Sci. Technol.* **2020**, *17*, 1200–1210. [CrossRef]
6. Francisco-Ortega, J.; Zona, S. Sweet sap from palms, a source of beverages, alcohol, vinegar, syrup, and sugar. *Vieraea* **2013**, *41*, 91–113. [CrossRef]
7. Yusoff, N.A.; Ahmad, M.; Al Hindi, B.; Widyawati, T.; Yam, M.F.; Mahmud, R.; Razak, K.N.; Asmawi, M.Z. Aqueous extract of *Nypa fruticans* Wurmb. vinegar alleviates postprandial hyperglycemia in normoglycemic rats. *Nutrients* **2015**, *7*, 7012–7026. [CrossRef] [PubMed]
8. Chatatikun, M.; Kwanhian, W. Phenolic profile of nipa palm vinegar and evaluation of its antilipidemic activities. *Evid. Based Complement. Alternat. Med.* **2020**, *2020*, 6769726. [CrossRef]
9. Santos, D.; Maurício, A.C.; Sencadas, V.; Santos, J.D.; Fernandes, M.H.; Gomes, P.S. Spray drying: An overview. *Biomater. Phys. Chem. -New Ed.* **2017**, 9–35. [CrossRef]
10. Shishir, M.R.I.; Chen, W. Trends of spray drying: A critical review on drying of fruit and vegetable juices. *Trends Food Sci. Technol.* **2017**, *65*, 49–67. [CrossRef]

11. Muhamad, I.I.; Hassan, N.D.; Mamat, S.N.; Nawi, N.M.; Rashid, W.A.; Tan, N.A. Extraction technologies and solvents of phyto-compounds from plant materials: Physicochemical characterization and identification of ingredients and bioactive compounds from plant extract using various instrumentations. In *Ingredients Extraction by Physicochemical Methods in Food*; Grumezescu, A.M., Holban, A.M., Eds.; Elsevier: Amsterdam, The Netherlands, 2017; pp. 523–560. [CrossRef]

12. Braga, V.; Guidi, L.R.; de Santana, R.C.; Zotarelli, M.F. Production and characterization of pineapple-mint juice by spray drying. *Powder Technol.* **2020**, *375*, 409–419. [CrossRef]

13. Mishra, P.; Mishra, S.; Mahanta, C.L. Effect of maltodextrin concentration and inlet temperature during spray drying on physicochemical and antioxidant properties of amla (*Emblica officinalis*) juice powder. *Food Bioprod. Process.* **2014**, *92*, 252–258. [CrossRef]

14. Abd Hashib, S.; Abd Rahman, N.; Suzihaque, M.; Ibrahim, U.K.; Hanif, N.E. Effect of slurry concentration and inlet temperature towards glass temperature of spray dried pineapple powder. *Procedia Soc. Behav. Sci.* **2015**, *195*, 2660–2667. [CrossRef]

15. Oberoi, D.P.S.; Sogi, D.S. Effect of drying methods and maltodextrin concentration on pigment content of watermelon juice powder. *J. Food Eng.* **2015**, *165*, 172–178. [CrossRef]

16. Chen, J.-C.; Zheng, B.-D.; Zhao, Y.-Y.; Xu, J.; Wu, J.-J.; Luo, C.; Pang, J.; He, G.-Q. Hypolipidemic effects on lipid metabolism and lipase inhibition by black vinegar powder. *J. Biobased Mater. Bioenergy* **2012**, *6*, 634–639. [CrossRef]

17. Chen, J.; Tian, J.; Ge, H.; Liu, R.; Xiao, J. Effects of tetramethylpyrazine from Chinese black vinegar on antioxidant and hypolipidemia activities in HepG2 cells. *Food Chem. Toxicol.* **2017**, *109*, 930–940. [CrossRef]

18. Othaman, M.A.; Sharifudin, S.A.; Mansor, A.; Kahar, A.; Long, K. Coconut water vinegar: New alternative with improved processing technique. *J. Eng. Sci. Technol.* **2014**, *9*, 293–302.

19. Yu, L.; Peng, Z.; Dong, L.; Wang, S.; Ding, L.; Huo, Y.; Wang, H. Bamboo vinegar powder supplementation improves the antioxidant ability of the liver in finishing pigs. *Livest. Sci.* **2018**, *211*, 80–86. [CrossRef]

20. Goula, A.M.; Adamopoulos, K.G. Spray drying of tomato pulp in dehumidified air: II. The effect on powder properties. *J. Food Eng.* **2005**, *66*, 35–42. [CrossRef]

21. AOAC International. *Official Methods of Analysis of AOAC International*, 21st ed.; AOAC International: Rockville, MD, USA, 2019.

22. AOAC International. *Official Methods of Analysis of AOAC International*, 17th ed.; AOAC International: Rockville, MD, USA, 2000.

23. Palachum, W.; Choorit, W.; Manurakchinakorn, S.; Chisti, Y. Guava pulp fermentation and processing to a vitamin B12-enriched product. *J. Food Process. Preserv.* **2020**, *44*, e14566. [CrossRef]

24. Chen, Z.; Chen, B.; Yao, S. High-performance liquid chromatography/electrospray ionization-mass spectrometry for simultaneous determination of taurine and 10 water-soluble vitamins in multivitamin tablets. *Anal. Chim. Acta* **2006**, *569*, 169–175. [CrossRef]

25. Sullivan, D.M.; Carpenter, D.E. *Methods of Analysis for Nutrition Labeling*; AOAC International: Rockville, MD, USA, 1993.

26. Palachum, W.; Choorit, W.; Chisti, Y. Nutritionally enhanced probioticated whole pineapple juice. *Fermentation* **2021**, *7*, 178. [CrossRef]

27. FDA. *Bacteriological Analytical Manual: Chapter 3: Aerobic Plate Count*; US Food and Drug Administration: Silver Spring, MD, USA, 2001. Available online: https://www.fda.gov/food/laboratory-methods-food/bam-chapter-3-aerobic-plate-count (accessed on 5 April 2022).

28. Tan, S.L.; Sulaiman, R.; Rukayadi, Y.; Ramli, N.S. Physical, chemical, microbiological properties and shelf life kinetic of spray-dried cantaloupe juice powder during storage. *LWT* **2021**, *140*, 110597. [CrossRef]

29. Saavedra-Leos, M.Z.; Leyva-Porras, C.; López-Martínez, L.A.; González-García, R.; Martínez, J.O.; Compeán Martínez, I.; Toxqui-Terán, A. Evaluation of the spray drying conditions of blueberry juice-maltodextrin on the yield, content, and retention of quercetin 3-D-galactoside. *Polymers* **2019**, *11*, 312. [CrossRef] [PubMed]

30. Şahin-Nadeem, H.; Torun, M.; Özdemir, F. Spray drying of the mountain tea (*Sideritis stricta*) water extract by using different hydrocolloid carriers. *LWT* **2011**, *44*, 1626–1635. [CrossRef]

31. Şahin-Nadeem, H.; Dinçer, C.; Torun, M.; Topuz, A.; Özdemir, F. Influence of inlet air temperature and carrier material on the production of instant soluble sage (*Salvia fruticosa* Miller) by spray drying. *LWT* **2013**, *52*, 31–38. [CrossRef]

32. Lee, K.-C.; Yoon, Y.S.; Li, F.-Z.; Eun, J.-B. Effects of inlet air temperature and concentration of carrier agents on physicochemical properties, sensory evaluation of spray-dried mandarin (*Citrus unshiu*) beverage powder. *Appl. Biol. Chem.* **2017**, *60*, 33–40. [CrossRef]

33. Ribeiro, C.M.C.M.; Magliano, L.C.d.S.A.; Costa, M.M.A.d.; Bezerra, T.K.A.; Silva, F.L.H.d.; Maciel, M.I.S. Optimization of the spray drying process conditions for acerola and seriguela juice mix. *Food Sci. Technol.* **2019**, *39*, 48–55. [CrossRef]

34. Both, E.; Boom, R.; Schutyser, M. Particle morphology and powder properties during spray drying of maltodextrin and whey protein mixtures. *Powder Technol.* **2020**, *363*, 519–524. [CrossRef]

35. Sinija, V.R.; Mishra, H.N. Moisture sorption isotherms and heat of sorption of instant (soluble) green tea powder and green tea granules. *J. Food Eng.* **2008**, *86*, 494–500. [CrossRef]

36. Solval, K.M.; Sundararajan, S.; Alfaro, L.; Sathivel, S. Development of cantaloupe (*Cucumis melo*) juice powders using spray drying technology. *LWT* **2012**, *46*, 287–293. [CrossRef]

37. Quek, S.Y.; Chok, N.K.; Swedlund, P. The physicochemical properties of spray-dried watermelon powders. *Chem. Eng. Process. Process Intensif.* **2007**, *46*, 386–392. [CrossRef]

38. Han, N.S.; Ahmad, W.; Ishak, W.R.W. Quality characteristics of *Pleurotus sajor-caju* powder: Study on nutritional compositions, functional properties and storage stability. *Sains Malays.* **2016**, *45*, 1617–1623.

Article

Effect of Fermented Camel Milk Containing Pumpkin Seed Milk on the Oxidative Stress Induced by Carbon Tetrachloride in Experimental Rats

Magdy Ramadan Shahein [1], El Sayed Hassan Atwaa [2], Barakat M. Alrashdi [3], Mahytab Fawzy Ramadan [2], El Sayed Abd El-Sattar [4], Adel Abdel Hameed Siam [5], Mohamed A. Alblihed [6] and Ehab Kotb Elmahallawy [7,*]

[1] Department of Food Science and Technology, Faculty of Agriculture, Tanta University, Tanta 31527, Egypt; magdrsh10@gmail.com

[2] Food Science Department, Faculty of Agriculture, Zagazig University, Zagazig 44511, Egypt; elsayedhassanattwa@gmail.com (E.S.H.A.); ahetabhassanien@hotmail.com (M.F.R.)

[3] Biology Department, College of Science, Jouf University, Sakaka 72388, Saudi Arabia; bmalrashdi@ju.edu.sa

[4] Department of Food and Dairy Technology, Faculty of Technology and Development, Zagazig University, Zagazig 44511, Egypt; e.abdelsattar82@gmail.com

[5] Department of Physiology, Faculty of Veterinary Medicine, Suez Canal University, Ismailia 41522, Egypt; adel_soliman@vet.suez.edu.eg

[6] Department of Microbiology, College of Medicine, Taif University, P.O.Box. 11099, Taif 21944, Saudi Arabia; mabulihd@tu.edu.sa

[7] Department of Zoonoses, Faculty of Veterinary Medicine, Sohag University, Sohag 82524, Egypt

* Correspondence: eehaa@unileon.es

Citation: Shahein, M.R.; Atwaa, E.S.H.; Alrashdi, B.M.; Ramadan, M.F.; Abd El-Sattar, E.S.; Siam, A.A.H.; Alblihed, M.A.; Elmahallawy, E.K. Effect of Fermented Camel Milk Containing Pumpkin Seed Milk on the Oxidative Stress Induced by Carbon Tetrachloride in Experimental Rats. *Fermentation* **2022**, *8*, 223. https://doi.org/10.3390/fermentation8050223

Academic Editor: Carlo Giuseppe Rizzello

Received: 30 March 2022
Accepted: 6 May 2022
Published: 13 May 2022

Publisher's Note: MDPI stays neutral with regard to jurisdictional claims in published maps and institutional affiliations.

Abstract: Oxidative stress can lead to chronic inflammation, nephrotoxicity, and renal damage. The consumption of plant-based dairy alternatives has increased rapidly worldwide due to their health effects. Bioactive components from natural sources, such as plants, are effective in protecting against oxidative stress. The present study evaluated the physicochemical and sensory properties of fermented camel milk made from camel milk mixed with pumpkin seed milk. Fermented camel milk consists of camel milk mixed with 25% and 50% pumpkin seed milk. This blend (fermented camel milk containing 50% pumpkin seed milk) was evaluated as an antioxidant agent in oxidative stress induced rats. A total of thirty-two male adult albino rats of Sprague Dawley® Rat strain weighing 150–180 g were randomly divided into four groups ($n = 8$). The first group was solely administered the standard diet and served as the negative control. The other rats ($n = 24$), received a basal diet, including being intraperitoneally injected with carbon tetrachloride, with a single dose at a rate of 2 mL/kg body weight) as a model for oxidative stress. The oxidative stress rats were divided into three groups; the first group did not receive any treatment and served as the positive control. The second and third groups were administered 10 g/day fermented camel milk and fermented camel milk containing 50% pumpkin seed milk. The results revealed that mixing the camel milk with pumpkin seed milk was more effective in increasing the total solids, protein, ash, fiber, acidity, viscosity, phenolic content, and antioxidant activity. These enhancements were proportional to the mixing ratio. Fermented camel milk containing 50% pumpkin seed milk exhibited the highest scores for sensory properties compared with the other fermented camel milk treatments. The group of rats with oxidative stress treated with fermented camel milk containing 50% pumpkin seed milk showed a significant decrease ($p \leq 0.05$) in the levels of malondialdehyde (MDA), low-density lipoprotein (LDL), cholesterol (CL), triglycerides (TGs), AST, ALT, creatinine, and urea, and increased ($p \leq 0.05$) high-density lipoprotein (HDL) and total protein and albumin compared with rats with oxidative stress. Consumption of fermented camel milk containing 50% pumpkin seed milk by the oxidative stress rat groups caused significant improvement in all of these factors compared with the positive control group. This study revealed that the administration of fermented camel milk containing 50% pumpkin seed milk to rats with oxidative stress prevented disorders related to oxidative stress compared with the untreated oxidative stress group. Thus, incorporating fermented camel milk might play a beneficial role in patients with oxidative stress.

Keywords: oxidative stress; pumpkin seed milk; camel milk; physicochemical; carbon tetrachloride; rats

1. Introduction

Milk and its products are considered essential foods for human development and can benefit in the oxidative defense of consumers. Furthermore, milk and dairy products with protective properties have the potential to act as coadjuvants in metabolic disorders, intestinal health, conventional therapies, addressing cardiovascular diseases, and chemopreventive properties [1]. Among others, camel milk has high nutritional and health value as it contains immune proteins such as lysozyme (which is an antioxidant and anti-inflammatory compound) and aminoglobulins, but no beta-lactoglobulin, which may cause allergic reactions in some people. It contains iron, potassium, and vitamins C, E, and A [2,3]. Most camel milk is traditionally consumed fresh as raw milk or when soured. However, recently, fresh pasteurized camel milk has been made available in the Middle East and Arab lands [4]. Importantly, Camel milk contains more free amino acids and peptides than cow's milk [5]. Moreover, non-protein-bound amino acids in camel milk are easily digested by microorganisms, thus, camel milk has higher metabolic activity when used in the starter culture preparation [6]. Fermented camel milk has several health benefits: antioxidant activity, angiotensin-converting enzyme inhibitory activity, a hypocholesterolemic effect, antimicrobial activity, anti-diarrhea activity, and anticancer activity [7]. There is increased attention to improving fermented dairy products as a result of their health benefits [8–16].

Oxidative stress is an imbalance between the production of reactive oxygen species and their elimination by protective mechanisms, which can lead to chronic inflammation [17]. Oxidative stress can activate various transcription factors, thus leading to the differential expression of various genes involved in inflammatory pathways [18,19]. The inflammation triggered by oxidative stress causes many chronic diseases [20]. Polyphenols might be useful as adjuvant therapy based on their potential anti-inflammatory effect, associated with their antioxidant activity and inhibition of enzymes involved in the production of eicosanoids [21,22]. Moreover, the kidneys are vital organs that are essential for the excretion of metabolic wastes and maintaining chemical homeostasis, among other functions [23]. Several studies have reported oxidative stress as a potential cause of different forms of renal damage and nephrotoxicity. Increased oxidative stress caused by free radical generation is a likely result of inflammatory responses [24]. Adding supplementary antioxidants from natural sources, such as plants, is efficient in protecting against oxidative stress. This may prompt more food manufacturers to replace synthetic antioxidants with ingredients containing natural antioxidative compounds. Therefore, natural additives have gained increased attention as they pose no health risk to consumers [25]. The antioxidant protection mechanism acting against the reactions of free radicals comprises enzymatic and non-enzymatic elements, part of which are synthesized in plants exclusively, and the body can obtain them only from foods [26]. Several antioxidants, e.g., flavonoids, phenolic acids, tannins, vitamin C, and vitamin E, interact synergistically with other reducing compounds and have diverse biological properties, such as anti-inflammatory, anti-carcinogenic, and anti-atherosclerotic effects [21]. Moreover, foods that are rich in bioactive compounds are advantageous as they boost the immune system.

It is noteworthy to state that cereals and their components have been accepted as functional foods as they provide the vitamins, dietary fiber, protein, antioxidants, energy, and minerals required for human health [8,10–16]. Moreover, cereals can be fermentable substances for the growth of probiotic bacteria [27,28]. Pumpkin (*Cucurbita pepo* L.) is a plump nutritious vegetable that belongs to the *Cucurbitaceae* family, with several varieties grown throughout the world, e.g., from South–Central America, to Mexico, Argentina, Chile, Europe, Asia (India and China), and Western America [29–31]. The pumpkin seed oil has been used since the end of the 19th century to treat urinary tract problems [32]. Pumpkin contains various biologically active components, such as sterols, proteins, peptides, polysaccharides, para-aminobenzoic acid [33], carotenoids, and γ-aminobutyric acid [34]. In addition, pumpkin seeds have a high protein content and essential fatty acids [35], mainly linoleic acid, stearic acid, oleic acid, and palmitic acid. Furthermore, it contains non-essential amino acids and phytosterols, e.g., sitosterol and stigmasterol [36]. Moreover, pumpkin seeds contain tocopherol (vitamin E), microelements (Na, K, and Cr), [37,38], and phenolic compounds, such as coumarins, flavonoids, pigments, pyrazine, and triterpenoids [39–41]. Importantly, a pumpkin seed extract promotes wound healing and has anti-arthritis, hair-growth-stimulating [42], anthelmintic [43], antitumor [44], hepatoprotective [45], and antioxidant effects [46]. In addition, the therapeutic activities of pumpkin seed extracts include the relief of symptoms associated with urinary bladder complications and prostate disorders [47]. A previous study [48] explained that pumpkin seeds and their different types are of high nutritional value as they are a good source of protein, fats, fiber, minerals, squalene, tocopherols and β-carotene, and therefore can be used as a good nutritional source in the food industry. Pumpkin seeds also have good antimicrobial properties and can be used for medicinal purposes. The chromatographic analysis showed Pumpkin seeds contain many phenolic compounds such as protocatechuic acid, myricetin, injectable acid, vanillin, quercetin, hydroxybenzoic acid, apigenin, etc., which are famous for their anti-inflammatory and anti-cancer activity. Revising the available literature, no information is available about the potential effect of adding pumpkin seed milk to fermented camel milk manufacturing. The present study aimed to investigate the impact of incorporating pumpkin seed milk into fermented camel milk manufacture, to fortify different proportions of fermented camel milk regarding its functional properties. Furthermore, the fortified fermented camel milk was tested for its therapeutic effect as a functional food in rats with high oxidative stress induced by carbon tetrachloride.

2. Material and Methods

2.1. Materials and Reagents

Fresh bulk camel milk was obtained from the Desert Research Center, Dokki, Egypt. Pumpkin seeds were obtained from the Crops Research Institute, Agricultural Research Center, Ministry of Agriculture Giza, Egypt. All solvents used for extraction and analyses were of analytical grade. Folin–Ciocalteu, gallic acid, and 2,2-diphenyl-1-pic-rylhydrazyl (DPPH) were purchased from Sigma (St. Louis, MO, USA). The kits that were used for the biochemical analyses were from Gamma Trade Company (Cairo, Egypt). The protocol reported by Reeves et al. 1993 [49] was used to prepare the standard basal diet. The yogurt cultures that were added included *Lactobacillus delbruekii* subsp. *bulgaricus* and *Streptococcus salivarius* subsp. *thermophilus* were purchased from the Ain Shams University, Agriculture Faculty, Microbiological Resources Center, Egypt.

2.2. Preparation of Pumpkin Seed Milk (PSM)

PSM was prepared as per Hassan et al., 2012, with some modifications [50]. Seeds (100 g) were cleaned, dehulled, and soaked in water at a 1:4 w/v ratio overnight at room temperature. The soaked seeds were blended in a grinder. The resulting emulsion was then filtered through a double-layer muslin cloth, boiled for 5 min with constant stirring, cooled to room temperature (22 °C), and kept at a refrigerator temperature before product formulation was used within 24 h.

2.3. Fermented Camel Milk Manufacture

Fermented camel milk was manufactured according to the method reported by Tamime and Robinson, 1999 [51]. Raw camel milk was divided into three equal portions; the first portion served as the control (C), the second portion was mixed with 25.0% PSM (T1), and the third portion was mixed with 50.0% PSM (T2). The milk used for all treatments was homogenized at 55 °C–60 °C for 2 min using a high-speed mixer (22,000 rpm), heat-treated in a thermostatically controlled water bath at 85 °C for 30 min, cooled to 42 °C in an ice bath, inoculated with 5% (w/v) yogurt culture, and incubated at 42 °C until a firm curd was obtained (around 12 h). The curd was then refrigerated at 4 °C overnight, stirred using the mixer, stored at 4 °C ± 1 °C, and analyzed 1 day after manufacture for its physicochemical and sensory properties.

2.4. Chemical Composition, Physicochemical Analysis, and Sensory Evaluation of the Fermented Camel Milk Treatments

In the fermented camel milk samples, the total solids, fat, protein, ash amounts, and titratable acidity were determined as described elsewhere [52]. The total solids of camel milk, pumpkin seed milk and fermented camel milk samples were determined using a drying oven for 24 h at 100 °C. The percentage of moisture content was calculated by the following formula.

$$\% \text{ moisture} = W1 - (W2 \times 100)/W1$$

where, W1 = initial weight of sample; W2 = weight of the dried sample

$$\text{Total solids } \% = 100 - \text{moisture content}$$

For fat content of camel milk and fermented camel milk samples, the Gerber method was applied, whereas for pumpkin seed milk, the Soxhelt apparatus method was used. The total nitrogen content (TN) was determined using the micro-Kjeldahl method, and protein content was calculated by multiplying the percentage of TN by 6.38 for milk components and 6.25 for pumpkin seed milk. To determine ash content, a 5 g sample was heated in a muffle furnace at 550 °C overnight. The titratable acidity, expressed as a percentage of lactic acid (%), was determined by mixing 10 g of fermented camel milk with 10 mL of distilled water and titrating with 0.1 N NaOH using phenolphthalein as an indicator to an end-point of faint pink color. Their pH values were monitored using a pH meter equipped with a glass electrode (HANNA, Instrument, Portugal). The viscosity of the fermented camel milk was determined according to Aryana, 2003 [53] using Rotational Viscometer Type Lab. Line Model 5437. Measurements were determined at a temperature of 30 °C after 15 s; the results were expressed as centipoise (cP). The total phenolic content (TPC; as mg GAE (gallic acid equivalents)/100 g) and antioxidant (AO) activity (%) of the prepared fermented camel milk treatments were assessed according to Maksimović et al. 2008 [54] and Apostolidis et al. 2007 [55], respectively. The sensory evaluation of the fermented camel milk treatments was carried out by ten trained panelists as described elsewhere [28].

2.5. Experimental Design of the Biological Study

A total of 32 male adult Sprague Dawley albino rats (weight range, 150–180 g) were provided by the Agricultural Research Center of Giza (Egypt) and housed in wire cages at 25 °C. After acclimation to a basal diet for 7 days, they were divided randomly into four groups (eight rats n each group), as follows: eight rats were kept as the normal control and were fed a standard diet as the negative control (group 1). Twenty-four rats were intraperitoneally injected with a single dose of carbon tetrachloride at a 2 mL/kg of body weight (oxidative stress rats), as described elsewhere [56]. Oxidative stress rats were then organized into three groups (eight rats in each group), as follows: oxidative stress rats as a positive control (group 2), oxidative stress rats fed a basal diet with 10 g/day plain fermented camel milk via an epigastric tube (group 3), and oxidative stress rats fed a basal diet with 10 g/day PSM-fermented camel milk (fermented camel milk manufactured from blended milk; i.e., 50% camel milk and 50% PSM) via a epigastric tube (group 4).

2.6. Biochemical Analysis

After five weeks, the rats were lightly anesthetized with diethyl ether. Blood samples were collected from the hepatic portal vein, submitted to centrifugation (to separate the serum) at 3000 rpm for 15 min, and stored at −40 °C. Caraway's method [57] was used to determine serum uric acid. In contrast, the serum creatinine level was measured using Bonsens' method [58], and serum urea was determined according to Marsch et al., 1965 [59].

Malondialdehyde (MDA) was determined in the serum according to Kei, 1978 [60]. The alanine amino transferase (ALT) and aspartate amino transferase (AST) enzymes were measured according to the methods described by Bergmeyer and Harder, 1986 [61]. The total cholesterol was determined according to the method of the Enzymatic Colorimeter [62]. Total lipids and triglycerides were determined according to the method of Devi and Sharma, 2004 [63]. LDL-cholesterol was calculated using the Friedewald formula [64], as follows:

$$\text{LDL-cholesterol} = \text{total cholesterol} - (\text{HDL-cholesterol}) - (\text{triglycerides}/5) \quad (1)$$

2.7. Statistical Analysis

The obtained data were statistically evaluated using analysis of variance, as reported by McClave and Benson, 1991 [65]. All data were subjected to one-way analysis of variance (ANOVA) software. Significant treatment means were separated by Duncan's new multiple range tests. Differences were considered significant at ($p \leq 0.05$).

3. Results and Discussion

3.1. Chemical Composition of Raw Camel Milk and PSM

Table 1 presents the approximate chemical composition of the PSM compared with the camel milk. The PSM had a higher protein content (7.84% vs. 3.94%), whereas it had a much higher fat content (6.90%) than camel milk (3.12%). The PSM was characterized by the unique presence of fibers and much higher levels of TPC (mg GAE/100 g) and DPPH inhibition % activity than present in the camel milk, i.e., 234.20 and 56.7 vs. 4.60 and 12.86, respectively. These results of approximate chemical composition are in line with those reported previously for PSM and camel milk [50,66]. The low TPC detected in camel milk was consistent with Bouhaddaoui et al., 2019 [67], who reported a level of 35.54 mg GAE/L. The high phenol content of PSM agreed with Peiretti et al., 2017 [68], who reported that pumpkin seeds contain a higher total phenolic level.

Table 1. Chemical composition of fresh camel milk and pumpkin seed milk.

Components (%)	Camel Milk	Pumpkin Seed Milk
Total Solids	14.14 ± 0.46 [b]	21.60 ± 1.04 [a]
Protein	3.94 ± 0.14 [b]	7.84 ± 0.82 [a]
Fat	3.12 ± 0.18 [b]	6.90 ± 0.55 [a]
Ash	0.82 ± 0.05 [b]	0.92 ± 0.04 [a]
Fiber	0.00 [b]	3.42 ± 0.24 [a]
Total phenolic content (mg/100 g)	3.62 ± 0.28 [b]	234.20 ± 5.62 [a]
DPPH inhibition % activity	12.86 ± 0.92 [b]	56.70 ± 2.14 [a]

Values (means ± SD) with different superscript letters are statistically significantly different ($p \leq 0.05$).

3.2. Chemical, Physicochemical, and Phytochemical Properties of Fermented Camel Milk Containing PSM

Table 2 shows the chemical composition of fermented camel milk containing PSM. Control fermented camel milk had a lower total solid (TS), protein, fat, ash, and fiber content. This difference was significant ($p \leq 0.05$) compared with the fermented camel milk containing PSM treatments. The TS, protein, fat, ash, and fiber content of fermented camel milk containing PSM increased gradually by increasing the mixing ratio, which was attributed to the high TS, protein, fat, ash, and fiber content of the prepared PSM compared with the camel milk used in this study (Table 1). These results are in agreement with those reported by Atwaa et al., 2020 [28], who found that partial replacement of camel milk with oat milk, of up to 40%, increased the TS, protein, ash, and carbohydrate content of the resultant fermented camel milk compared with camel milk yogurt. Moreover, the data illustrated in Table 2 indicated that the titratable acidity (TA) of the control fermented camel milk had the lowest value, which might be attributable to the high level of antimicrobial components, such as lysozyme, lactoferrin, and immunoglobulins in camel milk, which decreased the viability of the starter culture [69]. The acidity of fermented camel milk made from camel milk mixed with PSM increased gradually by increasing the mixing ratio; this may be attributed to the hypothesis that the PSM fermentable substance improves the viability of the starter culture [70].

Table 2. Chemical, physicochemical, and phytochemical properties of fermented camel milk containing pumpkin seed milk.

Items	Treatments		
	C	T_1	T_2
Chemical composition (%)			
Total Solids	14.22 ± 0.86 [c]	15.94 ± 0.74 [b]	16.54 ± 0.48 [a]
Protein	3.98 ± 0.32 [c]	4.92 ± 0. 48 [b]	5.86 ± 0.28 [a]
Fat	3.18 ± 0.24 [c]	4.04 ± 0.12 [b]	4.96 ± 0.32 [a]
Ash	0.85 ± 0.06 [c]	0.87 ± 0.04 [b]	0.90 ± 0.03 [a]
Fiber	0.00 [c]	0.82 ± 0.03 [b]	1.55 ± 0.09 [a]
Physicochemical properties			
Acidity [as lactic acid percentage (%)]	0.75 ± 0.04 [c]	0.80 ± 0.02 [b]	0.84 ± 0.03 [a]
pH values	4.82 ± 0.06 [a]	4.76 ± 0.04 [b]	4.68 ± 0.05 [c]
Viscosity (cP)	2160 ± 54 [c]	2420 ± 84 [b]	2590 ± 98 [a]
Phytochemical properties			
Total phenolic content (mg/100 g)	4.06 ± 0.22 [c]	58.54 ± 1.64 [b]	108.65 ± 4.42 [a]
DPPH inhibition % activity	13.40 ± 0.34 [c]	15.24 ± 0.56 [b]	18.12 ± 0.74 [a]

Values (means ± SD) with different superscript letters are statistically significantly different ($p \leq 0.05$). C: fermented camel milk made from camel milk as a control (C); T_1: fermented camel milk made from camel milk mixed with 25% pumpkin seed milk; T_2: fermented camel milk made from camel milk mixed with 50% pumpkin seed milk; cP:centipoise (Unit of dynamic viscosity).

Furthermore, the addition of vegetarian milk to camel milk reduces the concentration of the components of camel milk. The pH values of all treatments exhibited an opposite trend to that of TA. Similar results were obtained by Atwaa et al., 2020 [28], who found that partial replacement of camel milk with oat milk increased the TA and decreased the pH values of the resultant camel milk yogurt. The mixing of camel milk with PSM greatly increased the viscosity of fermented camel milk. Dabija et al., 2018 [70] and Johari et al., 2021 [71] reported similar results. They observed that the addition of pumpkin seed powder to milk caused an increase in the viscosity of the yogurt gel. Moreover, the addition of oat milk to camel milk caused an increase in the viscosity of fermented camel milk [28].

The TPC of fermented camel milk made from camel milk mixed with PSM was increased by increasing the mixing ratio compared with the control fermented camel milk. This might be attributed to the higher TPC of PSM vs. camel milk [72,73]. These results are consistent with those reported by Barakat and Hassan, 2017 [74], who found that the TPC and radical scavenging activity of yogurt increased when milk was fortified with different types of pumpkin. In addition, Atwaa et al., 2020 [28] found that adding oat milk to camel milk increased the TPC and radical scavenging activity of fermented camel milk.

3.3. Sensory Properties of Fermented Camel Milk Containing PSM

The data presented in Table 3 showed that mixing camel milk with PSM greatly increased the sensory attributes of the resultant fermented camel milk, particularly its flavor and, body and texture, compared with the control fermented camel milk. Moreover, this improvement was proportional to the mixing ratio. The control fermented camel milk had the lowest score for sensory properties, which may be attributed to the weak body and texture and inferior flavor of the curd produced from camel milk [73]. This finding is consistent with the observation reported by Atwaa et al., 2020 [28], who found that fortification of camel milk with oat milk increased the sensory attribute scores of the resultant fermented camel milk. Dabija et al., 2018 [70] and Johari et al., 2021 [71] observed that the addition of pumpkin seed powder to milk caused an improvement in the sensory attributes of the resultant yogurt.

Table 3. Sensory properties of fermented camel milk containing pumpkin seed milk.

Attributes	Treatments		
	C	T_1	T_2
Flavor (60)	40.7 ± 2.30 [c]	44.3 ± 2.60 [b]	46.2 ± 3.12 [a]
Body and Texture (30)	22.2 ± 1.18 [c]	25.4 ± 1.74 [b]	27.9 ± 1.58 [a]
Appearance (10)	7.8 ± 0.50 [c]	8.4 ± 0.82 [b]	9.2 ± 0.60 [a]
Total Scores (100)	70.7 ± 3.10 [c]	78.1 ± 3.24 [b]	83.3 ± 3.46 [a]

Values (means ± SD) with different superscript letters are statistically significantly different ($p \leq 0.05$).

3.4. Effect of Fermented Camel Milk Containing PSM on Final Weight and Body Weight Gain in Rats with Oxidative Stress

The data presented in Table 4 showed that final weight (FW) and body weight gain (BWG) was significantly ($p < 0.05$) affected by the treatments. Using 10 g/day fermented camel milk containing 50% PSM in oxidative stress rats induced the best values of FW (216.4 g) and BWG (21.80%) compared with the positive control group, which had an FW f 202.6 g and a BWG f 16.28%. This improvement may be ascribed to the high vitamin, mineral, and antioxidant content of PSM, which may protect the body's cells from the damage caused by free radicals [72]. Accordingly, Barakat and Mahmoud, 2011 [75], and Ghahremanloo et al., 2018 [76], found that adding pumpkin seed or pumpkin seed extract to the diet of obese rats promoted a significant increment ($p \leq 0.05$) in the BWG and enhanced their nutritional status compared with non-treated rats (positive control group). Dikhanbayeva et al., 2021 [77] found that the addition of camel milk curd masses to rat diets promoted a significant increment ($p \leq 0.05$) in the BWG and enhanced the nutritional status compared with the control group. Therefore, the group fed with fermented camel milk containing PSM exhibited the best outcomes as a result of the combined action of fermented camel milk and PSM.

Table 4. Final weight and body weight gain of oxidative stress rats treated with fermented camel milk containing pumpkin seed milk.

Group	Parameters			
	Initial Weight (g)	Final Weight (g)	Percentage of Increase in Body Weight (%)	BWG (%)
Group (1)	168.2 ± 3.8 [a]	222.3 ± 4.8 [a]	32.16 ± 1.3 [a]	24.33 ± 1.5 [a]
Group (2)	169.6 ± 2.9 [a]	202.6 ± 3.5 [d]	19.45 ± 0.96 [d]	16.28 ± 1.4 [d]
Group (3)	168.3 ± 4.3 [a]	210.5 ± 4.8 [c]	25.07 ± 1.7 [c]	20.04 ± 1.6 [c]
Group (4)	169.1 ± 4.6 [a]	216.4 ± 4.4 [b]	27.97 ± 1.6 [b]	21.80 ± 1.2 [b]

Mean values of six rats ± SD. a–d of the small letters in the same column are significantly different at $p \leq 0.05$). Group (1), non-treated non-oxidative stress rats (negative control). Group (2), oxidative stress rats (positive control). Group (3), oxidative stress rats treated with fermented camel milk. Group (4), oxidative stress rats treated with fermented camel milk containing 50% pumpkin seed milk.

3.5. Effect of Fermented Camel Milk Containing PSM on the Serum Lipid Profile of Oxidative Stress Rats

As depicted in Table 5, among the various groups of rats with oxidative stress, group 4 (treated with fermented camel milk containing PSM) had the lowest total cholesterol (75.6 mg/dL) compared with the positive control group, which had the highest total cholesterol (92.4 mg/dL). Regarding the triacylglyceride and LDL levels, the positive control group had the highest content of these components (103.4 and 43.12 mg/dL, respectively) compared with rats treated with fermented camel milk containing PSM, which exhibited a significant decrease in triacylglyceride and LDL content (78.50 and 23.50 mg/dL, respectively). In contrast, HDL content in the positive control group was the lowest (28.60 mg/dL) compared with the other groups. A diet supplemented with fermented camel milk containing PSM increased HDL levels significantly (33.40 mg/dL). Moreover, treatment with fermented camel milk containing PSM decreased total cholesterol, triglycerides, and LDL-c significantly compared with the positive control group. Pumpkin seed exerts a remarkable hypolipidemic effect by reducing total cholesterol and LDL, associated with a significant elevation of HDL [75,76,78]. Carbon tetrachloride causes oxidative stress, with a consequent increment in the formation of reactive oxygen species, which can promote the oxidation of pivotal cellular components (e.g., membrane lipids, proteins, and DNA), leading to cellular damage [79]. Therefore, the beneficial hypolipidemic effect on the lipid profile of the pumpkin seed observed here may be attributed to the phenolic compounds (as natural antioxidants) present in pumpkin seeds [68]. This hypolipidemic action has also been ascribed to the modulation of the lipid metabolism by phenolics and flavonoids, leading to a decrease in total cholesterol, triglycerides, and LDL (not associated with the increase in HDL levels) by upregulating the hepatic peroxisome proliferator-activated receptor α (PPAR-α) [80]. In addition, the high insulin concentration in camel milk can cause the activation of the lipoprotein lipase enzyme [81]. Ashraf et al., 2021 [82] reported that the high mineral content of camel milk (sodium, potassium, zinc, copper, and magnesium), as well as its high vitamin C level, may act as antioxidant agents by removing free radicals. Similar results were obtained by Ghahremanloo et al., 2018 [76] and Dikhanbayeva et al., 2021 [77], who found that pumpkin seed or fermented camel milk had a hypocholesterolemic effect. The group fed with fermented camel milk containing PSM exhibited the best outcomes regarding lipid profiles due to the combined action of fermented camel milk and PSM.

Table 5. Effect of fermented camel milk containing pumpkin seed milk on the lipid profile of experimental rats.

Groups	Parameters			
	Total Cholesterol (TC) (mg/dL)	Triglycerides (TG) (mg/dL)	HDL (mg/dL)	LDL (mg/dL)
Group (1)	70.6 ± 2.5 [d]	80.2 ± 2.2 [c]	39.4 ± 1.8 [a]	15.16 ± 1.1 [d]
Group (2)	92.4 ± 3.2 [a]	103.4 ± 3.2 [a]	28.6 ± 1.5 [d]	43.12 ± 1.3 [a]
Group (3)	85.8 ± 2.8 [b]	87.3 ± 2.4 [b]	32.9 ± 1.7 [c]	35.44 ± 1.2 [b]
Group (4)	75.6 ± 2.3 [c]	78.5 ± 2.6 [c]	36.4 ± 1.6 [b]	23.50 ± 1.6 [c]

Mean values of six rats ± SD. a–d of the small letters in the same column are significantly different at $p \leq 0.05$.

3.6. Effect of Fermented Camel Milk Containing PSM on Liver Function Parameters in Oxidative Stress Rats

The data illustrated in Table 6 showed that the untreated group (positive control) showed a significant ($p \leq 0.05$) increase in ALT and AST and a decrease in total protein and albumin compared with the normal control group. Conversely, the rats treated with fermented camel milk containing PSM showed a significant ($p \leq 0.05$) increase in total protein and albumin and decreased ALT and AST compared with the positive control group. This decrease in aminotransferase enzymes and restoration of various vital functions by hepatocytes can be attributed to the high content of phenolic and bioactive compounds in pumpkin seed, such as phenolic acids and flavonoids. They help the integrity of the plasma membrane in hepatocytes and protect it from damage and the release of the cytosol loaded with these enzymes [68]. Moreover, camel milk's high vitamin C and mineral content may act as antioxidants, restoring aminotransferase enzymes' values [83]. These results are in harmony with Ghahremanloo et al., 2018 [76] and Dikhanbayeva et al., 2021 [77], who reported that pumpkin seed or fermented camel milk significantly reduced the levels of ALT and AST and increased TP and albumin levels in the serum. Our results are also in agreement with a previous study [84], which found that the administration of pumpkin seed oil in sodium nitrate (NaNO3)-induced oxidative damage in rats restored most hematological and biochemical parameters, including ALT, AST, TP and albumin, to their normal level. Given the above findings, the rats fed with fermented camel milk containing PSM exhibited the best results regarding liver function parameters due to the combined action of fermented camel milk and PSM.

Table 6. Effect of fermented camel milk containing pumpkin seed milk on liver function parameters in experimental rats.

Group	Aspartate Aminotransferase (AST U/L)	Alanine Aminotransferase (ALT U/L)	Total Protein (g/dL)	Total Albumin (g/dL)
Group (1)	37.2 ± 1.14 [d]	44.2 ± 2.12 [d]	6.22 ± 0.48 [a]	3.94 ± 0.22 [a]
Group (2)	82.5 ± 2.30 [a]	85.6 ± 3.25 [a]	5.54 ± 0.62 [c]	2.82 ± 0.17 [c]
Group (3)	44.6 ± 1.62 [b]	52.9 ± 2.14 [b]	5.86 ± 0.35 [b]	3.18 ± 0.20 [b]
Group (4)	39.8 ± 1.26 [c]	46.8 ± 3.10 [c]	6.04 ± 0.32 [ab]	3.74 ± 0.24 [a]

Mean values of six rat's $\pm$ SD. a–d of the small letters in the same column are significantly different at ($p \leq 0.05$).

3.7. Effect of Fermented Camel Milk Containing PSM on Kidney Function and Oxidative Stress Markers in Oxidative Stress Rats

The data presented in Table 7 showed that the positive control group had a significant increase in creatinine and urea levels ($p < 0.05$) compared with the normal control group. In contrast, the group treated with fermented camel milk containing PSM showed a significant decrease in creatinine and urea compared with the positive control group. The creatinine and urea levels were reduced in the negative control groups, rats treated with fermented camel milk containing PSM, and rats treated with fermented camel milk compared with the positive control group. This conspicuous change may be partially attributed to the high content of bioactive components in pumpkin seeds, such as minerals, vitamins, and phenolic and flavonoid compounds, which may indirectly reduce uric acid levels and keep the kidney protected from the disorders potentially caused by oxidative stress [72]. These bioactive components act as superoxide scavengers, resulting in suppressing reactive oxygen species and uric acid formation [85]. Moreover, plant-based nutrition yielded a lower mean uric acid serum concentration than animal-based nutrition [86]. A previous study [87] reported that a higher plant protein intake was significantly associated with a lower risk of prevalent CKD. A 20 g increase in plant protein intake reduced the risk of developing CKD by 16%. In addition, the high mineral content of camel milk (sodium, potassium, zinc, copper, and magnesium) and its high vitamin C level may act as antioxidants, thereby affording an improvement in kidney function [77,81]. Furthermore, the positive control group exhibited the highest mean value of MDA (68.40 µmol/L) compared with the negative control, which showed the lowest value (43.60 µmol/L).

Table 7. Effect of fermented camel milk containing pumpkin seed milk on kidney function and oxidative stress markers in experimental rats.

Group	Creatinine (mg/dL)	Urea (mg/dL)	Malondialdehyde (MDA) (µmol/L)
Group (1)	0.48 ± 0.04 [d]	16.2 ± 0.42 [d]	43.60 ± 1.5 [d]
Group (2)	0.84 ± 0.06 [a]	25.8 ± 0.25 [a]	68.40 ± 2.2 [a]
Group (3)	0.66 ± 0.03 [b]	21.3 ± 0.34 [b]	51.92 ± 1.6 [b]
Group (3)	0.54 ± 0.02 [c]	18.2 ± 0.22 [c]	46.38 ± 1.2 [c]

Mean values of six rats ± SD. a–d of the small letters in the same column are significantly different at $p \leq 0.05$.

In addition, there was a significant decrease in MDA in the group treated with fermented camel milk containing PSM (46.38 µmol/L). MDA exerts an adverse effect as it alters the cell membranes structure and function [88]. The increase in the MDA level can lead to oxidative mechanisms, high cytotoxicity, and inhibitory actions. MDA acts as a tumor promoter and co-carcinogenic agent [89]. These results are in harmony with the finding of Barakat and Mahmoud, 2011 [75], who reported that pumpkin seed supplementation effectively decreased the level of creatinine, urea, and malondialdehyde. Previous studies [90,91] have found that fermented camel milk supplementation decreased creatinine, urea, and malondialdehyde compared with the positive control group. Therefore, the group fed with fermented camel milk containing PSM exhibited the best results regarding kidney function and decreased oxidative stress markers as a result of the combined action of fermented camel milk and PSM.

4. Conclusions

Based on the information provided above, the mixing of camel milk with PSM improved the chemical, antioxidant, viscosity, and sensory properties of fermented camel milk, and these improvements were proportional to a mixing ratio up to 50%, which improved the technological problems associated with camel milk such as weak texture and salty taste, and increased the content of phenolic components and antioxidants dietary fiber, thus increasing the nutritional value and healthy benefits to fermented camel milk. The consumption of fermented camel milk containing 50% PSM in oxidative stress rats caused a significant decrease in the levels of malondialdehyde (MDA), low-density lipoprotein (LDL), and cholesterol (CL), triglycerides (TGs), AST, ALT, creatinine, and urea. It increased high-density lipoprotein (HDL), total protein, and albumin compared with the non-treated group. The present findings suggest that more extensive studies of PSM are warranted. Incorporating it into dairy products may be beneficial, particularly in non-traditional milk, which has an undesirable taste.

Author Contributions: M.R.S., E.S.H.A., E.S.A.E.-S. and M.F.R. were involved in the conception of the research idea and methodology design, supervision, and performed data analysis and interpretation. B.M.A., A.A.H.S., M.A.A. and E.K.E. were involved in methodology and drafted and prepared the manuscript for publication and revision. The funders had no role in data collection and analysis, decision to publish, or preparation of the manuscript. All authors have read and agreed to the published version of the manuscript.

Funding: This research was funded by Saudi Ministry of Environment, Water, and Agriculture, facilitated by General Administration of Animal Productions.

Institutional Review Board Statement: This study was conducted with the approval of the approval of the Faculty of Veterinary Medicine, Suez Canal University and the institutional Review Board Number 2022003.

Informed Consent Statement: Not applicable.

Data Availability Statement: The data that support the findings of this study are available on request from the corresponding author.

Acknowledgments: The authors thank the Taif University Researchers supporting project number (TURSP-2020/93), Taif University, Taif, Saudi Arabia, for support.

Conflicts of Interest: The authors declare no conflict of interest.

References

1. Khan, I.T.; Nadeem, M.; Imran, M.; Ullah, R.; Ajmal, M.; Jaspal, M.H. Antioxidant properties of Milk and dairy products: A comprehensive review of the current knowledge. *Lipids Health Dis.* **2019**, *18*, 41. [CrossRef] [PubMed]
2. Salem, S.; Meead, G.; El-Rashody, F.M. Physicochemical and sensory properties of ice cream made from camel milk and fortified with dates products. *Int. J. Humanit. Arts Med. Sci.* **2017**, *5*, 29–40.
3. Khalesi, M.; Salami, M.; Moslehishad, M.; Winterburn, J.; Moosavi-Movahedi, A.A. Biomolecular content of camel milk: A traditional superfood towards future healthcare industry. *Trends Food Sci. Technol.* **2017**, *62*, 49–58. [CrossRef]
4. Kaskous, S. Importance of camel milk for human health. *Emir. J. Food Agric.* **2016**, *28*, 158–163. [CrossRef]
5. Meena, S.; Rajput, Y.; Sharma, R. Comparative fat digestibility of goat, camel, cow and buffalo milk. *Int. Dairy J.* **2014**, *35*, 153–156. [CrossRef]
6. Mudgil, P.; Kamal, H.; Yuen, G.C.; Maqsood, S. Characterization and identification of novel antidiabetic and anti-obesity peptides from camel milk protein hydrolysates. *Food Chem.* **2018**, *259*, 46–54. [CrossRef]
7. Solanki, D.; Hati, S. Fermented camel milk: A Review on its bio-functional properties. *Emir. J. Food Agric.* **2018**, *30*, 268–274.
8. Atwaa, E.S.H.; Shahein, M.R.; El-Sattar, E.S.A.; Hijazy, H.H.A.; Albrakati, A.; Elmahallawy, E.K. Bioactivity, Physicochemical and Sensory Properties of Probiotic Yoghurt Made from Whole Milk Powder Reconstituted in Aqueous Fennel Extract. *Fermentation* **2022**, *8*, 52. [CrossRef]
9. Shori, A.B. Camel milk and its fermented products as a source of potential probiotic strains and novel food cultures: A mini review. *PharmaNutrition* **2017**, *5*, 84–88. [CrossRef]
10. Shahein, M.R.; Atwaa, E.S.H.; El-Zahar, K.M.; Elmaadawy, A.A.; Hijazy, H.H.A.; Sitohy, M.Z.; Albrakati, A.; Elmahallawy, E.K. Remedial Action of Yoghurt Enriched with Watermelon Seed Milk on Renal Injured Hyperuricemic Rats. *Fermentation* **2022**, *8*, 41. [CrossRef]
11. Swelam, S.; Zommara, M.A.; Abd El-Aziz, A.E.-A.M.; Elgammal, N.A.; Baty, R.S.; Elmahallawy, E.K. Insights into Chufa Milk Frozen Yoghurt as Cheap Functional Frozen Yoghurt with High Nutritional Value. *Fermentation* **2021**, *7*, 255. [CrossRef]
12. Beltrán-Barrientos, L.; Hernández-Mendoza, A.; Torres-Llanez, M.; González-Córdova, A.; Vallejo-Córdoba, B. Invited review: Fermented milk as antihypertensive functional food. *J. Dairy Sci.* **2016**, *99*, 4099–4110. [CrossRef] [PubMed]
13. Shahein, M.R.; Atwaa, E.S.H.; Radwan, H.A.; Elmeligy, A.A.; Hafiz, A.A.; Albrakati, A.; Elmahallawy, E.K. Production of a Yogurt Drink Enriched with Golden Berry (*Physalispubescens* L.) Juice and Its Therapeutic Effect on Hepatitis in Rats. *Fermentation* **2022**, *8*, 112. [CrossRef]
14. Elkot, W.F.; Ateteallah, A.H.; Al-Moalem, M.H.; Shahein, M.R.; Alblihed, M.A.; Abdo, W.; Elmahallawy, E.K. Functional, Physicochemical, Rheological, Microbiological, and Organoleptic Properties of Synbiotic Ice Cream Produced from Camel Milk Using Black Rice Powder and Lactobacillus acidophilus LA-5. *Fermentation* **2022**, *8*, 187. [CrossRef]
15. Shahein, M.R.; Atwaa, E.S.H.; Elkot, W.F.; Hijazy, H.H.A.; Kassab, R.B.; Alblihed, M.A.; Elmahallawy, E.K. The Impact of Date Syrup on the Physicochemical, Microbiological, and Sensory Properties, and Antioxidant Activity of Bio-Fermented Camel Milk. *Fermentation* **2022**, *8*, 192. [CrossRef]
16. Shahein, M.R.A.; Atwaa, E.S.H.; Babalghith, A.O.; Alrashdi, B.M.; Radwan, H.A.; Umair, M.; Abdalmegeed, D.; Mahfouz, H.; Dahran, N.; Cacciotti, I.; et al. Impact of incorporation of Hawthorn (*C. oxyanatha*) leaves aqueous extract on yogurt properties and its therapeutic effects against oxidative stress in Rats induced by carbon tetrachloride. *Fermentation* **2022**, *8*, 200.
17. Hussain, T.; Tan, B.; Yin, Y.; Blachier, F.; Tossou, M.C.; Rahu, N. Oxidative stress and inflammation: What polyphenols can do for us? *Oxidative Med. Cell. Longev.* **2016**, *2016*, 7432797. [CrossRef]
18. Ďuračková, Z. Some current insights into oxidative stress. *Physiol. Res.* **2010**, *59*, 459–469. [CrossRef]
19. Kumar, S.; Pandey, A.K. Free radicals: Health implications and their mitigation by herbals. *Br. J. Med. Med. Res.* **2015**, *7*, 438–457. [CrossRef]
20. Kim, Y.S.; Young, M.R.; Bobe, G.; Colburn, N.H.; Milner, J.A. Bioactive food components, inflammatory targets, and cancer prevention. *Cancer Prev. Res.* **2009**, *2*, 200–208. [CrossRef]
21. Cosme, P.; Rodríguez, A.B.; Espino, J.; Garrido, M. Plant phenolics: Bioavailability as a key determinant of their potential health-promoting applications. *Antioxidants* **2020**, *9*, 1263. [CrossRef] [PubMed]
22. Vuolo, M.M.; Lima, V.S.; Junior, M.R.M. Phenolic compounds: Structure, classification, and antioxidant power. In *Bioactive Compounds*; Elsevier: Amsterdam, The Netherlands, 2019; pp. 33–50.
23. Samuel, S.A.; Francis, A.O.; Anthony, O.O. Role of the kidneys in the regulation of intra-and extra-renal blood pressure. *Ann. Clin. Hypertens.* **2018**, *2*, 48–58.
24. Tomsa, A.M.; Alexa, A.L.; Junie, M.L.; Rachisan, A.L.; Ciumarnean, L. Oxidative stress as a potential target in acute kidney injury. *PeerJ* **2019**, *7*, e8046. [CrossRef] [PubMed]
25. Shahidi, F.; Ambigaipalan, P. Phenolics and polyphenolics in foods, beverages and spices: Antioxidant activity and health effects—A review. *J. Funct. Foods* **2015**, *18*, 820–897. [CrossRef]

26. Butnariu, M.; Grozea, I. Antioxidant (Antiradical) Compounds. *J. Bioequivalence Bioavailab.* **2012**, *4*, 6. [CrossRef]
27. Charalampopoulos, D.; Wang, R.; Pandiella, S.; Webb, C. Application of cereals and cereal components in functional foods: A review. *Int. J. Food Microbiol.* **2002**, *79*, 131–141. [CrossRef]
28. Atwaa, E.; Hassan, M.; Ramadan, M.F. Production of probiotic stirred yoghurt from camel milk and oat milk. *J. Food Dairy Sci.* **2020**, *11*, 259–264. [CrossRef]
29. Dotto, J.M.; Chacha, J.S. The potential of pumpkin seeds as a functional food ingredient: A review. *Sci. Afr.* **2020**, *10*, e00575. [CrossRef]
30. Staff, T.P. *PDR for Herbal Medicines*; Physician's Desk Reference (PDR): New York, NY, USA, 2004.
31. Stovel, D. *Pumpkin, a Super Food for All 12 Months of the Year*; Storey Publishing: North Adams, MA, USA, 2012.
32. Csikós, E.; Horváth, A.; Ács, K.; Papp, N.; Balázs, V.L.; Dolenc, M.S.; Kenda, M.; Kočevar Glavač, N.; Nagy, M.; Protti, M. Treatment of Benign Prostatic Hyperplasia by Natural Drugs. *Molecules* **2021**, *26*, 7141. [CrossRef]
33. Caili, F.; Huan, S.; Quanhong, L. A review on pharmacological activities and utilization technologies of pumpkin. *Plant Foods Hum. Nutr.* **2006**, *61*, 70–77. [CrossRef]
34. Matus, Z.; Molnár, P.; Szabó, L.G. Main carotenoids in pressed seeds (*Cucurbitae semen*) of oil pumpkin (*Cucurbita pepo* convar. *pepo var. styriaca*). *Acta Pharm. Hung.* **1993**, *63*, 247–256. [PubMed]
35. Aktaş, N.; Uzlaşır, T.; Tuncil, Y.E. Pre-roasting treatments significantly impact thermal and kinetic characteristics of pumpkin seed oil. *Thermochim. Acta* **2018**, *669*, 109–115. [CrossRef]
36. Glew, R.; Glew, R.; Chuang, L.-T.; Huang, Y.-S.; Millson, M.; Constans, D.; Vanderjagt, D. Amino acid, mineral and fatty acid content of pumpkin seeds (*Cucurbita* spp.) and Cyperus esculentus nuts in the Republic of Niger. *Plant Foods Hum. Nutr.* **2006**, *61*, 49–54. [CrossRef] [PubMed]
37. Yadav, M.; Jain, S.; Tomar, R.; Prasad, G.; Yadav, H. Medicinal and biological potential of pumpkin: An updated review. *Nutr. Res. Rev.* **2010**, *23*, 184–190. [CrossRef] [PubMed]
38. Broznić, D.; Čanadi Jurešić, G.; Milin, Č. Involvement of α-, γ-and δ-tocopherol isomers from pumpkin (*Cucurbita pepo* L.) seed oil or oil mixtures in the biphasic DPPH disappearance kinetics. *Food Technol. Biotechnol.* **2016**, *54*, 200–210. [CrossRef] [PubMed]
39. Naziri, E.; Mitić, M.N.; Tsimidou, M.Z. Contribution of tocopherols and squalene to the oxidative stability of cold-pressed pumkin seed oil (*Cucurbita pepo* L.). *Eur. J. Lipid Sci. Technol.* **2016**, *118*, 898–905. [CrossRef]
40. Abou-Zeid, S.M.; AbuBakr, H.O.; Mohamed, M.A.; El-Bahrawy, A. Ameliorative effect of pumpkin seed oil against emamectin induced toxicity in mice. *Biomed. Pharmacother.* **2018**, *98*, 242–251. [CrossRef]
41. Aghaei, S.; Nikzad, H.; Taghizadeh, M.; Tameh, A.A.; Taherian, A.; Moravveji, A. Protective effect of Pumpkin seed extract on sperm characteristics, biochemical parameters and epididymal histology in adult male rats treated with Cyclophosphamide. *Andrologia* **2014**, *46*, 927–935. [CrossRef]
42. Cho, Y.H.; Lee, S.Y.; Jeong, D.W.; Choi, E.J.; Kim, Y.J.; Lee, J.G.; Yi, Y.H.; Cha, H.S. Effect of pumpkin seed oil on hair growth in men with androgenetic alopecia: A randomized, double-blind, placebo-controlled trial. *Evid. Based Complementary Altern. Med.* **2014**, *2014*, 549721. [CrossRef]
43. Ezea, B.O.; Ogbole, O.O.; Ajaiyeoba, E.O. In vitro anthelmintic properties of root extracts of three Musa species. *J. Pharm. Bioresour.* **2019**, *16*, 145–151. [CrossRef]
44. Jayaprakasam, B.; Seeram, N.P.; Nair, M.G. Anticancer and antiinflammatory activities of cucurbitacins from Cucurbita andreana. *Cancer Lett.* **2003**, *189*, 11–16. [CrossRef]
45. Balbino, S.; Dorić, M.; Vidaković, S.; Kraljić, K.; Škevin, D.; Drakula, S.; Voučko, B.; Čukelj, N.; Obranović, M.; Ćurić, D. Application of cryogenic grinding pretreatment to enhance extractability of bioactive molecules from pumpkin seed cake. *J. Food Process Eng.* **2019**, *42*, e13300. [CrossRef]
46. Fruhwirth, G.O.; Wenzl, T.; El-Toukhy, R.; Wagner, F.S.; Hermetter, A. Fluorescence screening of antioxidant capacity in pumpkin seed oils and other natural oils. *Eur. J. Lipid Sci. Technol.* **2003**, *105*, 266–274. [CrossRef]
47. Alhakamy, N.A.; Fahmy, U.A.; Ahmed, O.A. Attenuation of benign prostatic hyperplasia by optimized tadalafil loaded pumpkin seed oil-based self nanoemulsion: In vitro and in vivo evaluation. *Pharmaceutics* **2019**, *11*, 640. [CrossRef]
48. Singh, A.; Kumar, V. Nutritional, phytochemical, and antimicrobial attributes of seeds and kernels of different pumpkin cultivars. *Food Front.* **2022**, *3*, 182–193. [CrossRef]
49. Reeves, P.G. Purified diets for laboratory rodents: Final report of the American Institute of Nutrition ad hoc writing committee on the reformation of the AIN-76A rodent diet. *J. Nutr.* **1993**, *123*, 1939–1951. [CrossRef]
50. Hassan, A.A.; Aly, M.; El-Hadidie, S.T. Production of cereal-based probiotic beverages. *World Appl. Sci. J.* **2012**, *19*, 1367–1380.
51. Tamime, A.; Robinson, R. *Yoghurt. Science and Technology*; Woodhead Publishing Limited England: Cambridge, UK, 1999.
52. Chemists, A.; Horwitz, W. *Official Methods of Analysis*; Association of Official Analytical Chemists: Washington, DC, USA, 1975; Volume 222.
53. Aryana, K.J. Folic acid fortified fat-free plain set yoghurt. *Int. J. Dairy Technol.* **2003**, *56*, 219–222. [CrossRef]
54. Maksimović, Z.; Malenčić, Đ.; Kovačević, N. Polyphenol contents and antioxidant activity of Maydis stigma extracts. *Bioresour. Technol.* **2005**, *96*, 873–877. [CrossRef]
55. Apostolidis, E.; Kwon, Y.-I.; Shetty, K. Inhibitory potential of herb, fruit, and fungal-enriched cheese against key enzymes linked to type 2 diabetes and hypertension. *Innov. Food Sci. Emerg. Technol.* **2007**, *8*, 46–54. [CrossRef]

56. Kujawska, M.; Ignatowicz, E.; Murias, M.; Ewertowska, M.; Mikołajczyk, K.; Jodynis-Liebert, J. Protective effect of red beetroot against carbon tetrachloride-and N-nitrosodiethylamine-induced oxidative stress in rats. *J. Agric. Food Chem.* **2009**, *57*, 2570–2575. [CrossRef] [PubMed]

57. Caraway, W.T. Determination of uric acid in serum by a carbonate method. *Am. J. Clin. Pathol.* **1955**, *25*, 840–845. [CrossRef] [PubMed]

58. Bonsens, K.; Taussky, D. Determination of serum creatinine. *J. Chem. Inv.* **1984**, *27*, 648–660.

59. Marsh, W.H.; Fingerhut, B.; Miller, H. Automated and manual direct methods for the determination of blood urea. *Clin. Chem.* **1965**, *11*, 624–627. [CrossRef] [PubMed]

60. Kei, S. Serum lipid peroxide in cerebrovascular disorders determined by a new colorimetric method. *Clin. Chim. Acta* **1978**, *90*, 37–43. [CrossRef]

61. Bergmeyer, H.; Harder, M. A colorimetric method of the determination of serum glutamic oxaloacetic and glutamic pyruvic transaminase. *Clin. Biochem.* **1986**, *24*, 1–488.

62. Dougnon, T.V.; Bankolé, H.S.; Klotoé, J.R.; Sènou, M.; Fah, L.; Koudokpon, H.; Akpovi, C.; Dougnon, T.J.; Addo, P.; Loko, F. Treatment of hypercholesterolemia: Screening of Solanum macrocarpon Linn (Solanaceae) as a medicinal plant in Benin. *Avicenna J. Phytomed.* **2014**, *4*, 160–169. [CrossRef]

63. Devi, R.; Sharma, D. Hypolipidemic effect of different extracts of Clerodendron colebrookianum Walp in normal and high-fat diet fed rats. *J. Ethnopharmacol.* **2004**, *90*, 63–68. [CrossRef]

64. Friedewald, W.T.; Levy, R.I.; Fredrickson, D.S. Estimation of the concentration of low-density lipoprotein cholesterol in plasma, without use of the preparative ultracentrifuge. *Clin. Chem.* **1972**, *18*, 499–502. [CrossRef]

65. McClave, J.; Benson, P. *Statistical for Business and Economics*; Maxwell Macmillan International editions; Dellen Publishing Co.: New York, NY, USA, 1991; Volume 1991, pp. 272–295.

66. Fouzia, R.; Noureddine, S.; Mebrouk, K. Evaluation of the factors affecting the variation of the physicochemical composition of Algerian camel's raw milk during different seasons. *Adv. Environ. Biol.* **2013**, *7*, 4879–4884.

67. Bouhaddaoui, S.; Chabir, R.; Errachidi, F.; El Ghadraoui, L.; El Khalfi, B.; Benjelloun, M.; Soukri, A. Study of the biochemical biodiversity of camel milk. *Sci. World J.* **2019**, *2019*, 2517293. [CrossRef] [PubMed]

68. Peiretti, P.G.; Meineri, G.; Gai, F.; Longato, E.; Amarowicz, R. Antioxidative activities and phenolic compounds of pumpkin (*Cucurbita pepo*) seeds and amaranth (*Amaranthus caudatus*) grain extracts. *Nat. Prod. Res.* **2017**, *31*, 2178–2182. [CrossRef] [PubMed]

69. Galeboe, O.; Seifu, E.; Sekwati-Monang, B. Production of camel milk yoghurt: Physicochemical and microbiological quality and consumer acceptability. *Int. J. Food Stud.* **2018**, *7*, 51–63. [CrossRef]

70. Dabija, A.; Codină, G.G.; Gâtlan, A.-M.; Sănduleac, E.T.; Rusu, L. Effects of some vegetable proteins addition on yogurt quality. *Sci. Study Res. Chem. Chem. Eng. Biotechnol. Food Ind.* **2018**, *19*, 181–192.

71. Johari, S.; Hosseini Ghaboos, S.H.; Shahi, T. Investigation on the rheological properties of fortified yogurt containing pumpkin powder. *Food Sci. Technol.* **2021**, *18*, 349–358. [CrossRef]

72. Peng, M.; Lu, D.; Liu, J.; Jiang, B.; Chen, J. Effect of Roasting on the Antioxidant Activity, Phenolic Composition, and Nutritional Quality of Pumpkin (*Cucurbita pepo* L.) Seeds. *Front. Nutr.* **2021**, *8*, 647354. [CrossRef]

73. Soliman, T.N.; Shehata, S.H. Characteristics of fermented camel's milk fortified with kiwi or avocado fruits. *Acta Sci. Pol. Technol. Aliment.* **2019**, *18*, 53–63.

74. Barakat, H.; Hassan, M.F. Chemical, nutritional, rheological, and organoleptical characterizations of stirred pumpkin-yoghurt. *Food Nutr. Sci.* **2017**, *8*, 746–759. [CrossRef]

75. Barakat, L.A.; Mahmoud, R.H. The antiatherogenic, renal protective and immunomodulatory effects of purslane, pumpkin and flax seeds on hypercholesterolemic rats. *North Am. J. Med. Sci.* **2011**, *3*, 411–417. [CrossRef]

76. Ghahremanloo, A.; Hajipour, R.; Hemmati, M.; Moossavi, M.; Mohaqiq, Z. The beneficial effects of pumpkin extract on atherogenic lipid, insulin resistance and oxidative stress status in high-fat diet-induced obese rats. *J. Complementary Integr. Med.* **2018**, *15*, 20170051. [CrossRef]

77. Dikhanbayeva, F.; Zhaxybayeva, E.; Smailova, Z.; Issimov, A.; Dimitrov, Z.; Kapysheva, U.; Bansal, N. The effect of camel milk curd masses on rats blood serum biochemical parameters: Preliminary study. *PLoS ONE* **2021**, *16*, e0256661. [CrossRef] [PubMed]

78. Makni, M.; Fetoui, H.; Gargouri, N.; Garoui, E.M.; Jaber, H.; Makni, J.; Boudawara, T.; Zeghal, N. Hypolipidemic and hepatoprotective effects of flax and pumpkin seed mixture rich in ω-3 and ω-6 fatty acids in hypercholesterolemic rats. *Food Chem. Toxicol.* **2008**, *46*, 3714–3720. [CrossRef] [PubMed]

79. Yilmaz, S.; Kaya, E.; Kisacam, M.A. The effect on oxidative stress of aflatoxin and protective effect of lycopene on aflatoxin damage. *Aflatoxin Control Anal. Detect. Health Risks* **2017**, *30*, 67–90.

80. Zeni, A.L.B.; Moreira, T.D.; Dalmagro, A.P.; Camargo, A.; Bini, L.A.; Simionatto, E.L.; Scharf, D.R. Evaluation of phenolic compounds and lipid-lowering effect of Morus nigra leaves extract. *An. Acad. Bras. Ciências* **2017**, *89*, 2805–2815. [CrossRef] [PubMed]

81. Agrawal, R.P.; Saran, S.; Sharma, P.; Gupta, R.P.; Kochar, D.K.; Sahani, M.S. Effect of camel milk on residual β-cell function in recent onset type 1 diabetes. *Diabetes Res. Clin. Pract.* **2007**, *3*, 494–495. [CrossRef] [PubMed]

82. Ashraf, A.; Mudgil, P.; Palakkott, A.; Iratni, R.; Gan, C.-Y.; Maqsood, S.; Ayoub, M.A. Molecular basis of the anti-diabetic properties of camel milk through profiling of its bioactive peptides on dipeptidyl peptidase IV (DPP-IV) and insulin receptor activity. *J. Dairy Sci.* **2021**, *104*, 61–77. [CrossRef]

83. Mailam, A. A Comparative Study On Antidiabetic Effect Of Buffalo And Camel Fermented Milk In Induced Diabetic Rats. *Adv. Food Sci.* **2017**, *39*, 124–132.

84. Rouag, M.; Berrouague, S.; Djaber, N.; Khaldi, T.; Boumendjel, M.; Taibi, F.; Abdennour, C.; Boumendjel, A.; Messarah, M. Pumpkin seed oil alleviates oxidative stress and liver damage induced by sodium nitrate in adult rats: Biochemical and histological approach. *Afr. Health Sci.* **2020**, *20*, 413–425. [CrossRef]

85. Lin, S.; Zhang, G.; Liao, Y.; Pan, J.; Gong, D. Dietary flavonoids as xanthine oxidase inhibitors: Structure–affinity and structure–activity relationships. *J. Agric. Food Chem.* **2015**, *63*, 7784–7794. [CrossRef]

86. Chiu, T.H.; Liu, C.-H.; Chang, C.-C.; Lin, M.-N.; Lin, C.-L. Vegetarian diet and risk of gout in two separate prospective cohort studies. *Clin. Nutr.* **2020**, *39*, 837–844. [CrossRef]

87. Alvirdizadeh, S.; Yuzbashian, E.; Mirmiran, P.; Eghtesadi, S.; Azizi, F. A prospective study on total protein, plant protein and animal protein in relation to the risk of incident chronic kidney disease. *BMC Nephrol.* **2020**, *21*, 489. [CrossRef]

88. Nair, G.G.; Nair, C.K.K. Radioprotective effects of gallic acid in mice. *BioMed Res. Int.* **2013**, *2013*, 953079. [CrossRef] [PubMed]

89. Koc, M.; Taysi, S.; Buyukokuroglu, M.E.; Bakan, N. Melatonin protects rat liver against irradiation-induced oxidative injury. *J. Radiat. Res.* **2003**, *44*, 211–215. [CrossRef] [PubMed]

90. Hamed, H.; Gargouri, M.; Boulila, S.; Chaari, F.; Ghrab, F.; Kallel, R.; Ghannoudi, Z.; Boudawara, T.; Chaabouni, S.; El Feki, A. Fermented camel milk prevents carbon tetrachloride induced acute injury in kidney of mice. *J. Dairy Res.* **2018**, *85*, 251–256. [CrossRef] [PubMed]

91. El-Zahar, K.M.; Hassan, M.F.; Al-Qaba, S.F. Protective Effect of Fermented Camel Milk Containing Bifidobacterium longum BB536 on Blood Lipid Profile in Hypercholesterolemic Rats. *J. Nutr. Metab.* **2021**, *2021*, 1557945. [CrossRef]

 *fermentation*

Article

Impact of Incorporating the Aqueous Extract of Hawthorn (*C. oxyanatha*) Leaves on Yogurt Properties and Its Therapeutic Effects against Oxidative Stress Induced by Carbon Tetrachloride in Rats

Magdy Ramadan Shahein [1], El-Sayed H. Atwaa [2], Ahmad O. Babalghith [3], Barakat M. Alrashdi [4], Hanan A. Radwan [5], Muhammad Umair [6], Dyaaaldin Abdalmegeed [7,8], Hala Mahfouz [9], Naief Dahran [10], Ilaria Cacciotti [11] and Ehab Kotb Elmahallawy [12,*]

[1] Department of Food Science and Technology, Faculty of Agriculture, Tanta University, Tanta 31527, Egypt; magdrsh10@gmail.com
[2] Department of Food Science, Faculty of Agriculture, Zagazig University, Zagazig 44519, Egypt; elsayedhassanattwa@gmail.com
[3] Medical Genetics Department, College of Medicine, Umm Al Qura University, Makkah 21955, Saudi Arabia; aobabalghith@uqu.edu.sa
[4] Biology Department, College of Science, Jouf University, Sakaka 72388, Saudi Arabia; bmalrashdi@ju.edu.sa
[5] Department of Home Economics, Faculty of Specific Education, Zagazig University, Zagazig 44519, Egypt; drhananalsadeq81@gmail.com
[6] Department of Food Science and Engineering, College of Chemistry and Engineering, Shenzhen University, Shenzhen 518060, China; umair_uaf@hotmail.com
[7] Microbiology Section, Botany Department, Faculty of Science, Tanta University, Tanta 31527, Egypt; 2018116158@njau.edu.cn
[8] Key Laboratory of Food Processing and Quality Control, College of Food Science and Technology, Nanjing Agricultural University, Nanjing 210095, China
[9] Department of Medical Biochemistry and Molecular Biology, Faculty of Medicine, Kafrelsheikh University, Kafrelsheikh 33516, Egypt; hala_ali2015@med.kfs.edu.eg
[10] Department of Anatomy, Faculty of Medicine, University of Jeddah, Jeddah 21959, Saudi Arabia; ndahran@uj.edu.sa
[11] Department of Engineering, INSTM RU, University of Rome "Niccolò Cusano", 00166 Rome, Italy; ilaria.cacciotti@unicusano.it
[12] Department of Zoonoses, Faculty of Veterinary Medicine, Sohag University, Sohag 82524, Egypt
* Correspondence: eehaa@unileon.es

Citation: Shahein, M.R.; Atwaa, E.-S.H.; Babalghith, A.O.; Alrashdi, B.M.; Radwan, H.A.; Umair, M.; Abdalmegeed, D.; Mahfouz, H.; Dahran, N.; Cacciotti, I.; et al. Impact of Incorporating the Aqueous Extract of Hawthorn (*C. oxyanatha*) Leaves on Yogurt Properties and Its Therapeutic Effects against Oxidative Stress Induced by Carbon Tetrachloride in Rats. *Fermentation* **2022**, *8*, 200. https://doi.org/10.3390/fermentation8050200

Academic Editor: Carlo Giuseppe Rizzello

Received: 22 March 2022
Accepted: 26 April 2022
Published: 29 April 2022

Publisher's Note: MDPI stays neutral with regard to jurisdictional claims in published maps and institutional affiliations.

Abstract: The current study aimed to evaluate the chemical, phytochemical, and sensory properties; the nutritional value; and the antioxidant properties resulting from the incorporation of yogurt fortified with the aqueous extract of Hawthorn leaves in Sprague Dawley rats. The results revealed that the yogurt containing the aqueous extract from Hawthorn leaves exhibited no significant differences in terms of its protein, fat, and ash contents compared to control samples. Moreover, the highest total phenolic content (62.00 ± 1.70) and antioxidant activity (20.60 ± 0.74%) were detected in the yogurt containing 0.4% Hawthorn leaf extract compared to the other samples. The consumption of yogurt fortified with the aqueous extract from Hawthorn leaves by rats experiencing oxidative stress resulted in a significant decrease ($p \leq 0.05$) in the triglyceride, total cholesterol, low-density lipoprotein, aspartate aminotransferase, alanine aminotransferase, creatinine, urea, and malondialdehyde levels and a remarkable increase ($p \leq 0.05$) in the high-density lipoprotein, total protein, and albumin levels as well as in the total antioxidant potentials of serum compared to the positive control group, indicating that the extract from Hawthorn leaves can play a preventive role against oxidative stress. Collectively, our study concluded that the extract from Hawthorn leaves can provide health benefits to yogurt on the basis of its high bioactive components and can exert protective effects against oxidative stress in rats.

Keywords: yogurt; *Crataegus oxyacantha*; total phenolic; cholesterol; triglyceride; oxidative stress

1. Introduction

Oxidative stress is considered to be one of the main causes of chronic inflammation [1–3] and promotes the activation of various transcription factors, resulting in the consequent expression of genes that are related to inflammatory pathways [4,5]. Several previous studies have reported that oxidative stress could potentially lead to different forms of nephrotoxicity and renal damage, both of which have dramatic consequences, due to the pivotal role of the kidneys in excreting metabolic waste and in the chemical homeostasis balance [2,6]. The consumption of antioxidant-rich foods is considered to be pivotal for combating the oxidative stress. According to the available literature, several previous studies have reported that dairy products contain a low content of bioactive compounds, limiting their nutritional value. The addition of antioxidants, e.g., flavonoids, tannins, phenolic acids, vitamin E, and vitamin C, to foods has been proposed as an effective preventive adjuvant therapy, as antioxidants are able to synergistically interact with other reducing compounds, having anti-atherosclerotic, anti-inflammatory, and anti-carcinogenic actions [7]. The use of polyphenols has also been shown to have potent antioxidant activity, anti-inflammatory action, and the capability to inhibit the enzymes that are involved in the development of eicosanoids [7,8]. Recently, food producers have shown interest in substitution with synthetic additives (including nutritional and preservative agents, coloring, flavoring, and miscellaneous agents) [9–13] to improve the features of processed foods with natural ones [14–18]. Interestingly, the supplementary antioxidants from natural sources, such as plants, have proven to be efficient in providing protection against oxidative stress due to enzymatic and non-enzymatic elements, some of which are only synthesized in plants and can only be taken in by the body through food [19]. Moreover, foods that have been supplemented with bioactive compounds are able to enhance the immune system [20]. Yogurt is considered to be one of the most important fermented milk products and is produced by fermenting lactose to lactic acid via the action of a yogurt starter culture containing *Lactobacillus delbrueckii* ssp. *bulgaricus* and *Streptococcus thermophilus*. The yogurt starter culture interacts with the milk protein, which improves the body texture and sensory attributes of the product. It is widely known that synthetic additives, including nutritional and preservatives additives, coloring, and flavoring agents as well as other miscellaneous agents, are used to improve the characteristics and properties of processed foods [21]. Clearly, the incorporation of plant-based additives can be used to fortify dairy products and improve the nutritional value of these products [22]. Hawthorn *Crataegus oxyacantha* L. belongs to the Rosaceae family of spiny shrubs also known as the Hawthorn species and is medicinally recognized in European Pharmacopeia, having been used for a long time in folk medicine for the treatment of different diseases such as diarrhea, asthma, and insomnia [23] as well as for the treatment of angina pectoris, hypertension, and arrhythmia. It has been also proposed as an alternative anxiolytic, antihyperlipidemic, immunomodulatory, antihyperglycemic, and antimutagenic medicine [24]. Likewise, *C. oxyacantha* contains different active constituents, including oligomeric, procyanidins flavonoids, triterpenic acids, phenolic acids, fatty acids, triterpenes, sterols, and phenol carboxylic acids [24,25]. Moreover, it has a significant hypolipidemic effect since it improves lipid profiles and enhances rheological blood flow and immune function [26–28]. Little information is available about the effects of incorporating the aqueous extract of Hawthorn (*C. oxyanatha*) leaves on yogurt properties and its therapeutic effects. Given the above information, the present study investigated the impact of incorporating the aqueous extract of Hawthorn (*C. oxyacantha*) leaves during yogurt manufacture to determine whether it fortified different proportions of yogurt in terms of its functional properties. Yogurt is one of the most important types of fermented milk in the world and is produced as a consequence of the interaction of *Lactobacillus delbrueckii* ssp. *bulgaricus* and *Streptococcus thermophilus* with milk proteins, resulting in a product with improved texture and sensory properties [29]. Furthermore, the fortified yogurt was examined to determine its therapeutic effects as a functional food on the biomarkers of different liver and kidney functionalities, oxidative stress, and lipid profiles in rats with high oxidative stress induced by carbon tetrachloride.

2. Materials and Methods

2.1. Materials

Hawthorn (*Crataegus oxyacantha* L.) leaves were obtained from the Agricultural Research Center of Giza (Egypt), washed with tap water to eliminate toxic saponins and then dried at 45 °C for 12 h, ground into a fine powder using an electric blender, and then stored in a freezer in plastic bags until use. All of the solvents that were exploited for extraction and analyses were of analytical grade. Folin–Ciocalteu (FC), gallic acid, 2,2-diphenyl-1-picrylhydrazyl (DPPH), and 2,2-azinobis (3-ethylbenzothiazoline-6-sulphonate) were purchased from Sigma (St. Louis, MO, USA). The kits used for the biochemical analyses of the total cholesterol, triglycerides, high-density lipoprotein (HDL), low-density lipoprotein (LDL), aspartate transaminase (AST), alanine aminotransferase (ALT), total protein, total albumin, creatinine, urea, and malondialdehyde (MDA) were purchased from the Gamma Trade Company for Pharmaceutical and Chemicals (Cairo, Egypt). The total phenolic and radical scavenging activity of the Hawthorn leave extract were determined as described elsewhere [30] to prepare a standard basal diet consisting of 10% protein, 15% casein, 5% cellulose, 10% fat, and 65% corn starch. Fresh cow's milk (with a fat content of 3%) was provided by Zagazig University, Agriculture Faculty, Food Science Department, Dairy Technology Unit, Zagazig, Egypt. A yogurt culture containing *Lactobacillus delbruekii* subsp. *bulgaricus* EMCC1102 and *Streptococcus salivarius* subsp. *thermophilus* EMCC104 was purchased from the Microbiological Resources Center (MIRCEN), Faculty of Agriculture, Ain Shams University, Cairo, Egypt.

2.2. Preparation of Hawthorn Leaves Aqueous Extract

The dried Hawthorn leaves were reduced to powder (0.841 mm). The aqueous extracts were obtained by dipping 10 g of Hawthorn leaves in 100 mL of hot distilled water (95 °C) and by keeping them refrigerated at 4 °C overnight. Afterwards, the Hawthorn extract was mixed at 100 rpm for 1 h in a rotary shaker, filtered using a filter paper (Whatman No. 1), lyophilized, and stored at 4 °C.

2.3. Yogurt Fortified with Aqueous Hawthorn Leaves Extract Manufacture

Fresh cow's milk (3% fat) was heated for 15 min at 80 °C and cooled to 42–45 °C before inoculation with a 3% starter culture containing *Lactobacillus delbruekii* subsp. *bulgaricus* and *Streptococcus salivarius* subsp. *thermophilus*, which was prepared by adding 2.25×10^7 cells/mL. The inoculated flask contained 250 mL of sterile culture medium. The obtained milk was then divided into three aliquots: plain yogurt (PY), yogurt loaded with 0.2% *w/w* (HLEY0.2), and yogurt loaded with 0.4% *w/w* (HLEY0.4) Hawthorn leaf aqueous extract. The three portions were maintained in plastic containers (100 mL) and incubated at 43 °C until a pH value of 4.65 was reached. When the fermentation process was complete, all of containers were immediately cooled down and kept at 4 °C in a refrigerator overnight.

2.4. Chemical Compositions, Physicochemical Analysis, and Sensory Evaluation of Yogurt Fortified with Aqueous Hawthorn Leaves Extract

For the yogurt samples, the total protein, total solid, fat, and ash contents; peroxide value; acid value; and titratable acidity were determined according to the method outlined in [31]. Their pH values were monitored using a pH meter equipped with a glass electrode (HANNA, Instrument, Portugal). Total phenolic content (TPC) (expressed as mg GAE (gallic acid equivalents)/100 g) was assessed as described elsewhere [32] with minor modifications. Briefly, 100 μL of different concentrations of the test sample was mixed with 1 mL of diluted FC reagent (1:10). After 10 min, 1 mL of 7.5% (*w/v*) sodium carbonate solution was added to the mixture, and the mixture was incubated in the dark for 90 min. The absorbance was recorded at 725 nm. The TPC was calculated according to the calibration curve of gallic acid $Y = 0.2808X + 0.0301$; $R^2 = 0.9983$. The antioxidant (AO) activity (%) of the prepared yogurts was assessed as described elsewhere [33], and the absorbance was noted at 517 nm using a spectrophotometer (Thermo Scientific, Wilmington,

NC, USA). The scavenging activity was calculated with the formula to determine the DPPH radical scavenging % = (Absorbance of sample—Absorbance of blank)/Absorbance of control $\times$ 100. The yogurt samples underwent a sensory evaluation to determine its flavor (45), texture (30), acidity (15), and appearance (10) following the method reported by [34]. The sensory evaluation was carried out by a team of 10 trained professional panelists. The samples were packed and coded with a 3-digit code. The encoded samples were presented to the panelists on a tray. After testing each sample, the panelists were offered plain water to cleanse their palates before moving on to the next sample.

2.5. Animal Experimental Design

Forty-eight male adult Sprague Dawley albino strain rats weighing 180–220 g were purchased from the Agricultural Research Center of Giza, Giza, Egypt and housed in wire cages in a 25 °C environment. After acclimatization on a basal diet for seven days, they were divided into four groups (twelve rats for every group) as follows: Twelve rats represented the normal control group and were fed a standard diet as a negative control (group 1). Thirty six rats were intraperitoneally injected with a single dose of carbon tetrachloride at a rate of 2 mL/kg body weight (oxidative stress rats), as described elsewhere [35]. The oxidative stress rats were then organized into three groups (twelve rats in each group) as follows: oxidative stress rats as a positive control (Group 2), oxidative stress rats fed a basal diet with plain yogurt (10 g/day) with a epigastric tube (Group 3), and oxidative stress rats fed a basal diet with yogurt fortified with 0.4% Hawthorn leaf aqueous extract (10 g/day) with a epigastric tube (Group 4).

2.6. Biochemical Analysis

At the end of the 5-week trial period, the rats were lightly anesthetized with diethyl ether. Blood samples were collected from the hepatic portal vein and were submitted to centrifugation to separate the serum at 3000 rpm ($1.811\times g$) for 15 min and stored at -40 °C. After that, the biochemical parameters were determined using commercially available kits and related methods: total cholesterol (mg/dL) was evaluated using the enzymatic colorimeter method [36]; total triglycerides (mg/dL) were measured by following the method outlined in [37]; high-density lipoprotein (HDL, mg/dL) was measured by following the method outlined in [38]; and low-density lipoprotein (LDL, mg/dL) was measured using the Friedewald formula (LDL-cholesterol = Total cholesterol-HDL-cholesterol-(Triglycerides/5)) [39]. The alanine aminotransferase (ALT, U/L) and aspartate aminotransferase (AST, U/L) enzymes were evaluated by applying the methods described by Bergmeyer and Harder, 1986 [40]. The total albumin (g/dL), total protein (g/dL), and malondialdehyde (MDA, μmol/L) were measured on the basis of procedures reported elsewhere [41]. Moreover, the serum creatinine (mg/dL) and serum urea (mg/dL) levels were measured as described elsewhere [42,43]. The total antioxidant potentials (mM/L) of the serum were also determined [44].

2.7. Histopathological Examination

Specimens from the livers and kidneys of the rat were directly collected and weighed after r the rats were sacrificed at the end of the examined period, fixed in formalin (10%), dehydrated in ethyl alcohol, cleared in xylene, and embedded in paraffin wax. Thick sections were obtained and stained with hematoxylin and eosin [45].

2.8. Statistical Analysis

Statistical data analysis was performed using the one-way ANOVA test associated with the Duncan test, which was carried out using CoStat software (Monterey, CA, USA, version 6.4). Significant difference was determined at $p \leq 0.05$.

3. Results and Discussion

3.1. Total Phenolic Content and Radical Scavenging Activity of Hawthorn Leaves Extract

The analyses performed on the Hawthorn leaf extract revealed a TFC of 340 ± 6.53 mg GAE/100 g and a DPPH radical scavenging activity of $95.40 \pm 1.63\%$. A previous study [46] reported a total phenolic content of 343.50 mg GAE/g in the case of ethyl acetate extract from Hawthorn leaves. Another previous report [47] revealed a TPC for the dry extract from Hawthorn leaves and berry extracts that ranged from 77.4 to 94.2 mg GAE/g, showing evidence of a higher phenolic content in the case of the leaves with respect to the berries as well as a more effective DPPH radical scavenger activity (IC50 = 29.7 µg/g vs. IC50 = 111.9 µg/g, for leaves and berries extracts, respectively).

3.2. Chemical Compositions, Physicochemical, and Sensory Characteristics of Yogurt Fortified with Aqueous Hawthorn Leaves Extract

The data presented in Table 1 reveal that the total solids, protein, and fat contents of the yogurt did not seem to be affected by the fortification from the Hawthorn leaf extract with the two ratios (HLEY0.2 and HLEY0.4). This result agrees with a previous study [48] that found that fortifying yogurt with different types of herbal extracts did not affect the total solids, protein, and fat contents of the resultant yogurt. On the other hand, comparing the pH values of the PY, HLEY0.2, and HLEY0.4 samples in Table 1, a significantly lower pH value ($p \leq 0.05$) was detected in PY compared to (HLEY0.2) and (HLEY0.4), which might be attributed to the effect of aqueous extract from Hawthorn leaves on the growth of microorganism and pH values. On the contrary, a previous study evidenced slightly higher pH values in plain yogurt than those in essential oil-treated yogurts [49]. The titratable acidity values in PY significantly ($p \leq 0.05$) increased compared to (HLEY0.2) and (HLEY0.4) due to the ability of Hawthorn leaf extract to inhibit/decrease bacteria growth in the yogurt. These results agree with those reported in a previous study [50], which indicated that the addition of Cinnamon herb extract addition influenced the titratable acidity in yogurt, leading to a slight pH increase. Furthermore, adding Hawthorn leaf extract to the yogurt in the present study increased its TPC and AO activity as a function of the concentration of the Hawthorn lead extract. This experimental evidence is in agreement with the findings reported in a previous study [51] that demonstrated how an increasing the content of Hawthorn extract in yogurt samples was positively correlated with the TPC and AO. Additionally, the data collected in Table 1 show that the yogurt containing Hawthorn leaf extract had lower peroxide and acid values than the plain yogurt. The obtained results confirm previous investigations on the presence of natural antioxidants [47,52]. The results in Figure 1 are related to the results of the sensory evaluation conducted on PY, HLEY0.2, and HLEY0.4, and all three yogurt samples demonstrate acceptable sensory features. However, PY was more acceptable compared to the other two treatments, with an overall less desirable flavor being achieved in HLEY0.2 and HLEY0.4. These results confirm those that were reported previously [53], which showed that yogurt supplemented with Cinnamon herb extract had a more undesirable flavor profile overall compared to plain yogurt. Furthermore, a previous study [54] found that the addition of Hawthorn leaf powder reduced the organoleptic properties of cupcakes. In addition, another study [55] found that adding olive leaf extract at the concentrations of 0.2 and 0.4% had an unfavorable effect on the organoleptic properties of yogurt samples. In addition, another previous report [56] found that adding walnut leaf extract to yogurt reduced the organoleptic properties of the product. A possible explanation for this finding might be attributed to the fact that most herbs contain a distinctive richness and a diverse population of metabolites that is responsible for their taste and flavor [53].

Table 1. Chemical compositions and physicochemical characteristics of yogurt fortified with aqueous Hawthorn leaf extract.

Treatments	Chemical Compositions					Physicochemical Characteristics				
	Total Solids	Fat	Protein	Ash	Acidity %	pH	TPC	AO Activity	PV	AV
PY	10.24 ± 0.16 [A]	3.12 ± 0.08 [A]	3.94 ± 0.08 [A]	0.72 ± 0.02 [B]	0.82 ± 0.02 [A]	4.68 ± 0.02 [C]	18.00 ± 2.45 [C]	7.70 ± 0.57 [C]	1.20 ± 0.05 [A]	0.50 ± 0.02 [A]
HLEY 0.2	10.42 ± 0.20 [A]	3.14 ± 0.08 [A]	3.98 ± 0.13 [A]	0.76 ± 0.03 [AB]	0.76 ± 0.02 [B]	4.75 ± 0.02 [B]	38.00 ± 2.45 [B]	12.40 ± 0.49 [B]	0.90 ± 0.03 [B]	0.42 ± 0.02 [B]
HLEY 0.4	10.66 ± 0.24 [A]	3.18 ± 0.10 [A]	4.02 ± 0.16 [A]	0.80 ± 0.03 [A]	0.70 ± 0.02 [C]	4.82 ± 0.03 [A]	62.00 ± 3.27 [A]	20.60 ± 0.82 [A]	0.60 ± 0.02 [C]	0.34 ± 0.02 [C]

[A,B,C] Letters in the same column are significantly different at ($p \leq 0.05$). Mean values $\pm$ standard deviation, $n = 3$. PY, plain yogurt; HLEY0.2, yogurt containing 0.2% Hawthorn leaf extract; HLEY0.4, yogurt containing 0.4% Hawthorn lead extract; TPC, total phenolic compounds (mg GAE/100 g); AO, antioxidant activity (%); PV, peroxide value (meq/Kg oil); AV, acid value (mg KOH/g oil).

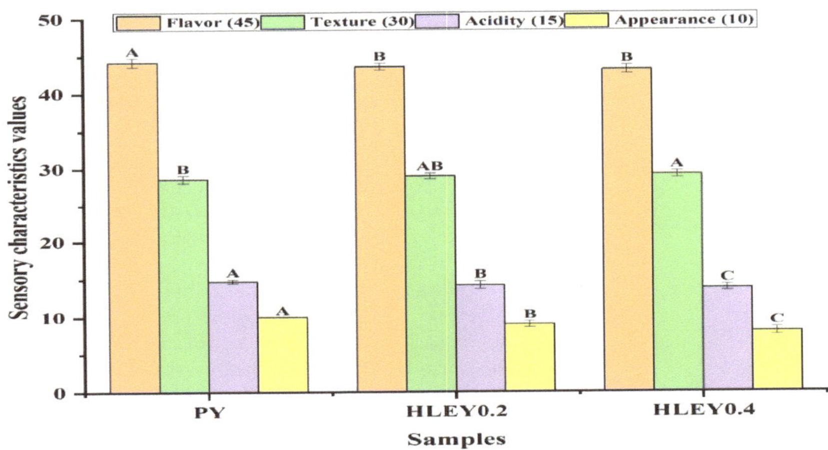

Figure 1. Sensory characteristics of yogurt fortified with aqueous Hawthorn leaf extract. Values with the different superscript letters are significantly different at ($p \leq 0.05$). Mean values $\pm$ standard deviation, $n = 10$. PY, plain yogurt; HLEY0.2, yogurt containing 0.2% Hawthorn leaf extract; HLEY0.4, yogurt containing 0.4% Hawthorn leaf extract.

3.3. Final Weight and Body Weight Gain in All Groups of Rats

As shown in Table 2, the negative control (Group 1) achieved the highest weight parameter values, whereas the treatments (Groups 3 and 4) had a remarkable ($p \leq 0.05$) effect on the final weight and on body weight gain. The best results were achieved in the case of Group 4, of the group in which the rats were yogurt fortified with 0.4% aqueous Hawthorn leaf extract, reaching higher final weight (275.48 g) and body weight gain (21.09%) values compared to the positive control group (final weight of 265.40 g and body weight gain of 15.49%). This improvement may be able to be ascribed to the high vitamin, mineral, and antioxidant contents in Hawthorn, which may prevent damage to body cells caused by free radicals [23,24]. Accordingly, a previous study [57] found that the addition of Hawthorn leaf extract in rat diets promoted a significant increase in body weight gain and enhanced their nutritional status.

Table 2. Final weight (g) and body weight gain (%) in rats treated with yogurt fortified with 0.4% aqueous Hawthorn leaf extract.

Groups	Initial Weight (g)	Final Weight (g)	Body Weight Gain (BWG (%))
Group (1)	227.40 ± 2.49 [A]	287.07 ± 1.72 [A]	26.25 ± 1.18 [A]
Group (2)	229.80 ± 1.54 [A]	265.40 ± 1.66 [D]	15.49 ± 0.30 [D]
Group (3)	228.60 ± 1.41 [A]	271.20 ± 2.62 [C]	18.63 ± 0.67 [C]
Group (4)	227.50 ± 1.78 [A]	275.48 ± 2.14 [B]	21.09 ± 0.56 [B]

(BWG %) = Final weight − Initial weight ÷ Initial Weight × 100. Group (1) non-treated rats (negative control). Group (2) treated oxidative stress rats (positive control). Group (3) oxidative stress rats fed a basal diet with plain yogurt (10 g/day). Group (4) oxidative stress rats fed a basal diet with yogurt fortified with 0.4% aqueous Hawthorn leaf extract (10 g/day). [A,B,C,D] Letters in the same column are significantly different at ($p \leq 0.05$). Mean values of ± standard deviation.

3.4. Effect of Yogurt Fortified with Hawthorn Leaves Extract on the Serum Lipid Profile, Liver Function Parameters, Kidney Function Parameters, Liver Weight, Kidney Weight, and Total Antioxidant Potentials of Serum of Oxidative Stress Rats

The detected lipid profiles for all of the investigated groups are compared and shown in Figure 2a. It should be noticed that the non-treated rats (negative control) showed lower total cholesterol triglycerides and LDL contents (67.40, 77.50, and 15.70 mg/dL, respectively) and a higher HDL content (36.20 mg/dL) compared to the other groups. Among the oxidative stress rat groups, Group 4, which was treated with yogurt fortified with 0.4% aqueous Hawthorn leaf extract, presented with lower total cholesterol (72.50 mg/dL) as well as significantly ($p \leq 0.05$) lower triglyceride and LDL contents (79.20 and 22.96 mg/dL, respectively) compared to the positive control group (total cholesterol, triglycerides, and LDL of 135.20, 119.20, and 85.96 mg/dL, respectively). Concerning the HDL content, the positive control group showed the lowest value (25.40 mg/dL) compared to the rats who had been fed a normal diet and yogurt fortified with 0.4% aqueous Hawthorn leaf extract, with a significant increase (33.70 mg/dL) being recorded. Clearly, comparing the present results, the treatment with yogurt fortified with 0.4% aqueous Hawthorn leaf extract resulted in a significant reduction in the total cholesterol, triglyceride, and LDL levels with respect to the positive control. It is noteworthy to state that *C. oxyacantha* Hawthorn demonstrates remarkable hypolipidemic and hypocholesterolemic effects [57,58] through the reduction of ApoB synthesis, total cholesterol, and LDL, and these affects are associated with a significant increase in the HDL [26]. Meanwhile, the positive control condition, in which oxidative stress was induced by carbon tetrachloride, demonstrated a consequent increase in the formation of reactive oxygen species and was able to promote the oxidation of pivotal cellular components (e.g., membrane lipids, proteins, and DNA), leading to cellular damage [59]. The reported beneficial hypolipidemic effect of the Hawthorn leaf extract on the lipid profile may be due to phenolic compounds the natural antioxidants that are present in Hawthorn leaves [25,28]. This hypolipidemic action can also be ascribed to the lipid metabolism modulation caused by phenolics such as chlorogenic acid and flavonoids such as quercetin, leading to a decrease in the total cholesterol, triglycerides, and LDL and was not associated with the increase in HDL levels because it up-regulated hepatic peroxisome proliferator-activated receptor (PPAR-α) expression [60].

Concerning the liver function parameters, as illustrated in Figure 2b, a significant increase in the AST and ALT and a decrease in the total albumin were revealed in the untreated group (positive control, Group 2) compared to the negative control group (Group 1). On the other hand, the group treated that received the yogurt that had been fortified with 0.4% aqueous Hawthorn leaf extract (Group 4) presented a significant increase in the total albumin and a decrease in AST and ALT compared to the positive control group, which is consistent with results that have been described elsewhere [57]. This decrease in the values of aminotransferase enzymes and the restoration of some vital functions by the hepatocytes can be ascribed to the high contents of phenolic and bioactive components in Hawthorn leaves, which preserve the plasma membrane in hepatocytes and protect it from the rupture and the exit of the cytosol that is loaded with these enzymes [25,28].

Regarding the kidney function parameters, as shown in Figure 2c, a significant increase in creatinine and urea ($p \leq 0.05$) was revealed in the positive control group compared to in the normal control group (negative group), whereas the group treated with yogurt fortified with 0.4% aqueous Hawthorn leaf extract showed a significant decrease in creatinine and urea. This conspicuous change could be partially attributed to the bioactive components present within Hawthorn leaves, such as phenolic compounds, flavonoids, minerals, and vitamins [24], as these may indirectly reduce uric acid levels, keeping the kidneys safe from the damage that can potentially be caused by oxidative stress. These bioactive components act as superoxide scavengers, resulting in the suppression of reactive oxygen species; thus, in the body cells, oxidative stress and inflammation are reduced. Moreover, they can inhibit the formation of uric acid through the direct uricosuric potential or increase the glomerular filtration rate, resulting in a consequent decrease in the uric acid levels in the blood [61,62]. Concerning MDA, the highest mean value was achieved in the positive control group (67.77 µmol/L), whereas the lowest one was observed in the case of the negative control group (46.34 µmol/L). A significant decrease of up to 47.18 µmol/L was revealed in the rats treated with yogurt fortified with 0.4% aqueous Hawthorn leaf extract. These findings were corroborated those by Qi et al., 2019 [58], who found that supplementation with Hawthorn leaf extract was effective in decreasing the creatinine, urea, and MDA levels compared to the positive control group. It should be borne in mind that MDA plays a very negative role and is able to alter the structure and function of the cell membrane [63]. The formation of MDA and increasing its levels can lead to inhibitory actions, oxidative mechanisms, high cytotoxicity, and tumor development, as it can act as a co-carcinogenic agent [64].

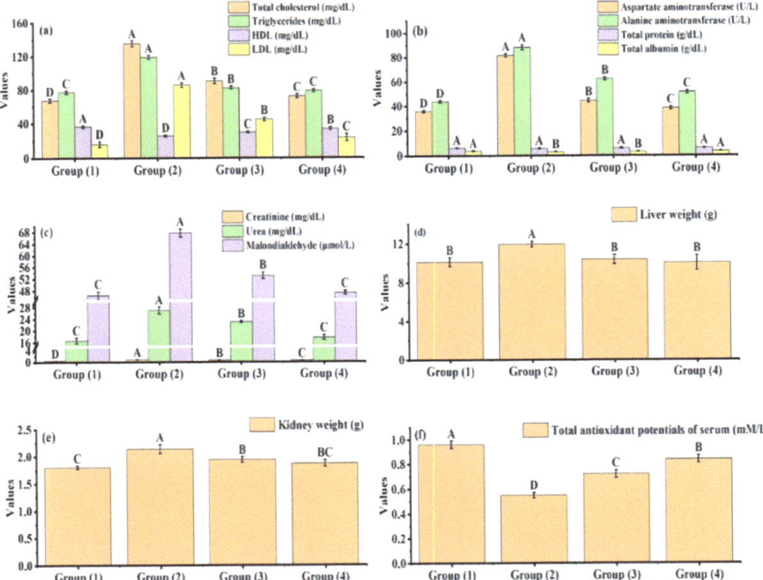

Figure 2. Effect of yogurt fortified with 0.4% aqueous Hawthorn leaf extract on serum: lipid profile (**a**), liver function parameters (**b**), kidney function parameters (**c**), liver weight (**d**), kidney weight (**e**), and total antioxidant potentials of serum (**f**) of the oxidative stress rats. Group (1) non-treated rats (negative control). Group (2) treated oxidative stress rats (positive control). Group (3) oxidative stress rats fed a basal diet with plain yogurt (10 g/day). Group (4) oxidative stress rats fed on basal diet with yogurt fortified with 0.4% aqueous Hawthorn leaf extract (10 g/day). Values with different superscript letters are significantly different at ($p \leq 0.05$). Mean values ± standard deviation.

When comparing the liver and kidney weights, the results of which are shown in Figure 2d,e, respectively, a significant ($p \leq 0.05$) increase in the liver and kidney weights of the positive control group was revealed compared to the other groups. This experimental evidence suggests that the liver weight was affected by oxidative stress, which caused spindle cell proliferation, collagen fibers admixed with erythrocytes, and inflammatory cells among the blood inside the bile duct lumen in the liver, something that is evident in the histological sections (Figure 3). Additionally, these results indicate the influence of oxidative stress on the kidney weight, causing the degeneration of collected tubules and desquamated lining epithelium (Figure 4b, histological section). It seems that Hawthorn leaf extract has the ability to improve liver and kidney function (Groups 3 and 4) due to its antioxidant content, which is in harmony with what has been reported elsewhere [58,65]. As shown in Figure 2f, the oxidative stress rat groups (2, 3, and 4) presented significantly lower serum antioxidant potential values compared to the negative control (Group 1), whereas the serum antioxidant potential in Group 4 (yogurt fortified with Hawthorn leaf extract) was significantly ($p \leq 0.05$) higher than it was in the positive control (Group 2). This experimental evidence confirms that the yogurt fortified with Hawthorn leaf extract has antioxidative and beneficial effects when the liver and kidneys are recovering from carbon tetrachloride injury [57,66].

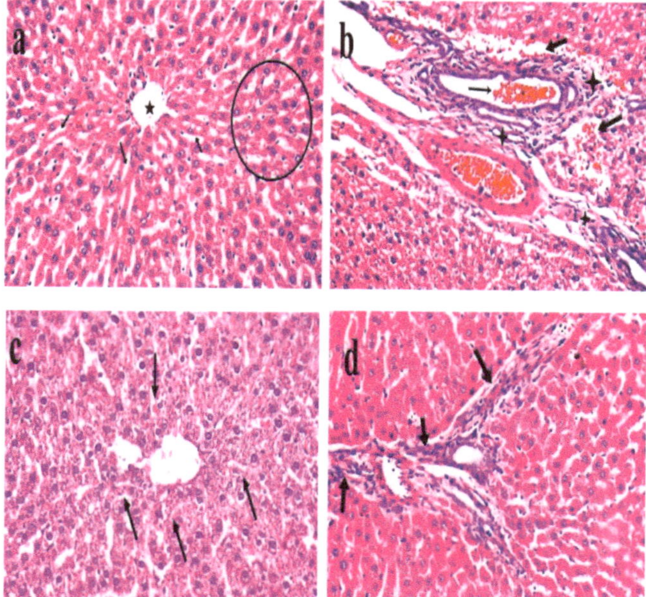

Figure 3. Representative photomicrographs of rat livers. (**a**) Group 1, negative control: normal hepatic parenchymal structures of the central vein (star), hepatic cords (circle), and sinusoids (arrows) containing Kupffer cells. (**b**) Group 2, positive control: portal trade fibrosis (stars) represented by spindle cells proliferation, collagen fibers ad-mixed with erythrocytes, and inflammatory cells (thick arrows) among the blood inside the bile duct lumen (thin arrow). (**c**) Group 3, plain yogurt (10 g/day): massive distribution of slightly vacuolated hepatocytes (arrows). (**d**) Group 4, yogurt with 0.4% aqueous Hawthorn leaf extract (10 g/day): interlobular fibrous bridge (arrows) with normal hepatic cords and sinusoids. Hematoxylin and eosin staining, magnification ×400.

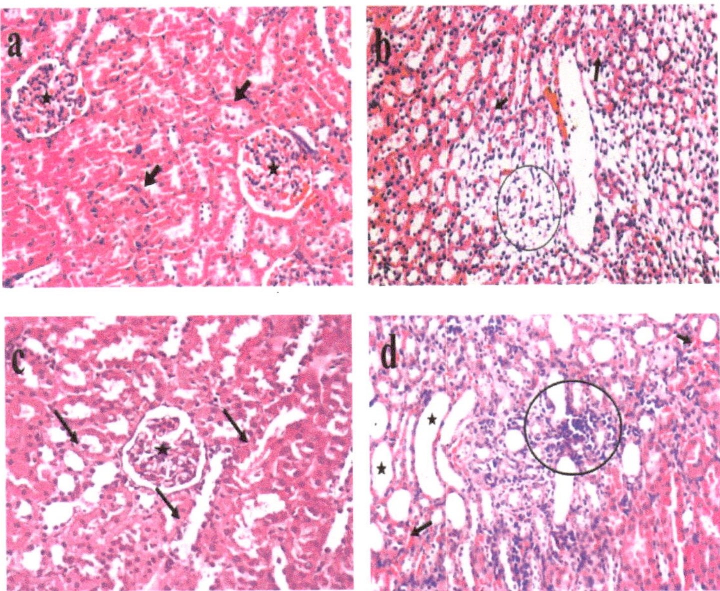

Figure 4. Representative photomicrographs of rat kidneys. (**a**) Group 1, negative control: normal renal tubules (arrows) and glomeruli (stars). (**b**) Group 2, positive control: degeneration collected tubules (circle) with desquamated lining epithelium (arrows). (**c**) Group 3, plain yogurt (10 g/day): vacuolation of the glomerular epithelium (arrow) and focally distributed degeneration of some renal tubules (star). (**d**) Group 4, yogurt with 0.4% aqueous Hawthorn leaf extract (10 g/day): focal interstitial lymphocytic infiltrations (circle), cystic dilated renal tubules (stars), and slightly congested interstitial capillaries (arrows). Hematoxylin and eosin staining, magnification × 400.

3.5. Light Microscopic Histological Study

In the present work, hepatic damage was assessed by means of histological organ examination, and the degree of histological changes in the different groups of rats were determined (Figure 3). The histological examinations of the livers in the normal rats (negative control group) showed the appearance of normally preserved normal hepatocytes (Figure 3a), whereas the liver tissue sections of the rats submitted to oxidative stress during the experiment (Group 2) were characterized by portal trade fibrosis represented by spindle cells proliferation, collagen fibers admixed with erythrocytes, and inflammatory cells among the blood inside the bile duct lumen (Figure 3b). On the other hand, the histological examination of the liver tissue sections taken from the rats treated with plain yogurt showed a mass distribution of slightly vacuolated hepatocytes (Figure 3c). The rats treated with yogurt fortified with aqueous Hawthorn leaf extract showed a consistent decrease in hepatocyte necrosis processes and a more regular liver structure. Furthermore, slight fibrosis in the portal area as well as the restoration of the presence of a fibrous bridge between the portal vein cavity was reported, and the cells and hepatic lobes were restored to normal (Figure 3d). These findings corroborate those reported elsewhere [26,67] and highlight no histological changes in the liver, kidney, and heart in the groups treated with wild plant extracts. Similarly, the renal damage was also evaluated through histological organ examinations to determine the degree of histological changes, as shown in Figure 4. Figure 4a reveals the normal appearance of renal tubules and glomeruli in the negative control group (normal). Meanwhile, the renal tissue samples from the Group 2 rats showed degeneration in the collected tubules and a desquamated lining epithelium, as shown in Figure 4b and represent positive control group. In the case of the rats treated with plain yogurt, the kidney sections were characterized by glomerular epithelium vacuolation and the focally distributed degeneration of some of the renal tubules (Figure 4c). On the other hand, the histological examination of kidney sections taken from the rats treated with the yogurt fortified with aqueous Hawthorn leaf extract presented normal glomerular structures and slight congestion in the cellular capillaries (Figure 4d), which is in harmony with several previous reports [58].

4. Conclusions

In summary, Hawthorn (*Crataegus oxyacantha*) leaf extract might exert protective effects against oxidative stress in rats due to their high bioactive components. Clearly, it can be used in some vital food products such as yogurt. Using *Crataegus oxyacantha* leaf aqueous extract at concentrations of up to 0.4% during the manufacture of yogurt greatly affected its physicochemical, phytochemical, and sensory properties and imparted health benefits to the yogurt on the basis of its high bioactive components. The present findings also concluded that the addition of *Crataegus oxyacantha* leaf extract as a bioactive supplement to yogurt is useful to maintain good oxidative status, which was positively reflected in the general health of the oxidative stress rats.

Author Contributions: M.R.S., E.-S.H.A. and H.A.R. were involved in the conception of the research idea and the design of the methodology, and supervision, and performed data analysis and interpretation. A.O.B., M.U., D.A., H.M., N.D., I.C., B.M.A. and E.K.E. were involved in the methodology and drafted and prepared the manuscript for publication and revision. All authors have read and agreed to the published version of the manuscript.

Funding: This research received no external funding.

Institutional Review Board Statement: The experimental protocol was approved by the Ethics Committee of Suez Canal University, Suez, Egypt (ref. no.: 2021047).

Informed Consent Statement: Not applicable.

Data Availability Statement: Not applicable.

Conflicts of Interest: The authors declare no conflict of interest.

References

1. Hussain, T.; Tan, B.; Yin, Y.; Blachier, F.; Tossou, M.C.; Rahu, N. Oxidative stress and inflammation: What polyphenols can do for us? *Oxidative Med. Cell. Longev.* **2016**, *2016*, 7432797. [CrossRef] [PubMed]
2. Tomsa, A.M.; Alexa, A.L.; Junie, M.L.; Rachisan, A.L.; Ciumarnean, L. Oxidative stress as a potential target in acute kidney injury. *PeerJ* **2019**, *7*, e8046. [CrossRef] [PubMed]
3. Kim, Y.S.; Young, M.R.; Bobe, G.; Colburn, N.H.; Milner, J.A. Bioactive food components, inflammatory targets, and cancer prevention. *Cancer Prev. Res.* **2009**, *2*, 200–208. [CrossRef] [PubMed]
4. Ďuračková, Z. Some current insights into oxidative stress. *Physiol. Res.* **2010**, *59*, 459–469. [CrossRef]
5. Kumar, S.; Pandey, A.K. Free radicals: Health implications and their mitigation by herbals. *J. Adv. Med. Med. Res.* **2015**, *7*, 438–457. [CrossRef]
6. Samuel, S.A.; Francis, A.O.; Anthony, O.O. Role of the kidneys in the regulation of intra-and extra-renal blood pressure. *Ann. Clin. Hypertens.* **2018**, *2*, 048–058.
7. Cosme, P.; Rodríguez, A.B.; Espino, J.; Garrido, M. Plant phenolics: Bioavailability as a key determinant of their potential health-promoting applications. *Antioxidants* **2020**, *9*, 1263. [CrossRef]
8. Vuolo, M.M.; Lima, V.S.; Junior, M.R.M. Phenolic compounds: Structure, classification, and antioxidant power. In *Bioactive Compounds*; Elsevier: Amsterdam, The Netherlands, 2019; pp. 33–50.
9. Carocho, M.; Morales, P.; Ferreira, I.C.F.R. Natural food additives: Quo vadis? *Trends Food Sci. Technol.* **2015**, *45*, 284–295. [CrossRef]
10. Hosseini-Ara, R.; Mokhtarian, A.; Karamrezaei, A.H.; Toghraie, D. Computational analysis of high precision nano-sensors for diagnosis of viruses: Effects of partial antibody layer. *Math. Comput. Simul.* **2022**, *192*, 384–398. [CrossRef]
11. Elkot, W.F.; Ateteallah, A.H.; Al-Moalem, M.H.; Shahein, M.R.; Mohamed AAlblihed, M.A.; Abdo, W.; Elmahallawy, E.K. Functional, Physicochemical, Rheological, Microbiological, and Organoleptic Properties of Synbiotic Ice cream Produced from Camel Milk Using Black Rice Powder and Lactobacillus acidophilus LA-5. *Fermentation* **2022**, *8*, 187. [CrossRef]

12. Shahein, M.R.; Atwaa, E.S.H.; Elkot, W.F.; Hijazy, H.H.A.; Kassab, R.B.; Alblihed, M.A.; Elmahallawy, E.K. The impact of date syrup on the physicochemical, microbiological, and sensory properties, and antioxidant activity of bio-fermented camel milk. *Fermentation* **2022**, *8*, 192. [CrossRef]
13. Shahein, M.R.; Atwaa, E.S.H.; Radwan, H.A.; Elmeligy, A.A.; Hafiz, A.A.; Albrakati, A.; Elmahallawy, E.K. Production of a Yogurt Drink Enriched with Golden Berry (*Physalis pubescens* L.) Juice and Its Therapeutic Effect on Hepatitis in Rats. *Fermentation* **2022**, *8*, 112. [CrossRef]
14. Shahidi, F.; Ambigaipalan, P. Phenolics and polyphenolics in foods, beverages and spices: Antioxidant activity and health effects–A review. *J. Funct. Foods* **2015**, *18*, 820–897. [CrossRef]
15. Atwaa, E.S.H.; Shahein, M.R.; El-Sattar, E.S.A.; Hijazy, H.H.A.; Albrakati, A.; Elmahallawy, E.K. Bioactivity, Physicochemical and Sensory Properties of Probiotic Yoghurt Made from Whole Milk Powder Reconstituted in Aqueous Fennel Extract. *Fermentation* **2022**, *8*, 52. [CrossRef]
16. Shahein, M.R.; Atwaa, E.S.H.; El-Zahar, K.M.; Elmaadawy, A.A.; Hijazy, H.H.A.; Sitohy, M.Z.; Albrakati, A.; Elmahallawy, E.K. Remedial Action of Yoghurt Enriched with Watermelon Seed Milk on Renal Injured Hyperuricemic Rats. *Fermentation* **2022**, *8*, 41. [CrossRef]
17. Swelam, S.; Zommara, M.A.; Abd El-Aziz, A.E.-A.M.; Elgammal, N.A.; Baty, R.S.; Elmahallawy, E.K. Insights into Chufa Milk Frozen Yoghurt as Cheap Functional Frozen Yoghurt with High Nutritional Value. *Fermentation* **2021**, *7*, 255. [CrossRef]
18. Beltrán-Barrientos, L.; Hernández-Mendoza, A.; Torres-Llanez, M.; González-Córdova, A.; Vallejo-Córdoba, B. Invited review: Fermented milk as antihypertensive functional food. *J. Dairy Sci.* **2016**, *99*, 4099–4110. [CrossRef]
19. Butnariu, M.; Grozea, I. Antioxidant (antiradical) compounds. *J. Bioequiv. Availab.* **2012**, *4*, 17–19. [CrossRef]
20. Galanakis, C.M. The food systems in the era of the coronavirus (COVID-19) pandemic crisis. *Foods* **2020**, *9*, 523. [CrossRef]
21. Al-Shawi, S.G.; Ali, H.I.; Al-Younis, Z.K. The effect of adding thyme extacts on microbiological, chemical and sensory characteristics of yogurt. *J. Pure Appl. Microbiol* **2020**, *14*, 1367–1376. [CrossRef]
22. Bertolino, M.; Belviso, S.; Dal Bello, B.; Ghirardello, D.; Giordano, M.; Rolle, L.; Gerbi, V.; Zeppa, G. Influence of the addition of different hazelnut skins on the physicochemical, antioxidant, polyphenol and sensory properties of yogurt. *LWT-Food Sci. Technol.* **2015**, *63*, 1145–1154. [CrossRef]
23. Benabderrahmane, W.; Lores, M.; Benaissa, O.; Lamas, J.P.; de Miguel, T.; Amrani, A.; Benayache, F.; Benayache, S. Polyphenolic content and bioactivies of *Crataegus oxyacantha* L. (Rosaceae). *Nat. Prod. Res.* **2021**, *35*, 627–632. [CrossRef] [PubMed]
24. Orhan, I.E. Phytochemical and pharmacological activity profile of *Crataegus oxyacantha* L. (hawthorn)-A cardiotonic herb. *Curr. Med. Chem.* **2018**, *25*, 4854–4865. [CrossRef]
25. Venskutonis, P.R. Phytochemical composition and bioactivities of hawthorn (*Crataegus* spp.): Review of recent research advances. *J. Food Bioact.* **2018**, *4*, 69–87. [CrossRef]
26. Martínez-Rodríguez, J.L.; Reyes-Estrada, C.A.; Gutiérrez-Hernández, R.; Lopez, J.A. Antioxidant, hypolipidemic and preventive effect of Hawthorn (*Crataegus oxyacantha*) on alcoholic liver damage in rats. *J. Pharmacogn. Phytother.* **2016**, *8*, 193–202.
27. Yang, T.; Zhang, X.; Pan, S.; Zhang, X.; Dang, Z. Effect of hawthorn Jiangzhi powder on blood lipids in patients with hyperlipidemia: A pathological analysis of 484 cases. *Chin. J. Integr. Tradit. West. Med. Intensive Crit. Care* **2017**, *24*, 166–169.
28. Rasheed, H.A.; Hussien, N.R.; Al-Naimi, M.S.; Al-Kuraishy, H.M.; Al-Gareeb, A.I. Fenofibrate and *Crataegus oxyacantha* is an effectual combo for mixed dyslipidemia. *Biomed. Biotechnol. Res. J. BBRJ* **2020**, *4*, 259.
29. Pakseresht, S.; Mazaheri Tehrani, M.; Razavi, S.M.A. Optimization of low-fat set-type yoghurt: Effect of altered whey protein to casein ratio, fat content and microbial transglutaminase on rheological and sensorial properties. *J. Food Sci. Technol.* **2017**, *54*, 2351–2360. [CrossRef]
30. Reeves, P.G.; Nielsen, F.H.; Fahey, G.C., Jr. *AIN-93 Purified Diets for Laboratory Rodents: Final Report of the American Institute of Nutrition ad Hoc Writing Committee on the Reformulation of the AIN-76A Rodent Diet*; Oxford University Press: Oxford, UK, 1993.
31. Chemists, A.O.O.A.; Horwitz, W. *Official Methods of Analysis*; Association of Official Analytical Chemists: Washington, DC, USA, 1975; Volume 222.
32. Maksimović, Z.; Malenčić, Đ.; Kovačević, N. Polyphenol contents and antioxidant activity of Maydis stigma extracts. *Bioresour. Technol.* **2005**, *96*, 873–877. [CrossRef]
33. Apostolidis, E.; Kwon, Y.I.; Shetty, K. Inhibitory potential of herb, fruit, and fungal-enriched cheese against key enzymes linked to type 2 diabetes and hypertension. *Innov. Food Sci. Emerg. Technol.* **2007**, *8*, 46–54. [CrossRef]
34. Ali, H.I.; Dey, M.; Alzubaidi, A.K.; Alneamah, S.J.A.; Altemimi, A.B.; Pratap-Singh, A. Effect of rosemary (*Rosmarinus officinalis* L.) supplementation on probiotic yoghurt: Physicochemical properties, microbial content, and sensory attributes. *Foods* **2021**, *10*, 2393. [CrossRef] [PubMed]
35. Kujawska, M.; Ignatowicz, E.; Murias, M.; Ewertowska, M.; Mikołajczyk, K.; Jodynis-Liebert, J. Protective effect of red beetroot against carbon tetrachloride-and N-nitrosodiethylamine-induced oxidative stress in rats. *J. Agric. Food Chem.* **2009**, *57*, 2570–2575. [CrossRef] [PubMed]
36. Dougnon, T.V.; Bankolé, H.S.; Klotoé, J.R.; Sènou, M.; Fah, L.; Koudokpon, H.; Akpovi, C.; Dougnon, T.J.; Addo, P.; Loko, F. Treatment of hypercholesterolemia: Screening of Solanum macrocarpon Linn (Solanaceae) as a medicinal plant in Benin. *Avicenna J. Phytomedicine* **2014**, *4*, 160. [CrossRef]
37. Devi, R.; Sharma, D. Hypolipidemic effect of different extracts of Clerodendron colebrookianum Walp in normal and high-fat diet fed rats. *J. Ethnopharmacol.* **2004**, *90*, 63–68. [CrossRef]

38. Bachorik, P.S.; Walker, R.E.; Kwiterovich, P.O. Determination of high density lipoprotein-cholesterol in human plasma stored at −70 degrees C. *J. Lipid Res.* **1982**, *23*, 1236–1242. [CrossRef]

39. Friedewald, W.T.; Levy, R.I.; Fredrickson, D.S. Estimation of the concentration of low-density lipoprotein cholesterol in plasma, without use of the preparative ultracentrifuge. *Clin. Chem.* **1972**, *18*, 499–502. [CrossRef]

40. Bergmeyer, H.; Harder, M. A colorimetric method of the determination of serum glutamic oxaloacetic and glutamic pyruvic transaminase. *Clin. Biochem.* **1986**, *24*, 458–481.

41. Koracevic, D.; Koracevic, G.; Djordjevic, V.; Andrejevic, S.; Cosic, V. Method for the measurement of antioxidant activity in human fluids. *J. Clin. Pathol.* **2001**, *54*, 356–361. [CrossRef]

42. Hemieda, F.A.-K.E.-S. Influence of gender on tamoxifen-induced biochemical changes in serum of rats. *Mol. Cell. Biochem.* **2007**, *301*, 137–142. [CrossRef]

43. Marsh, W.H.; Fingerhut, B.; Miller, H. Automated and manual direct methods for the determination of blood urea. *Clin. Chem.* **1965**, *11*, 624–627. [CrossRef]

44. Han, K.-H.; Sekikawa, M.; Shimada, K.-i.; Hashimoto, M.; Hashimoto, N.; Noda, T.; Tanaka, H.; Fukushima, M. Anthocyanin-rich purple potato flake extract has antioxidant capacity and improves antioxidant potential in rats. *Br. J. Nutr.* **2007**, *96*, 1125–1134. [CrossRef]

45. Hosni, M.; El Maghrbi, A.; Ganghish, K. Occurrence of Trichinella spp. in wild animals in northwestern Libya. *Open Vet. J.* **2013**, *3*, 85–88. [PubMed]

46. Öztürk, N.; Tunçel, M. Assessment of phenolic acid content and in vitro antiradical characteristics of hawthorn. *J. Med. Food* **2011**, *14*, 664–669. [CrossRef]

47. Šavikin, K.P.; Krstić-Milošević, D.B.; Menković, N.R.; Beara, I.N.; Mrkonjić, Z.O.; Pljevljakušić, D.S. Crataegus orientalis leaves and berries: Phenolic profiles, antioxidant and anti-inflammatory activity. *Nat. Prod. Commun.* **2017**, *12*, 1934578X1701200204. [CrossRef]

48. Jafari, S.M.; Vakili, S.; Dehnad, D. Production of a Functional Yogurt Powder Fortified with Nanoliposomal Vitamin D Through Spray Drying. *Food Bioprocess Technol.* **2019**, *12*, 1220–1231. [CrossRef]

49. Azizkhani, M.; Tooryan, F. Antimicrobial activities of probiotic yogurts flavored with peppermint, basil, and zataria against Escherichia coli and Listeria monocytogenes. *J. Food Qual. Hazards Control* **2016**, *3*, 79–86.

50. Suliman, A.; Ahmed, K.; Mohamed, B.; Babiker, E. Potential of cinnamon (*Cinnamomum cassia*) as an anti-oxidative and anti-microbial agent in sudanese yoghurt (Zabadi). *J. Dairy Vet. Sci.* **2019**, *12*, 555833.

51. Dabija, A.; Codină, G.G.; Ropciuc, S.; Gâtlan, A.-M.; Rusu, L. Assessment of the antioxidant activity and quality attributes of yogurt enhanced with wild herbs extracts. *J. Food Qual.* **2018**, *2018*, 5329386. [CrossRef]

52. Olmedo, R.H.; Nepote, V.; Grosso, N.R. Preservation of sensory and chemical properties in flavoured cheese prepared with cream cheese base using oregano and rosemary essential oils. *LWT-Food Sci. Technol.* **2013**, *53*, 409–417. [CrossRef]

53. Yadav, K.; Shukla, S. Microbiological, physicochemical analysis and sensory evaluation of herbal yogurt. *Pharma Innov.* **2014**, *3*, 1–4.

54. Shelbaya, L.A.; Ghoneim, G.A. Biological Evaluation of Cupcake Treated with Hawthorn Leaves on Diabetic Rats. *J. Food Dairy Sci.* **2017**, *8*, 25–30. [CrossRef]

55. Cho, W.-Y.; Kim, D.-H.; Lee, H.-J.; Yeon, S.-J.; Lee, C.-H. Quality characteristic and antioxidant activity of yogurt containing olive leaf hot water extract. *CyTA-J. Food* **2020**, *18*, 43–50. [CrossRef]

56. Shiravani, M.; Ansari, S. Yogurt fortification with walnut leaf extract and investigation of its physicochemical and sensory properties. *J. Innov. Food Sci. Technol.* **2020**, *12*, 1–17.

57. Al Shammari, A.M.N. Biological Effects of Hawthorn Leaves Powder and its Extract on Biological and Biochemical Change of Induced Obese Rats. *Life Sci. J.* **2020**, *17*, 62–66.

58. Qin, C.; Xia, T.; Li, G.; Zou, Y.; Cheng, Z.; Wang, Q. Hawthorne leaf flavonoids prevent oxidative stress injury of renal tissues in rats with diabetic kidney disease by regulating the p38 MAPK signaling pathway. *Int. J. Clin. Exp. Pathol.* **2019**, *12*, 3440–3446. [PubMed]

59. Yilmaz, S.; Kaya, E.; Kisacam, M.A. The effect on oxidative stress of aflatoxin and protective effect of lycopene on aflatoxin damage. *Aflatoxin-Control Anal. Detect. Health Risks* **2017**, *30*, 67–90.

60. Zeni, A.L.B.; Moreira, T.D.; Dalmagro, A.P.; Camargo, A.; Bini, L.A.; Simionatto, E.L.; Scharf, D.R. Evaluation of phenolic compounds and lipid-lowering effect of Morus nigra leaves extract. *An. Da Acad. Bras. De Ciências* **2017**, *89*, 2805–2815. [CrossRef]

61. Yu, Z.; Fong, W.P.; Cheng, C.H. The dual actions of morin (3,5,7,2′,4′-pentahydroxyflavone) as a hypouricemic agent: Uricosuric effect and xanthine oxidase inhibitory activity. *J. Pharmacol. Exp. Ther.* **2006**, *316*, 169–175. [CrossRef]

62. Lin, S.; Zhang, G.; Liao, Y.; Pan, J.; Gong, D. Dietary flavonoids as xanthine oxidase inhibitors: Structure–affinity and structure–activity relationships. *J. Agric. Food Chem.* **2015**, *63*, 7784–7794. [CrossRef]

63. Nair, G.G.; Nair, C.K.K. Radioprotective effects of gallic acid in mice. *BioMed Res. Int.* **2013**, *2013*, 953079. [CrossRef] [PubMed]

64. Koc, M.; Taysi, S.; Buyukokuroglu, M.E.; Bakan, N. Melatonin protects rat liver against irradiation-induced oxidative injury. *J. Radiat. Res.* **2003**, *44*, 211–215. [CrossRef]

65. Li, Z.; Xu, J.; Zheng, P.; Xing, L.; Shen, H.; Yang, L.; Zhang, L.; Ji, G. Hawthorn leaf flavonoids alleviate nonalcoholic fatty liver disease by enhancing the adiponectin/AMPK pathway. *Int. J. Clin. Exp. Med.* **2015**, *8*, 17295–17307. [PubMed]

66. Tatsimo, S.J.N.; Tamokou, J.d.D.; Havyarimana, L.; Csupor, D.; Forgo, P.; Hohmann, J.; Kuiate, J.-R.; Tane, P. Antimicrobial and antioxidant activity of kaempferol rhamnoside derivatives from Bryophyllum pinnatum. *BMC Res. Notes* **2012**, *5*, 158. [CrossRef] [PubMed]
67. Sadek, M.A.; Mousa, S.; El-Masry, S.; Demain, A.M. Influence of Hawthorn (*Crataegus oxyacantha*) Leaves Extract Administration on Myocardial Infarction Induced by Isoproterenol in Rats. *Cardiol. Angiol. Int. J.* **2018**, 1–12. [CrossRef]

fermentation

Article

Changes in Phenolic Profiles and Inhibition Potential of Macrophage Foam Cell Formation during Noni (*Morinda citrifolia* Linn.) Fruit Juice Fermentation

Kun Cai [1,2,3], Rong Dou [1], Xue Lin [1,2,3], Xiaoping Hu [1,2,3], Zhulin Wang [1], Sixin Liu [2,3,4], Congfa Li [1,2,3,*] and Wu Li [1,2,3,*]

[1] School of Food Science and Engineering, Hainan University, Haikou 570228, China; caikun@hainanu.edu.cn (K.C.); lori@hainanu.edu.cn (R.D.); 993048@hainanu.edu.cn (X.L.); huxiaoping03@hainanu.edu.cn (X.H.); 18085231210024@hainanu.edu.cn (Z.W.)

[2] Key Laboratory of Food Nutrition and Functional Food of Hainan Province, Haikou 570228, China; sixinliu@126.com

[3] Key Laboratory of Tropical Agricultural Products Processing Technology of Haikou, Haikou 570228, China

[4] School of Sciences, Hainan University, Haikou 570228, China

[*] Correspondence: congfa@hainanu.edu.cn (C.L.); leewuu@163.com (W.L.); Tel./Fax: +86-898-66278326 (C.L.); +86-898-66278326 (W.L.)

Abstract: The dynamic changes in phenolic composition and antioxidant activity, and the potential effect on foam cell formation and cholesterol efflux during noni (*Morinda citrifolia* Linn.) fruit juice fermentation were investigated in this study. The composition of phenolic compounds was significantly different at various fermentation times. Rutin, quercetin, and isoquercitrin were the major phenolics in fermented noni fruit juice based on a quantitative analysis of representative phenolics. The contents of caffeic acid, 2,4-dihydroxybenzoic acid, p-coumaric acid, rutin, and quercetin tended to increase, while those of isoquercitrin decreased during the fermentation process. Fermented noni juice extracts showed high antioxidant activities against 2,2-diphenyl-1-picrylhydrazyl and 2,2′-azino-bis(3-ethylbenzothiazoline-6-sulphonic acid), hydroxyl radical scavenging activity, and ferric reducing antioxidant power. Notably, the highest antioxidant activity was observed after 28 days of fermentation. Furthermore, the treatment of fermented noni juice extracts was shown to reduce foam cell formation, intracellular cholesterol level, and the cholesterol esterification ratio. A correlation analysis indicated a strong positive relationship between the phenolic composition, antioxidant activity, and the ratio of cholesterol ester and total cholesterol. This study may provide a theoretical basis for the quality improvement and standardized production of fermented noni fruit juice, thus promoting the development of the noni food industry.

Keywords: noni fruit juice; fermentation; phenolic; antioxidant activity; macrophages; atherosclerosis

Citation: Cai, K.; Dou, R.; Lin, X.; Hu, X.; Wang, Z.; Liu, S.; Li, C.; Li, W. Changes in Phenolic Profiles and Inhibition Potential of Macrophage Foam Cell Formation during Noni (*Morinda citrifolia* Linn.) Fruit Juice Fermentation. *Fermentation* **2022**, *8*, 201. https://doi.org/10.3390/fermentation8050201

Academic Editor: Carlo Giuseppe Rizzello

Received: 27 March 2022
Accepted: 27 April 2022
Published: 29 April 2022

Publisher's Note: MDPI stays neutral with regard to jurisdictional claims in published maps and institutional affiliations.

1. Introduction

Polyphenols are a class of widely distributed, natural polyhydroxylated phenolic compounds which are characterized by their strong antioxidant properties [1]. Based on their chemical structure, polyphenols can be categorized into flavonoids, phenolic acids, and lignans. Numerous phenolic compounds exhibit significant biological activities, such as anti-inflammatory [2], antitumor [3], and antidepressant [4,5] effects, as well as cardiovascular protection [6]. Epidemiological and clinical studies have shown that the consumption of foods and beverages rich in phenolic compounds correlates with a reduced risk of several chronic diseases, such as obesity [7], type 2 diabetes [8], cardiovascular diseases [9], and neurodegenerative disorders [10]. The main sources of phenolic compounds are fruits and vegetables [11]. In the past two decades, various pharmacological effects and health benefits of dietary polyphenols have been revealed [12]. In addition, previous studies have shown that the fermentation process may have a significant effect on the compositions and bioactivities of phenolic compounds. For instance, the gentisic acid content of curly kale significantly increased after fermentation [13]. Gong X et al. [14] reported that the composition of phenolics of *Dendrobium candidum* substantially changed after fermentation, and the content of syringic acid, 4-hydroxybenzoic acid, and p-hydroxycinnamic acid increased significantly.

Morinda citrifolia Linn. (Noni) is an evergreen, fruit-bearing plant which is widely grown in tropical and subtropical areas. Noni has been used to treat headaches, arthritis, and diabetes in Polynesia for more than a thousand years [15]. This plant contains many valuable chemical and biological compounds, such as phenolic compounds, polysaccharides, organic acids, vitamins, and minerals [16]. Phenolic compounds play a key role in the plant's therapeutic properties; for example, rutin, b-sitosterol, asperuloside, and andursolic acid are important biologically active compounds present in noni [17]. The flavonoids isolated from noni fruit, such as kaempferol, quercetin, catechin, and epicatechin, have shown antidepressant and antioxidant activities [18]. With ongoing investigations into the health-promoting properties of noni, noni-related products, especially fermented noni juice, have become remarkably popular worldwide. Fermented noni juice is a beverage with high nutritional value [19]. Traditionally, this juice is produced by fermenting noni fruits in sealed jars or barrels for approximately 2 months or longer and then recovering the juice through drip extraction and/or mechanical pressure [20,21]. It is generally produced by natural fermentation processes, i.e., it is fermented by its intrinsic enzymes and natural mutualistic microorganisms. Deng et al. [22] compared the phytochemical fingerprints of 13 commercial noni fermented juices in the global market and found that they all contained scopolamine, rutin, and quercetin.

Most commercial fermented noni juice is currently produced through a natural fermentation process. Due to the lack of scientific evaluation methods and unified quality standards, the marketplace is full of various forms of fermented noni juice, with differences in quality and price [23]. Available data on the content of phenolic compounds during noni fermentation are limited, and the composition and activity of phenolic compounds remain unclear. Therefore, the content and composition of phenolic extracts were analyzed in this study using ultraperformance liquid chromatography-quadrupole-time-of-flight tandem mass (UPLC–Q–TOF–MS/MS). Meanwhile, their antioxidant activities and inhibition potential of macrophage foam cell formation were also evaluated. This study may provide basic data for product quality improvement and standardized production of fermented noni juice.

2. Materials and Methods

2.1. Chemical and Reagents

Phenolic acid and flavonoid standards (caffeic acid, 2,4-dihydroxybenzoic acid, p-coumaric acid, rutin, quercetin, and isoquercitrin) were purchased from Sigma-Aldrich (Beijing, China). 2,2-diphenyl-1-picrylhydrazyl (DPPH), 2,4,6-tripyridyl-s-triazine (TPTZ), and 2,2′-azino-bis [3-ethylbenzothiazoline-6-sulfonic acid] (ABTS) were purchased from Solarbio Technology Co., Ltd. (Beijing, China). Formic acid and acetonitrile was LC/MS grade and obtained from Fisher Scientific (Hampton, NH, USA). Ethyl acetate, ethanol, $K_2S_2O_8$, methanol, o-phenanthroline, $FeSO_4$, $FeCl_3$, formaldehyde and isopropanol of analytical grade were obtained from Sinopharm Chemical Reagent Co., Ltd. (Shanghai, China). DMEM (Dulbecco's modified Eagle's medium) was bought from Hyclone Co. (Shanghai, China). The cholesterol assay kit was purchased from Pulilai Gene Technology Co., Ltd. (Beijing, China). Oil Red O were purchased from Sigma-Aldrich (Louis, MO, USA).

2.2. Sample Preparation

Mature yellow or white Noni fruits (cultivars "Kuke") were provided by the Hainan Wanning Noni Industrial Base (Hainan, China). The noni fruits were cleaned with running water, soaked with 60 mg/L chlorine dioxide for 30 min, and then rinsed with sterile water. A total of 15 sterile fermenters with each containing about 5 kg noni fruits were fermented at room temperature (25–28 °C) and kept from light. The fermented juice was collected on days 0, 7, 28, and 63 and stored at −80 °C for use.

2.3. Extraction of Phenolic Compounds

Fresh or fermented noni juice was centrifuged at 8000× g for 10 min (4 °C). The supernatants were extracted four times with ethyl acetate (1:1, v/v). The ethyl acetate fractions were combined and evaporated to dryness at 40 °C and redissolved in 10 mL 50% ethanol. The phenolic extracts were filtered through a Nylon filter (0.22 µm, 25 mm) and stored at −80 °C until further analysis. The phenolic extracts were marked as 0-FJP (from fresh noni juice), 7-FJP (after 7-day fermentation), 28-FJP (after 28-day fermentation), and 63-FJP (after 63-day fermentation).

2.4. Phenolic Composition Analysis by UPLC–Q-TOF–MS/MS

The phenolic compounds were analyzed on an Agilent 1290 UPLC coupled to a 6540 Q-TOF MS system with a dual ESI source (Agilent Technologies, Santa Clara, CA, USA). The column used was an Agilent ZORBAX Eclipse Plus C18 (3.0 × 100 mm, 1.8 µm). Elution was undertaken at a flow rate of 0.5 mL/min with solvent A (acetonitrile) and solvent B (0.15% formic acid) in a linear gradient manner: 10–20% A (0–6 min), 20–25% A (6–10 min), 60–95% A (15–20 min), and 95% A (20–25 min). DAD detector was used between 200–400 nm. The injection volume was 20 µL, and the temperature was 30 °C. MS spectrometry testing conditions were as follows: ionization mode, electrospray ionization, and the accurate mass data correction using electrospray ionization-L low concentration tuning mix (G1969-85000). MS operation conditions were as follows: drying gas temperature, 300 °C; drying gas flow rate, 8 mL/min; nebulizer pressure, 45 psi; sheath gas temperature, 400 °C; sheath gas flow, 12 L/min; capillary voltages, 3.5 kV; nozzle voltage, 175 V; and the mass range was scanned as m/z 100–1100. The LC-MS data were collected and analyzed using Mass Hunter B.03.01 (Agilent Technologies). Negative ion analysis and MRE scan modes were also adopted.

2.5. Quantitative Analysis by UPLC

The quantification of targeted phenolic compounds was performed in accordance with the method proposed by Wang, R.M. et al. [24] and conducted on an Agilent 1290 series UPLC (Agilent Technologies, Santa Clara, CA, USA) equipped with a photodiode array detector. The same column and conditions were described as above.

2.6. Antioxidant Activity Assay
2.6.1. DPPH Radical Scavenging Activity Assay

The DPPH radical scavenging activity was measured in accordance with the previously reported method [25]. The sample (1.0 mL) was mixed with a freshly prepared methanol solution of DPPH (1.0 mL, 80 µg/mL). The absorbance was measured at 517 nm after the mixture was incubated for 30 min at 25 °C. The radical scavenging capability of DPPH was calculated using the following equation:

$$\text{DPPH scavenging activity}(\%) = \left[\frac{\left(A_{control} - A_{sample} \right)}{A_{control}} \right] \times 100 \tag{1}$$

where $A_{control}$ is the absorbance of the control and A_{sample} is the absorbance of the extract.

2.6.2. ABTS$^+$ Radical Scavenging Activity Assay

The ABTS$^+$ radical cation decolorization assay was performed using the previously reported method with slight modifications [26]. Briefly, 5 mL of 7 mM ABTS$^+$ and 140 mM $K_2S_2O_8$ aqueous solutions were mixed in equal volumes and reacted for 12–16 h in the dark. The stock solution was diluted with methanol to an absorbance value of 0.7 ± 0.02 at 734 nm to prepare an ABTS$^+$ solution. Then, 0.4 mL of FJP was mixed with 2.6 mL of

$ABTS^+$ solution and reacted for 6 min in the dark. The absorbance at 734 nm was measured using an ultraviolet spectrophotometer with absolute ethyl alcohol as a blank control. The radical scavenging activity was calculated using the following equation:

$$ABTS^+ \text{ scavenging activity}(\%) = \left[\frac{\left(A_{control} - A_{sample}\right)}{A_{control}}\right] \times 100 \tag{2}$$

where $A_{control}$ is the absorbance of the control and A_{sample} is the absorbance of the extract.

2.6.3. OH·(Hydroxyl) Radical Scavenging Activity Assay

The OH radical scavenging activity was determined in accordance with the method proposed by Ren et al. [27] with minor modifications. The reaction mixture contained 1.0 mL of phosphate buffer saline (PBS, 0.02 mol/L, pH 7.4), 0.5 mL o-phenanthroline (2.5 mmol/L), 0.5 mL $FeSO_4$ (2.5 mmol/L), 0.5 mL H_2O_2 (2.0 mmol/L), and 0.5 mL sample. The reaction mixture was centrifuged at $4000 \times g$ for 10 min after reaction for 1 h in a water bath at 37 °C. Optical density (OD) at 536 nm of the supernatant was measured as A_1. OD 536 nm of the blank, in which the sample was replaced with distilled water, was measured as A_0. Distilled water was used to replace H_2O_2 as the control group to measure OD 536 nm, which was recorded as A_2.

$$OH \cdot \text{scavenging activity}(\%) = \left[1 - \frac{(A_1 - A_2)}{A_0}\right] \times 100 \tag{3}$$

where A_0 is the absorbance of the control, A_1 is the absorbance of the extract and A_2 is the absorbance of the reagent blank without sodium salicylate.

2.6.4. Ferric Reducing Antioxidant Power (FRAP) Assay

The FRAP was determined in accordance with the method proposed in a previous report [27]. The FRAP dye was prepared by mixing sodium acetate solution (300 mM), TPTZ solution (10 mM), and $FeCl_3$ solution (20 mM) in a 10:1:1 (v/v) ratio, respectively. A total of 200 µL of extract (50-fold dilution) or standard was added to 2800 µL of prepared FRAP dye solution. The mixture reacted for 30 min under darkness for 10 min (37 °C). The absorbance was measured at 593 nm. The antioxidant capacity was expressed in µg Trolox equivalents per 100 mL of extract (µmol TE/100 mL). The standard curve, which is plotted at different concentrations of standard, ranges from 0–500 µg/mL.

2.7. Cell Experiments

2.7.1. Cell Culture and Treatment

The cell culture was performed following the method proposed by Mueller et al. [28] with slight modifications. RAW264.7 macrophages in the logarithmic growth phase were digested into a single-cell suspension, inoculated into 96-well plates (2×10^5 cells per well), and cultured in DMEM at 37 °C in a humidified atmosphere with 5% CO_2. The medium was replaced with serum-free DMEM for 12 h starving culture after cells were adherent.

Experimental samples were divided into five groups: control group (DMEM), oxidized low-density lipoprotein (ox-LDL) group (100 µg/mL ox-LDL), and sample groups (0-FJP, 7-FJP, 28-FJP, and 63-FJ) were all diluted with DMEM solution under the necessary concentrations: 10, 50, and 100 ng/mL, respectively. The cells were further incubated for 24 h.

2.7.2. Oil Red O Staining and Determination of the Intracellular lipid

Intracellular lipid droplets were determined by oil red O staining. The cells were washed three times in ice-cold PBS (pH7.4) after incubation, followed by 10% formaldehyde fixation for 60 min at room temperature (25 °C). Neutral lipids were stained using 0.5% oil red O in isopropanol for 30 min, and cell morphology was observed using microscopy

(Olympus IX81, Tokyo, Japan). The plate was washed three times with PBS, and 100 µL cell lysis solution (ethanol: acetic acid = 1:1 v/v) was added into each well. The absorbance was acquired at 520 nm after complete lysis.

2.7.3. Determination of the Cholesterol Content

Total cholesterol (TC) and free cholesterol (FC) contents were quantified using the cholesterol assay kit (Pulilai Gene Technology Co., Ltd. Beijing, China) according to the manufacturer's instructions.The esterified cholesterol (CE) content was determined by subtracting FC from TC. CE/TC ratio levels were calculated as follows: CE/TC ratio = (TC − FC)/TC.

2.8. Statistical Analysis

All experiments were performed in triplicate with independent replicates. The data were reported as the mean ± standard deviation (SD) of three replicates. The results were compared by one-way analysis of variance (ANOVA), and Tukey's test was used to identify significant differences. Differences of $p < 0.05$ were considered significant. The data and correlations were analyzed using Excel and SPSS software (version 18.0, SPSS, Inc. (Chicago, IL, USA)), respectively.

3. Results

3.1. Identification of Phenolic Compositions

Table 1 lists the detected phenolic compounds with retention time, observed m/z, generated molecular formula, and proposed compound detected in the extracts of fermented noni juice. A total of 21 compounds, including 12 phenolic acids, 7 flavonoid compounds, and 2 other compounds, were tentatively identified in the current study. The phenolic compounds changed in composition and quantity during the entire fermentation. Ten phenolic compounds were detected in the fresh noni juice (0-FJP), while 15, 15, and 14 phenolics were found in 7-FJP, 28-FJP, and 63-FJP, respectively.

The parent ion at m/z 151.03 [M−H]$^-$ indicated that compound 1 was likely to be 2-hydroxy-2-phenylacetic acid with the molecular formula $C_8H_8O_3$, which has also been tentatively characterized in palm fruits [29]. Compounds 2, 3, and 4 were hydroxycoumarins derivatives and were respectively identified as esculin [30], 4-hydroxycoumarin [31], and scopoletin [32] based on the parent ion at m/z 339.07 [M−H]$^-$, m/z 161.02 [M−H]$^-$, m/z 191.03 [M−H]$^-$, and the related MS^2 data. Three hydroxybenzoic acids (compounds 5, 6, and 7) were characterized. Among them, compound 5 with [M−H]$^-$ ion at m/z 153.01 was identified as protocatechuic acid consistent with previous literature [29], which was recognized in all the samples. Compound 6 with [M−H]$^-$ at m/z 137.02 was tentatively characterized as 3,4-dihydroxybenzaldehyde, while compound 7 with [M−H]$^-$ at m/z 135.04 was tentatively identified as 2-dydroxy-4-methylbenzaldehyde. Only one hydroxyphenylpropanoic acid (compound 8) was detected in 0-FJP, 7-FJP, and 28-FJP. Compound 8 showed [M−H]$^-$ at m/z 165.05, and the molecular formula $C_9H_{10}O_3$ was tentatively characterized as 3-hydroxyphenylpropionic acid [33]. Three kinds of hydroxycinnamic acid were also characterized (compounds 9, 10, and 11). Compound 9, which exhibited the parent ion at m/z 147.04 [M−H]$^-$, was tentatively identified as cinnamic acid; this compound is found in a variety of dietary plant materials, such as fruits, tea, vegetables, and cereals [34,35]. Compound 10 was identified as caffeic acid based on the parent ion at m/z 179.04 [M−H]$^-$ and the similar fragmentation pattern reported by Li et al. [31]. p-coumaric acid (compound 11 with [M−H]$^-$ ion at m/z 163.03) present in 0-FJP and 28-FJP was easily identified and confirmed by MS^2 data and previous literature [36]. Compound 12 was detected in all fermented noni juice phenolic extracts (7-FJP, 28-FJP, and 63-FJP) with [M−H]$^-$ ion at m/z 212.06 tentatively identified as 2,3-dihydroxy-1-guaiacylpropanone. Three flavones (compounds 13, 14, and 15) were also identified during the entire fermentation process. Compound 13 was tentatively characterized as quercetin 3-O-glucosyl-xyloside, because it was detected in negative ionization modes with [M−H]$^-$ at m/z 595.13 and yielded aglycone at m/z 300, which is coherent with that derived from a cross-ring cleavage of its sugar moiety (quercetin moiety) [37]. In addition, compound 14, found in all samples, demonstrated a molecular ion at m/z 609.14 $[M-H]^-$ and fragment ions at m/z 301 and 151, and was easily identified as rutin based on the characterization and literature data [38,39]. Compound 15 showed $[M-H]^-$ at m/z 593.15 with the molecular formula $C_{27}H_{30}O_5$ and was temporarily regarded as luteolin 7-neohesperidoside. Regarding isoflavonoids (derivatives), four compounds (compounds 16, 17, 18, and 19) were tentatively detected and identified. Compound 16, determined by comparing with results from a previous study, was characterized in *Dalbergia odorifera* using LC-MS/MS and was tentatively identified as sativanone [40]. Compounds 17 and 18 with respective precursor ions at $[M-H]^-$ m/z 463.08 and $[M-H]^-$ m/z 301.03 were detected in all samples, tentatively representing isoquercitrin and quercetin, which were identified on the basis of their characterization and literature data [41]. Compound 19, which has a precursor ion $[M-H]^-$ m/z at 313.07, was tentatively characterized as 2′,7-dihydroxy-4′,5′-dimethoxyisoflavone and was present in 0-FJP, 7-FJP, and 63-FJP. Two other phenolic derivatives, including isoscutellarein 7-xyloside (compound 20, $[M-H]^-$ m/z 417.08) and mucronulatol ($\pm$) (compound 21, $[M-H]^-$ m/z 301.10), were also tentatively detected (compounds 20 and 21).

Table 1. UPLC–ESI–QTOF–MS/MS characterization of phenolic compounds in different fermentation times.

No.	Proposed Compounds	Molecular Formula	RT (min)	m/z $[M-H]^-$	m/z Fragments	FIor (ppm)	0-FJP	7-FJP	28-FJP	63-FJP
	Phenolic acids (12)									
1	2-Hydroxy-2-phenylacetic acid	$C_8H_8O_3$	1.84	151.03	91, 65	−0.172	−	−	+	−
2	Esculin	$C_{15}H_{16}O_9$	2.67	339.07	177, 159, 133	−0.06	−	+	+	−
3	4-Hydroxycoumarin	$C_9H_6O_3$	7.99	161.02	133, 89	−0.026	−	−	+	+
4	Scopoletin	$C_{10}H_8O_4$	9.42	191.03	120, 104	−0.033	+	−	−	+
5	2,4-Dihydroxybenzoic acid	$C_7H_6O_4$	3.13	153.01	153	−0.006	+	+	+	+
6	3,4-Dihydroxybenzaldehyde	$C_7H_6O_3$	4.80	137.02	109, 81	0.136	−	+	−	+
7	2-Hydroxy-4-methylbenzaldehyde	$C_8H_8O_2$	4.06	135.04	135, 89	0.064	−	+	+	−
8	3-Hydroxyphenylpropionic acid	$C_9H_{10}O_3$	3.22	165.05	119	−0.851	+	+	+	−
9	Cinnamic acid	$C_9H_8O_2$	3.23	147.04	147, 103	−0.008	+	−	−	−
10	Caffeic acid	$C_9H_8O_4$	3.9	179.04	1, 358, 979	0.076	+	+	+	+
11	p-Coumaric acid	$C_9H_8O_3$	6.308	163.03	133, 92	−0.081	+	+	−	−
12	2,3-Dihydroxy-1-guaiacylpropanone	$C_{10}H_{12}O_5$	4.834	211.06	137, 109	−0.038	−	+	+	+
	Flavonols (7)									
13	Quercetin 3-O-glucosyl-xyloside	$C_{26}H_{28}O_{16}$	13.10	595.13	300, 271	−0.112	−	+	+	+
14	Rutin	$C_{27}H_{30}O_{16}$	13.548	609.14	301, 151	0.021	+	+	+	+
15	Luteolin 7-neohesperidoside	$C_{27}H_{30}O_{15}$	15.92	593.15	285, 151	−0.039	−	+	+	−
16	Sativanone	$C_{17}H_{16}O_5$	13.58	299.09	271, 255	−0.091	−	−	+	+
17	Isoquercitrin	$C_{21}H_{20}O_{12}$	13.67	463.08	300, 271, 151	−0.007	+	+	+	+
18	Quercetin	$C_{15}H_{10}O_7$	20.45	301.03	273, 229, 121	0.028	+	+	+	+
19	2′,7-Dihydroxy-4′,5′-dimethoxyisoflavone	$C_{17}H_{14}O_6$	34.84	313.07	270, 211	0.005	+	+	−	+
	Other phenolics (2)									
20	Isoscutellarein 7-xyloside	$C_{20}H_{18}O_{10}$	9.05	417.08	243, 175, 149	−0.052	−	−	−	+
21	Mucronulatol ($\pm$)	$C_{17}H_{18}O_5$	14.34	301.10	149, 123	−0.022	−	+	+	+
Total		21					10	15	15	14

Notes: RT, Retention Time; (+), Detected; (−), Not detected.

Many phenolics in the fermented noni juice did not come directly from fresh noni fruit, because phenolics were mostly bound with sugars and other compounds in the form of glycosidic bonds. Acidity increased and the glycosidic bonds could be hydrolyzed as fermentation continued; thus, significant amounts of phenolics were released [42]. The phenolics composition changed in the *Momordica charantia* juice after fermentation, demonstrating a reduction in caffeic and p-coumaric acids and an increase in dihydrocaffeic and phloretic acids [43]. Li et al. [44] reported that the phenolics composition also changed during apple beverage fermentation with *Lactobacillus plantarum* ATCC14917, in which sugar moieties were removed and galloyl moieties of a variety of phenolic compounds hydrolyzed, resulting in changes in phenolic profiles [45].

3.2. Quantification of Predominant Individual Phenolic Compounds

Table 2 provides the contents of predominant individual phenolic compounds, including caffeic acid, 2, 4-dihydroxybenzoic acid, p-coumaric acid, rutin, quercetin, and isoquercitrin. The amounts of the individual phenolic compounds were between 0.15–26.69 mg/L. The contents of individual phenolic substantially changed in the fermentation process. The major phenolic compounds identified were rutin, quercetin, and isoquercitrin. The result was similar to previous studies, which reported that rutin and quercetin are characteristic compounds in commercial noni juice fermentation [22]. The content of caffeic and 2,4-dihydroxybenzoic acids gradually increased from 0 to 28 days of the fermentation process and then slightly decreased on day 63. A similar trend was observed on the content of rutin and quercetin, which reached the highest value on the 28th day. By contrast, isoquercitrin content showed a marked decrease over time. The isoquercitrin concentration dropped from 8.25 ± 0.34 mg/L to 4.47 ± 0.05 mg/L at the end of the 63-day fermentation.

Table 2. Contents of predominant individual phenolic compounds of noni juice extracts during fermentation.

No.	Phenolic Compounds	0-FJP (mg/L)	7-FJP (mg/L)	28-FJP (mg/L)	63-FJP (mg/L)
1	caffeic acid	0.96 ± 0.14 c	1.21 ± 0.02 bc	1.68 ± 0.04 a	1.37 ± 0.18 b
2	2,4-dihydroxybenzoic acid	0.75 ± 0.13 c	1.47 ± 0.11 ab	1.7 ± 0.08 a	1.26 ± 0.11 b
3	p-coumaric acid	1.55 ± 0.06 c	3.48 ± 0.02 ab	ND	ND
4	rutin	17.67 ± 1.56 b	20.45 ± 1.32 b	26.69 ± 2.25 a	9.25 ± 0.62 c
5	quercetin	3.03 ± 0.35 c	6.4 ± 0.8 b	10.55 ± 1.36 a	8.16 ± 1.04 ab
6	isoquercitrin	8.25 ± 0.34 a	6.57 ± 0.36 ab	5.14 ± 2.10 b	4.47 ± 0.05 b

Data expressed as mean $\pm$ SD of three replicates; Different letter (a–c) indicate that the data are significantly different from the other data of the same row ($p < 0.05$). The significant difference was calculated through a one-way analysis of variance (ANOVA) and Tukey's test. ND, not detected.

To date, over 200 compounds, especially phenolic compounds and alkaloids, have been identified in noni and related products [22]. Fermented noni juice contains high quantities of total phenolic compounds, flavonoids, condensed tannin, and scopoletin [46]. Previous studies have shown that fermentation generally changes the content of phenolics. Several studies have reported that fermentation influences the phenolic content during the fermentation of plant sources. Li et al. [47] reported that the levels of vanillic acid considerably increased during wampee leaf fermentation, reaching their highest values in the 6th day (4.68 mg/100 g), while 7-hydroxycoumarin almost remained stable. Liu et al. [48] also showed that the catechin content from the beginning (0 days) to the end (27 days) of fermentation ranged from 16.0 ± 4.2 mg/L to 478.5 ± 3.5 mg/L, while quercetin content showed a fluctuating change during fermentation and reached its highest values on day 12, which is consistent with the current results.

3.3. Changes in Antioxidant Activity during Fermentation

The antioxidation activity of noni juice during fermentation was monitored in this study. Table 3 shows that fermented noni juice had higher antioxidant activity compared with unfermented noni juice extracts. The highest antioxidant activity during fermentation was found in the 28-FJP considering DPPH and $ABTS^+$ assay. The DPPH and $ABTS^+$ radical scavenging activity in 28-FJP ($68 \pm 0.29\%$ and $72.05 \pm 1.67\%$, respectively) were significantly higher than that in 7-FJP ($64.49 \pm 0.22\%$ and $67.59 \pm 1.05\%$, respectively) and 63-FJP ($63.92 \pm 0.53\%$ and $67.61 \pm 1.06\%$, respectively). Compared with the 0-FJP, the DPPH and $ABTS^+$ radical scavenging activity in 28-FJP increased by 13.13% and 7.99%, respectively. However, the DPPH and $ABTS^+$ radical scavenging activity of 7-FJP was insignificantly different compared with 63-FJP. The same trend was also observed with OH· and FRAP assays, in which 28-FJP ($57.41 \pm 0.63\%$ and 1144.01 ± 6.86 μmol TE/100 mL) exhibited significantly higher antioxidant activities than those in 63-FJP (49.71 ± 0.57% and 927.10 ± 51.14 μmol TE/100 mL) and 7-FJP (52.45 ± 0.95% and 1013.33 ± 11.98 μmol TE/100 mL).

Table 3. Antioxidant activity of noni juice extracts during fermentation.

Samples	DPPH Radical Scavenging Rate (%)	ABTS⁺ Radical Scavenging Rate(%)	OH· Radical Scavenging Rate (%)	FRAP Assay (μmol TE/100 mL)
0-FJP	60.11 ± 0.02 c	66.72 ± 1.27 b	45.22 ± 0.96 d	877.7 ± 12.87 c
7-FJP	64.49 ± 0.22 b	67.59 ± 1.05 b	52.45 ± 0.95 b	1013.33 ± 11.98 b
28-FJP	68 ± 0.29 a	72.05 ± 1.67 a	57.41 ± 0.63 a	1144.01 ± 6.86 a
63-FJP	63.92 ± 0.53 b	67.61 ± 1.06 b	49.71 ± 0.57 c	927.1 ± 51.14 c

Data expressed as mean ± SD of three replicates; Different letters (a–d indicate that the data are significantly different from the other data of the same column ($p < 0.05$). The significant difference was calculated through a one-way analysis of variance (ANOVA) and Tukey's test.

All four antioxidant indexes (Table 3) showed significant improvements after fermentation, probably due to their strong ability to hydrolyze the polyphenols present in the noni juice. Furthermore, the relatively high diversity and content of phenolic compounds in 7-FJP and 28-FJP may explain the relatively high antioxidant activity compared with 0-FJP. Therefore, an increase in antioxidant activities was associated with the release of phenolic compounds during fermentation. Similar to the current study, Yang et al. [46] reported that fermented noni juice exhibited high reductive activity and improved superoxide anion and H_2O_2 scavenging activities.

3.4. Oil Red O Staining and Analysis of Lipid Accumulation

Furthermore, oil red O staining was performed to evaluate the protective effect of FNJ against Ox-LDL-induced lipid accumulation in macrophages. Figure 1 shows that treatment with FJP of different fermentation times for 24 h lowered the Ox-LDL-induced cellular accumulation of lipid droplets in macrophages.

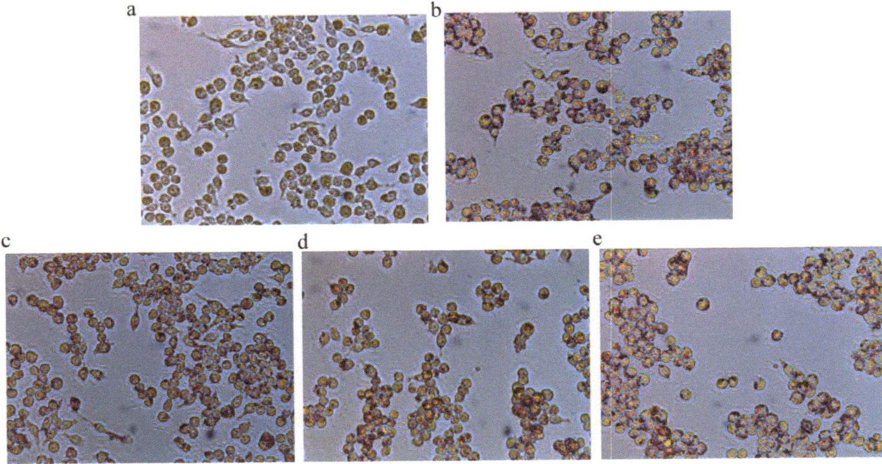

Figure 1. Representative microscopy images of oil red O-stained lipid droplets at a magnification of ×400. (**a**) control group, (**b**) ox-LDL group, (**c**) 100 ng/mL 0-FJP + ox-LDL group, (**d**) 100 ng/mL 28-FJP + ox-LDL group, and (**e**) 100 ng/mL 63-FJP + ox-LDL group.

Figure 2 shows that the OD value in the ox-LDL group reached 0.24, which revealed a significant difference from the control group (0.12). The OD values of experimental groups were significantly different from that of the ox-LDL group ($p < 0.05$) under the concentration range from 10–100 ng/mL. 0-FJP and 28-FJP significantly inhibited lipid accumulation in macrophages ($p < 0.05$) and showed a dose-dependent manner relationship. The inhibition rates of 0-FJP and 28-FJP groups were 42.62% and 45.70%, respectively, under the 100 ng/mL concentration, showing the strongest inhibitory effect on macrophage lipids. Maximal inhibition was observed for group 63-FJP under the 10 ng/mL concentration, revealing an inhibition rate of 31.05% ($p < 0.05$). The 28-FJP exhibited the strongest inhibitory effect on lipid droplets compared with other groups under the concentration range of 10 to 100 ng/mL. These findings suggest that appropriate FJP concentrations can enhance the ability of macrophages to clear pathological lipids and inhibit lipid aggregation to a certain extent.

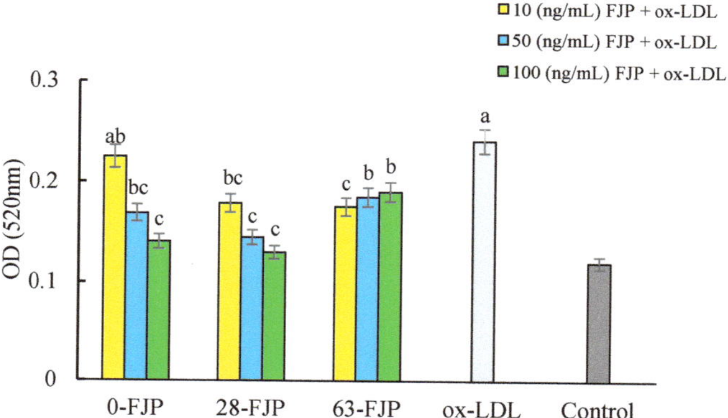

Figure 2. Effect of FJP on lipid accumulation in macrophage foam cells. Data expressed as mean $\pm$ SD of three replicates; Different letters (a–c) indicate that the data are significantly different from the other data of the same column ($p < 0.05$). The significant difference was calculated through a one-way analysis of variance (ANOVA) and Tukey's test.

3.5. Effect of FJP on Intracellular Cholesterol Content in Foam Cells

Additionally, the cholesterol levels were analyzed. Table 4 shows that the percentage of cholesterol ester was 54.79 $\pm$ 2.72% after treating with 80 ng/mL OX-LDL for 24 h, indicating the establishment of the foam cell model. Compared with the ox-LDL group, the total cholesterol and cholesterol ester content reduced in cells in the sample groups, and the foam phenomenon of cells demonstrated reduction ranges of 10–100 ng/mL. The 0-FJP and 28-FJP decreased the intracellular CE/TC ratio in a dose-dependent manner. The decrease in intracellular CE/TC value was the strongest and the CE/TC value was 39.8 $\pm$ 0.29% and 44.22 $\pm$ 0.28%, respectively ($p < 0.01$) when the concentration was 100 ng/mL in 0-FJP and 28-FJP. However, 63-FJP had a negative dose-dependent relationship when the extract concentration at 10 ng/mL the CE/TC ratio was 40.38 $\pm$ 0.07% ($p < 0.01$), showing the strongest effect on the decrease in intracellular CE/TC ratio. These findings confirm that 0-FJP, 28-FJP, and 63-FJP could inhibit ox-LDL-induced macrophage lipid deposition. Moreover, fermented noni juice regulated cholesterol metabolism in macrophages, increased the flow of cholesterol from the inside to the outside of macrophages, and reduced the accumulation of intracellular cholesterol.

The circulating level of ox-LDL is considered to be one of the most important biomarkers in the progression of atherosclerosis. A considerable amount of evidence shows that dietary antioxidants (e.g., phenolic and flavonoid) can potentially protect against LDL oxidation [49]. For example, olive (*Olea europaea* L.) extracts show anti-inflammatory and anti-atherogenic functions, such as the inhibition of LDL oxidation, platelet aggregation, and other factors involved in the development of atherosclerosis [50,51]. Grape seed phenolics can slow down the formation of foam cells by regulating key genes in cholesterol efflux and inflammation [52]. Zhang et al. [53] reported that procyanidin A2 and its metabolites significantly reduced the ox-LDL-induced intracellular CE/TC ratio in megalocytes.

Table 4. Effects of noni juice extracts on cholesterol efflux from macrophages during fermentation.

Samples	Extract Concentration (ng/mL)	TC	FC	CE/TC (%)
Control	0	124.69 ± 0.53 h	71.98 ± 0.48 f	42.27 ± 0.17 e
ox-LDL	80	231.08 ± 0.94 a	105.57 ± 1.26 a	54.31 ± 0.43 a
0-FJP	10	158.31 ± 1.95 d	91.29 ± 2.06 c	42.33 ± 0.59 e
	50	148.79 ± 0.53 e	89.13 ± 0.21 c	40.09 ± 0.35 f
	100	140.79 ± 0.37 f	84.75 ± 0.48 d	39.8 ± 0.29 f
28-FJP	10	182.58 ± 0.47 b	90.35 ± 0.58 c	50.51 ± 0.37 b
	50	150.63 ± 0.56 e	81.54 ± 0.42 e	45.86 ± 0.46 c
	100	130.56 ± 0.69 g	72.82 ± 0.68 f	44.22 ± 0.28 d
63-FJP	10	158.58 ± 0.55 d	94.53 ± 0.44 b	40.38 ± 0.07 f
	50	160.33 ± 0.93 d	86.31 ± 0.94 d	46.16 ± 0.27 c
	100	178.54 ± 0.66 c	95.19 ± 0.5 b	46.68 ± 0.09 c

Data expressed as mean ± SD of three replicates; Different letters (a–h) indicate that the data are significantly different from the other data of the same column ($p < 0.05$). The significant difference was calculated through a one-way analysis of variance (ANOVA) and Tukey's test.

3.6. Correlation Analysis

A correlation study between the results obtained in different antioxidant activity assays, CE/TC ratios, and determined phenolic contents was performed (Table 5) to elucidate the influence of phenolics on the antioxidant activity and cholesterol efflux from macrophages. Among these findings, the contents of caffeic acid, 2,4-dihydroxybenzoic acid, and quercetin showed a highly significant positive correlation ($p < 0.05$) with antioxidant activity. The contents of p-coumaric acid with DPPH, OH·, and FRAP assay demonstrated a significant positive correlation ($p < 0.05$). The contents of rutin with ABTS$^+$·, OH, and FRAP assay showed a significant positive correlation ($p < 0.05$), while the isoquercitrin revealed a negative correlation with DPPH·assay ($p < 0.05$). Moreover, the contents of caffeic and 2,4-dihydroxybenzoic acids showed a positive correlation with CE/TC ratio ($p < 0.05$). Many studies have demonstrated the correlations between bioactive compounds and antioxidant activities in numerous fruits and vegetables [54,55]. Zhu et al. [56] found the significant positive correlations of gallic acid, rutin, quercetin, and kaempferol-3-O-rutinoside contents with DPPH, ABTS$^+$, OH, and FRAP values. These results are consistent with the present study.

Table 5. Pearson correlation analysis between phenolics and antioxidant activities.

	Caffeic Acid	2,4-Dihydroxybenzoic Acid	p-Coumaric Acid	Rutin	Quercetin	Isoquercitrin
DPPH	0.876 **	0.923 **	0.998 **	0.52	0.905 **	−0.604 *
ABTS$^+$	0.698 *	0.675 *	0.440	0.656 *	0.783 **	−0.284
OH·	0.860 **	0.951 **	0.973 **	0.637 *	0.856 **	−0.521
FRAP	0.831 **	0.872 **	0.984 **	0.747 **	0.774 **	−0.402
CE/TC	0.684 *	0.692 *	0.662	0.252	0.827 **	−0.342

* Significant correlation in 0.05 level (bilateral); ** Significant correlation in 0.01 level (bilateral); DPPH: Diphenylpicrylhydrazyl Assay; ABTS: 2,2′-Azino-bis-3ethylbenzothiazoline-6-sulfonic Acid Assay; OH·: Hydroxyl radical scavenging rate; FRAP: Ferric reducing antioxidant potential; CE/TC: Cholesterol ester/Total cholesterol ratio.

4. Conclusions

The characteristics and contents of FJP at different fermentation times and their corresponding in vitro biological activities (especially antioxidant and inhibited lipid deposition) were reported in this study. Twenty-one phenolic compounds, comprising 12 phenolic acids, 7 flavonoids, and 2 others, were tentatively identified. Rutin, quercetin, and isoquercitrin were the most abundant components in FJP. The 28-FJP showed the highest antioxidant activities of DPPH, ABTS$^+$, OH·, and FRAP, among other samples. FJP treatment significantly inhibited lipid deposition and reduced ox-LDL-induced intracellular CE/TC ratio. Furthermore, significant positive correlations between phenolics, flavonoids, antioxidant activities, and CE/TC ratio were observed by Pearson's correlation coefficient analysis. The present study suggests that the optimal fermentation period to achieve maximal yields of phenolic compounds is around 28 days,. The obtained findings provide useful information for processing fermented noni juice, which is rich in phenolics with upgraded bioactivities.

Author Contributions: K.C.: Conceptualization, Methodology, Writing-original manuscript; R.D.: Investigation, Methodology; X.L., X.H. and Z.W.: Investigation, Data Curation; S.L.: Supervision; C.L.: Resources, Writing-review and editing, Funding acquisition; W.L.: Supervision, Writing-review and editing. All authors have read and agreed to the published version of the manuscript.

Funding: This research was funded by [the National Natural Science Foundation of China] grant number [31760456], [the National Natural Science Foundation of China] grant number [32160559], [the Hainan Provincial Natural Science Foundation of China] grant number [321QN191] and [the Hainan Provincial Natural Science Foundation of China] grant number [320RC511]. The APC was funded by [the National Natural Science Foundation of China].

Institutional Review Board Statement: Not applicable.

Informed Consent Statement: Not applicable.

Data Availability Statement: The original contributions presented in the study are included in the article, further inquiries can be directed to the corresponding author.

Acknowledgments: This research was funded by the National Natural Science Foundation of China, grant number 31760456, the National Natural Science Foundation of China, grant number 32160559, the Hainan Provincial Natural Science Foundation of China, grant number 321QN191 and the Hainan Provincial Natural Science Foundation of China, grant number 320RC511.

Conflicts of Interest: The authors declare no conflict of interest.

References

1. Zhang, H.; Tsao, R. Dietary polyphenols, oxidative stress and antioxidant and anti-inflammatory effects. *Curr. Opin. Food Sci.* **2016**, *8*, 33–42. [CrossRef]
2. Abbasi-Parizad, P.; De Nisi, P.; Adani, F.; Pepe Sciarria, T.; Squillace, P.; Scarafoni, A.; Iametti, S.; Scaglia, B. Antioxidant and Anti-Inflammatory Activities of the Crude Extracts of Raw and Fermented Tomato Pomace and Their Correlations with Aglycate-Polyphenols. *Antioxidants* **2020**, *9*, 179. [CrossRef] [PubMed]
3. Navarro, M.; Arnaez, E.; Moreira, I.; Quesada, S.; Azofeifa, G.; Wilhelm, K.; Vargas, F.; Chen, P. Polyphenolic Characterization, Antioxidant, and Cytotoxic Activities of Mangifera indica Cultivars from Costa Rica. *Foods* **2019**, *8*, 384. [CrossRef] [PubMed]
4. Huang, X.; Li, W.; You, B.; Tang, W.; Gan, T.; Feng, C.; Li, C.; Yang, R. Serum Metabonomic Study on the Antidepressant-like Effects of Ellagic Acid in a Chronic Unpredictable Mild Stress-Induced Mouse Model. *J. Agric. Food Chem.* **2020**, *68*, 9546–9556. [CrossRef]
5. Lira, S.M.; Dionisio, A.P.; Holanda, M.O.; Marques, C.G.; Silva, G.S.D.; Correa, L.C.; Santos, G.B.M.; de Abreu, F.A.P.; Magalhaes, F.E.A.; Reboucas, E.L.; et al. Metabolic profile of pitaya (Hylocereus polyrhizus (F.A.C. Weber) Britton & Rose) by UPLC-QTOF-MS(E) and assessment of its toxicity and anxiolytic-like effect in adult zebrafish. *Food Res. Int.* **2020**, *127*, 108701. [CrossRef]
6. Adriouch, S.; Lampure, A.; Nechba, A.; Baudry, J.; Assmann, K.; Kesse-Guyot, E.; Hercberg, S.; Scalbert, A.; Touvier, M.; Fezeu, L.K. Prospective Association between Total and Specific Dietary Polyphenol Intakes and Cardiovascular Disease Risk in the Nutrinet-Sante French Cohort. *Nutrients* **2018**, *10*, 1587. [CrossRef]
7. Guo, X.; Tresserra-Rimbau, A.; Estruch, R.; Martinez-Gonzalez, M.A.; Medina-Remon, A.; Fito, M.; Corella, D.; Salas-Salvado, J.; Portillo, M.P.; Moreno, J.J.; et al. Polyphenol Levels Are Inversely Correlated with Body Weight and Obesity in an Elderly Population after 5 Years of Follow Up (The Randomised PREDIMED Study). *Nutrients* **2017**, *9*, 452. [CrossRef]
8. Burton-Freeman, B.M.; Sandhu, A.K.; Edirisinghe, I. Red Raspberries and Their Bioactive Polyphenols: Cardiometabolic and Neuronal Health Links. *Adv. Nutr.* **2016**, *7*, 44–65. [CrossRef]
9. Mattera, R.; Benvenuto, M.; Giganti, M.G.; Tresoldi, I.; Pluchinotta, F.R.; Bergante, S.; Tettamanti, G.; Masuelli, L.; Manzari, V.; Modesti, A.; et al. Effects of Polyphenols on Oxidative Stress-Mediated Injury in Cardiomyocytes. *Nutrients* **2017**, *9*, 523. [CrossRef]
10. Ravi, S.K.; Narasingappa, R.B.; Vincent, B. Neuro-nutrients as anti-alzheimer's disease agents: A critical review. *Crit. Rev. Food Sci. Nutr.* **2019**, *59*, 2999–3018. [CrossRef]

11. Nowak, D.; Goslinski, M.; Przygonski, K.; Wojtowicz, E. The antioxidant properties of exotic fruit juices from acai, maqui berry and noni berries. *Eur. Food Res. Technol.* **2018**, *244*, 1897–1905. [CrossRef]

12. Tresserra-Rimbau, A.; Lamuela-Raventos, R.M.; Moreno, J.J. Polyphenols, food and pharma. Current knowledge and directions for future research. *Biochem. Pharmacol.* **2018**, *156*, 186–195. [CrossRef] [PubMed]

13. Michalak, M.; Szwajgier, D.; Paduch, R.; Kukula-Koch, W.; Wasko, A.; Polak-Berecka, M. Fermented curly kale as a new source of gentisic and salicylic acids with antitumor potential. *J. Funct. Foods* **2020**, *67*, 103866. [CrossRef]

14. Gong, X.; Jiang, S.; Tian, H.; Xiang, D.; Zhang, J. Polyphenols in the Fermentation Liquid of Dendrobium candidum Relieve Intestinal Inflammation in Zebrafish Through the Intestinal Microbiome-Mediated Immune Response. *Front. Immunol.* **2020**, *11*, 1542. [CrossRef] [PubMed]

15. Dixon, A.R.; McMillen, H.; Etkin, N.L. Ferment this: The transformation of Noni, a traditional Polynesian medicine (*Morinda citrifolia*, Rubiaceae). *Econ. Bot.* **1999**, *53*, 51–68. [CrossRef]

16. Wall, M.M.; Miller, S.; Siderhurst, M.S. Volatile changes in Hawaiian noni fruit, *Morinda citrifolia* L., during ripening and fermentation. *J. Sci. Food Agric.* **2018**, *98*, 3391–3399. [CrossRef] [PubMed]

17. Pawlus, A.D.; Kinghorn, D.A. Review of the ethnobotany, chemistry, biological activity and safety of the botanical dietary supplement *Morinda citrifolia* (noni). *J. Pharm. Pharmacol.* **2007**, *59*, 1587–1609. [CrossRef] [PubMed]

18. Mohd Zin, Z.; Abdul Hamid, A.; Osman, A.; Saari, N.; Misran, A. Isolation and Identification of Antioxidative Compound from Fruit of Mengkudu (*Morinda citrifolia* L.). *Int. J. Food Prop.* **2007**, *10*, 363–373. [CrossRef]

19. Chan-Blanco, Y.; Vaillant, F.; Mercedes Perez, A.; Reynes, M.; Brillouet, J.-M.; Brat, P. The noni fruit (*Morinda citrifolia* L.): A review of agricultural research, nutritional and therapeutic properties. *J. Food Compos. Anal.* **2006**, *19*, 645–654. [CrossRef]

20. Yang, J.; Paulino, R.; Janke-Stedronsky, S.; Abawi, F. Free-radical-scavenging activity and total phenols of noni (*Morinda citrifolia* L.) juice and powder in processing and storage. *Food Chem.* **2007**, *102*, 302–308. [CrossRef]

21. Cheng, Y.L.; Li, P.Z.; Hu, B.; Xu, L.; Liu, S.N.; Yu, H.; Guo, Y.H.; Xie, Y.F.; Yao, W.R.; Qian, H. Correlation analysis reveals the intensified fermentation via Lactobacillus plantarum improved the flavor of fermented noni juice. *Food Biosci.* **2021**, *43*, 101234. [CrossRef]

22. Deng, S.X.; West, B.J.; Jensen, C.J. A quantitative comparison of phytochemical components in global noni fruits and their commercial products. *Food Chem.* **2010**, *122*, 267–270. [CrossRef]

23. Motshakeri, M.; Ghazali, H.M. Nutritional, phytochemical and commercial quality of Noni fruit: A multi-beneficial gift from nature. *Trends Food Sci. Technol.* **2015**, *45*, 118–129. [CrossRef]

24. Wang, R.M.; Yao, L.L.; Lin, X.; Hu, X.P.; Wang, L. Exploring the potential mechanism of Rhodomyrtus tomentosa (Ait.) Hassk fruit phenolic rich extract on ameliorating nonalcoholic fatty liver disease by integration of transcriptomics and metabolomics profiling. *Food Res. Int.* **2022**, *151*, 110824. [CrossRef]

25. Wang, Z.; Feng, Y.; Yang, N.; Jiang, T.; Xu, H.; Lei, H. Fermentation of kiwifruit juice from two cultivars by probiotic bacteria: Bioactive phenolics, antioxidant activities and flavor volatiles. *Food Chem.* **2022**, *373*, 131455. [CrossRef] [PubMed]

26. Re, R.; Pellegrini, N.; Proteggente, A.; Pannala, A.; Yang, M.; Rice-Evans, C. Antioxidant activity applying an improved ABTS radical cation decolorization assay. *Free Radic. Biol. Med.* **1999**, *26*, 1231–1237. [CrossRef]

27. Ren, B.B.; Chen, C.; Li, C.; Fu, X.; You, L.; Liu, R.H. Optimization of microwave-assisted extraction of Sargassum thunbergii polysaccharides and its antioxidant and hypoglycemic activities. *Carbohyd. Polym.* **2017**, *173*, 192–201. [CrossRef]

28. Mueller, M.; Hobiger, S.; Jungbauer, A. Anti-inflammatory activity of extracts from fruits, herbs and spices. *Food Chem.* **2010**, *122*, 987–996. [CrossRef]

29. Ma, C.; Dunshea, F.R.; Suleria, H.A.R. LC-ESI-QTOF/MS Characterization of Phenolic Compounds in Palm Fruits (Jelly and Fishtail Palm) and Their Potential Antioxidant Activities. *Antioxidants* **2019**, *8*, 483. [CrossRef]

30. Adnan, M.; Patel, M.; Deshpande, S.; Alreshidi, M.; Siddiqui, A.J.; Reddy, M.N.; Emira, N.; De Feo, V. Effect of Adiantum philippense Extract on Biofilm Formation, Adhesion with Its Antibacterial Activities against Foodborne Pathogens, and Characterization of Bioactive Metabolites: An in vitro-in silico Approach. *Front. Microbiol.* **2020**, *11*, 823. [CrossRef]

31. Li, W.; Yang, R.L.; Ying, D.Y.; Yu, J.W.; Sanguansri, L.; Augustin, M.A. Analysis of polyphenols in apple pomace: A comparative study of different extraction and hydrolysis procedures. *Ind. Crops Prod.* **2020**, *147*, 112250. [CrossRef]

32. Kramberger, K.; Barlic-Maganja, D.; Bandelj, D.; Baruca Arbeiter, A.; Peeters, K.; Miklavcic Visnjevec, A.; Jenko Praznikar, Z. HPLC-DAD-ESI-QTOF-MS Determination of Bioactive Compounds and Antioxidant Activity Comparison of the Hydroalcoholic and Water Extracts from Two Helichrysum italicum Species. *Metabolites* **2020**, *10*, 403. [CrossRef] [PubMed]

33. Tang, J.F.; Dunshea, F.R.; Suleria, H.A.R. LC-ESI-QTOF/MS Characterization of Phenolic Compounds from Medicinal Plants (Hops and Juniper Berries) and Their Antioxidant Activity. *Foods* **2020**, *9*, 7. [CrossRef]

34. Chou, O.; Ali, A.; Subbiah, V.; Barrow, C.J.; Dunshea, F.R.; Suleria, H.A.R. LC-ESI-QTOF-MS/MS Characterisation of Phenolics in Herbal Tea Infusion and Their Antioxidant Potential. *Fermentation* **2021**, *7*, 73. [CrossRef]

35. Guzman, J.D. Natural Cinnamic Acids, Synthetic Derivatives and Hybrids with Antimicrobial Activity. *Molecules* **2014**, *19*, 19292–19349. [CrossRef]

36. Zhong, B.M.; Robinson, N.A.; Warner, R.D.; Barrow, C.J.; Dunshea, F.R.; Suleria, H.A.R. LC-ESI-QTOF-MS/MS Characterization of Seaweed Phenolics and Their Antioxidant Potential. *Mar. Drugs* **2020**, *18*, 331. [CrossRef]

37. Castro-Lopez, C.; Bautista-Hernandez, I.; Gonzalez-Hernandez, M.D.; Martinez-Avila, G.C.G.; Rojas, R.; Gutierrez-Diez, A.; Medina-Herrera, N.; Aguirre-Arzola, V.E. Polyphenolic Profile and Antioxidant Activity of Leaf Purified Hydroalcoholic Extracts from Seven Mexican Persea americana Cultivars. *Molecules* **2019**, *24*, 173. [CrossRef]

38. Pierson, J.T.; Monteith, G.R.; Roberts-Thomson, S.J.; Dietzgen, R.G.; Gidley, M.J.; Shaw, P.N. Phytochemical extraction, characterisation and comparative distribution across four mango (*Mangifera indica* L.) fruit varieties. *Food Chem.* **2014**, *149*, 253–263. [CrossRef]

39. Li, S.J.; Wang, R.M.; Hu, X.P.; Li, C.F.; Wang, L. Bio-affinity ultra-filtration combined with HPLC-ESI-qTOF-MS/MS for screening potential alpha-glucosidase inhibitors from Cerasus humilis (Bge.) Sok. leaf-tea and in silico analysis. *Food Chem.* **2022**, *373*, 131528. [CrossRef]

40. Zhao, X.S.; Zhang, S.H.; Liu, D.; Yang, M.H.; Wei, J.H. Analysis of Flavonoids in Dalbergia odorifera by Ultra-Performance Liquid Chromatography with Tandem Mass Spectrometry. *Molecules* **2020**, *25*, 389. [CrossRef]

41. Guo, J.J.; Zhang, D.; Yu, C.; Yao, L.; Chen, Z.; Tao, Y.D.; Cao, W.G. Phytochemical Analysis, Antioxidant and Analgesic Activities of Incarvillea compacta Maxim from the Tibetan Plateau. *Molecules* **2019**, *24*, 1692. [CrossRef]

42. Huynh, N.T.; Van Camp, J.; Smagghe, G.; Raes, K. Improved Release and Metabolism of Flavonoids by Steered Fermentation Processes: A Review. *Int. J. Mol. Sci.* **2014**, *15*, 19369–19388. [CrossRef] [PubMed]

43. Gao, H.; Wen, J.J.; Hu, J.L.; Nie, Q.X.; Chen, H.H.; Nie, S.P.; Xiong, T.; Xie, M.Y. Momordica charantia juice with Lactobacillus plantarum fermentation: Chemical composition, antioxidant properties and aroma profile. *Food Biosci.* **2019**, *29*, 62–72. [CrossRef]

44. Li, Z.X.; Teng, J.; Lyu, Y.L.; Hu, X.Q.; Zhao, Y.L.; Wang, M.F. Enhanced Antioxidant Activity for Apple Juice Fermented with Lactobacillus plantarum ATCC14917. *Molecules* **2019**, *24*, 51. [CrossRef] [PubMed]

45. Dueñas, M.; Fernández, D.; Hernández, T.; Estrella, I.; Muñoz, R. Bioactive phenolic compounds of cowpeas (*Vigna sinensis* L). Modifications by fermentation with natural microflora and with Lactobacillus plantarum ATCC 14917. *J. Sci. Food Agric.* **2005**, *85*, 297–304. [CrossRef]

46. Yang, S.C.; Chen, T.I.; Li, K.Y.; Tsai, T.C. Change in Phenolic Compound Content, Reductive Capacity and ACE Inhibitory Activity in Noni Juice during Traditional Fermentation. *J. Food Drug Anal.* **2007**, *15*, 290–298. [CrossRef]

47. Li, Q.; Chang, X.; Guo, R.; Wang, Q.; Guo, X. Dynamic effects of fermentation on phytochemical composition and antioxidant properties of wampee (*Clausena lansium* (Lour.) Skeel) leaves. *Food Sci. Nutr.* **2019**, *7*, 76–85. [CrossRef]

48. Liu, Y.; Cheng, H.; Liu, H.; Ma, R.; Ma, J.; Fang, H. Fermentation by Multiple Bacterial Strains Improves the Production of Bioactive Compounds and Antioxidant Activity of Goji Juice. *Molecules* **2019**, *24*, 3519. [CrossRef]

49. Aviram, M.; Fuhrman, B. Wine flavonoids protect against LDL oxidation and atherosclerosis. *Ann. N.Y. Acad. Sci.* **2002**, *957*, 146–161. [CrossRef]

50. Tejada, S.; Pinya, S.; Bibiloni, M.D.; Tur, J.A.; Pons, A.; Sureda, A. Cardioprotective Effects of the Polyphenol Hydroxytyrosol from Olive Oil. *Curr. Drug Targets* **2017**, *18*, 1477–1486. [CrossRef]

51. Kim, Y.; Keogh, J.B.; Clifton, P.M. Benefits of Nut Consumption on Insulin Resistance and Cardiovascular Risk Factors: Multiple Potential Mechanisms of Actions. *Nutrients* **2017**, *9*, 1271. [CrossRef] [PubMed]

52. Terra, X.; Pallares, V.; Ardevol, A.; Blade, C.; Fernandez-Larrea, J.; Pujadas, G.; Salvado, J.; Arola, L.; Blay, M. Modulatory effect of grape-seed procyanidins on local and systemic inflammation in diet-induced obesity rats. *J. Nutr. Biochem.* **2011**, *22*, 380–387. [CrossRef]

53. Zhang, Y.Y.; Li, X.L.; Li, T.Y.; Li, M.Y.; Huang, R.M.; Li, W.; Yang, R.L. 3-(4-Hydroxyphenyl)propionic acid, a major microbial metabolite of procyanidin A2, shows similar suppression of macrophage foam cell formation as its parent molecule. *Rsc. Adv.* **2018**, *8*, 6242–6250. [CrossRef]

54. Belwal, T.; Pandey, A.; Bhatt, I.D.; Rawal, R.S.; Luo, Z. Trends of polyphenolics and anthocyanins accumulation along ripening stages of wild edible fruits of Indian Himalayan region. *Sci. Rep.* **2019**, *9*, 5894. [CrossRef]

55. Afanas'ev, I.B.; Dorozhko, A.I.; Brodskii, A.V.; Kostyuk, V.A.; Potapovitch, A.I. Chelating and free radical scavenging mechanisms of inhibitory action of rutin and quercetin in lipid peroxidation. *Biochem. Pharmacol.* **1989**, *38*, 1763–1769. [CrossRef]

56. Zhu, H.L.; Liu, S.X.; Yao, L.L.; Wang, L.; Li, C.F. Free and Bound Phenolics of Buckwheat Varieties: HPLC Characterization, Antioxidant Activity, and Inhibitory Potency towards alpha-Glucosidase with Molecular Docking Analysis. *Antioxidants* **2019**, *8*, 606. [CrossRef] [PubMed]

MDPI AG
Grosspeteranlage 5
4052 Basel
Switzerland
Tel.: +41 61 683 77 34

Fermentation Editorial Office
E-mail: fermentation@mdpi.com
www.mdpi.com/journal/fermentation